Mobile and Wireless
Design Essentials

Martyn Mallick

WILEY

Wiley Publishing, Inc.

Publisher: Robert Ipsen
Editor: Carol Long
Developmental Editor: Scott Amerman
Managing Editor: Micheline Frederick
Text Design & Composition: Wiley Composition Services

This book is printed on acid-free paper. ⊗

Library of Congress Cataloging-in-Publication Data:

ISBN: 0-471-21419-1

Printed in the United States of America

10 9 8 7 6 5 4 3 2 1

To my beautiful wife Catherine, for her support, understanding, and encouragement.

Contents

Acknowledgments

When I started writing *Mobile and Wireless Design Essentials* a year and a half ago, I never imagined what a monumental effort it would require. Without the support and assistance from my family, friends, and colleagues it would not have been possible to write on such a broad, evolving topic.

In particular, I must thank my wife Cathy for being patient and understanding when I wrote late into the night, and during most weekends.

I would also like to thank the rest of my family, including my parents Adrian and JoAnn, and my siblings David, Anita, Bryan, and Marissa for helping me stay focused throughout the writing process. Thanks also goes to Carol Long and her team at John Wiley & Sons including Scott, Micheline, Felicia, Janet, Erica, Adaobi, and Holly for offering their insight into the publishing process and for keeping me on schedule. Also deserving my gratitude are many of my colleagues at iAnywhere Solutions including David, Bharat, Alex, Liam, Matt, Chris, and Eric, who allowed me to bounce ideas off of them, and provided technical editing during the latter phases of writing. Additionally, a special mention goes to Michelle Cheng, a very gifted individual, who worked endless hours putting together all of the figures you see throughout the book. And finally I would like to thank my employer for providing an environment that is conducive to expanding my knowledge about both the business and technical aspects of the mobile and wireless industry. I hope the knowledge I have gained, and the lessons I have learned, prove to be valuable as you develop your mobile and wireless solutions.

Martyn Mallick
January 2003

Introduction

Mobile and wireless application development has come a long way in the past few years. It has progressed beyond the hype of wireless Web applications for consumers to the reality of high-value mobile applications for corporate users. Opportunities abound for creating new mobile and wireless applications that provide vital benefits to any business. A sampling of these benefits includes increased worker productivity, reduced processing costs, heightened accuracy, and competitive advantage. In contrast is the concern that developing mobile and wireless applications will involve many new technologies and concepts that many corporate developers are still learning to use.

One of the challenges in the mobile application space is the variety of application architectures available. Though many by now are now familiar with Wireless Application Protocol (WAP) applications, they are not familiar with smart client and messaging application architectures. (Note: WAP is a specific protocol, but is commonly used to describe any type of thin client wireless application. For a detailed discussion of thin client applications, see Chapter 11, "Thin Client Overview," and Chapter 12, "Thin Client Development.") *Thin client* refers to server-based applications that make it possible to browse the Internet on a wireless device. All of the business logic and data access logic is located on the server. The only software required on the client is a micro-browser, which is often preinstalled on wireless devices. Thin client applications are attractive because they can build upon existing Internet applications and do not require deployment to the client device. They can be viewed by anyone with a wireless Web-enabled device and can be updated at any time simply by changing the software on the enterprise server.

Thin client applications have one fundamental shortcoming, however: They require a wireless network connection to be effective. Without a connection, information cannot be retrieved from the server, essentially making the application useless. Even when a connection is available, unreliable wireless network coverage, slow data transfer rates, and cost also impact the success of thin client applications.

Consequently, a movement is growing toward *smart client* applications. These applications allow corporations to deploy an application to the mobile device so the user can continue to interact with the application even when a wireless data connection is unavailable. (For more detailed information on smart client applications and technology, see Chapter 7, "Smart Client Overview," and Chapter 8, "Smart Client Development.") These applications commonly include a form of persistent data storage that communicates with enterprise systems using data synchronization. This combination enables applications to have sophisticated user interfaces and high-performance data access, making them suitable for offline computing.

The third mobile application architecture of interest is *messaging*. (For more detailed information on messaging technology, see Chapter 5, "Mobile and Wireless Messaging.") Messaging technology can be used either on its own or to enhance existing applications. Adding notification capabilities to an application can increase its effectiveness dramatically. Mobile users can have important data "pushed" to them, as opposed to constantly requesting it from an enterprise server. Information notifications can be applied to both thin client and smart client applications. Messaging applications can also be developed on their own using messaging as the data delivery mechanism. In these applications, message queues are present on both the client and the server, allowing for information to be stored when a user is not connected to the network. Once the user connects, the stored messages are automatically forwarded to him or her. This type of messaging is commonly referred to as store-and-forward.

The technologies available to companies that want to extend their enterprise systems to their mobile workforce are covered in depth in the chapters that follow. All three mobile applications architectures are covered in some depth, as is related information on mobile devices, wireless networks, mobile and wireless security, mobile information management, and location-based services. This book provides all of the information you require to build highly successful mobile and wireless applications. Though the content is mainly focused on the creation of enterprise applications, you will find information relevant to developing consumer applications as well.

Overview of the Book and Technology

Developing successful mobile applications requires the integration of many technologies. Handset manufacturers and wireless network operators lay the basic groundwork for many applications, but they are only one component of the solution. Mobile middleware software, in conjunction with client-side technology, rounds out the solution. Developers are responsible for putting the pieces together to form effective mobile and wireless applications.

In order to be successful with this task, a broad knowledge of the mobile and wireless industry as a whole is required. Obtaining this knowledge can be a difficult quest. Although many books have been written on mobile and wireless computing, they typically focus on a very specific topic, meaning that developers would have to read several of these books to learn what is required for a single project. Obviously, this is too time-consuming. Moreover, due to the specific nature of the content being covered, many of these books simply rehash industry specifications or product manuals and do not cover the important issues for designing and developing enterprise mobile solutions. To fill in these gaps, readers then have to spend still more time doing their own research on the Internet before they can be productive.

The lack of comprehensive books on mobile and wireless computing was the motivating factor behind this book. *Mobile and Wireless Design Essentials* was written to make things easier for developers. It is a single resource whose objective is to bring you up to speed on the full spectrum of technologies and issues related to mobile and wireless computing. Every chapter provides insightful information on its respective topic, beginning with an overview of the subject matter and followed by in-depth analysis of how it is useful to application developers. Each chapter also contains helpful Web links where you can go for more information on any of the topics covered. These Web links alone will save you countless hours of searching for information.

To meet this goal — to be an all-in-one resource—a great number of topics are covered in *Mobile and Wireless Design Essentials*. The book starts with an overview of the mobile and wireless landscape, to ensure that all readers are "on the same page," before tackling the topics of developing smart client and thin client applications. Each of these application architectures has several chapters dedicated to related design and development issues. The final parts of the book focus on related technologies that are being used today—such as mobile email, mobile device management, and location-based services—and technologies that you will come across in the near future, including mobile Web services, M-Services and BREW.

How This Book Is Organized

Mobile and Wireless Design Essentials is divided into 4 parts and 18 chapters. Part I provides core information, making it essential to read first, to get the most from the rest of the book. After that, feel free to read the rest of the material in any order, as each part is self-contained. The same is true for most of the chapters: each is complete in itself so you can easily reference any one to get what you need on the subject covered. The chapters were written to give you a concise overview of the technology under discussion; you will not be overwhelmed by needless details.

Part Summary

Part I, "Introduction to the Mobile and Wireless Landscape," lays the foundation on which the other parts of the book are based. To that end, it covers mobile devices, wireless networks, mobile application architectures, mobile and wireless security, and messaging technology. These are core topics that should be well understood by anyone involved in a mobile application project. Subsequently, each of these topics is covered in its own chapter to provide a concise introduction to mobile and wireless computing.

Part II, "Building Smart Client Applications," provides an in-depth look at the concepts and technologies pertinent to developing smart client applications. It starts with an overview of the smart client architecture and proceeds to the development process, persistent data technology, and enterprise data synchronization. This part gives you everything you need to know in respect to the design and development of smart client applications.

Part III, "Building Wireless Internet Applications," provides an in-depth look at the concepts and technologies inherent to developing thin client applications. Similar to Part II, this part starts with an overview of the thin client architecture and proceeds to discuss the concepts and technologies involved in thin client development. A complete overview of the thin client markup languages—HDML, WML, cHTML, and XHTML—is provided, along with the techniques that can be used to generate this technology for the wide range of wireless devices being used today. Wireless Internet technology and voice application development using VoiceXML are the final topics discussed in Part III.

Part IV, "Beyond Enterprise Data," takes a look at technologies that are not core to the leading mobile application architectures but that are significant in the adoption and deployment of mobile applications. The first topic is mobile information management, which includes both personal information management (PIM) and mobile device management capabilities. These technologies are becoming increasingly important as mobile devices proliferate and become the responsibility of enterprise IT staffs. The second topic in Part IV is location-based services (LBS). Much of the hype around LBS refers to its use in the consumer market, but many corporate applications can also benefit from location information. Part IV finishes with information on four technologies that gaining momentum in mobile computing: Mobile Web Services, BREW, SALT, and M-Services.

Chapter Summary

Chapter 1, "Welcome to Mobile and Wireless," is a nontechnical introduction to mobile wireless computing. It starts with an overview of key terms (e.g., mobile, wireless, m-commerce, and m-business) and moves on to a discussion of the benefits and challenges that surround mobile computing. An overview is also provided of the main mobility enablers, including wireless networks, mobile devices, and software infrastructure. This chapter will help organizations understand the risks and rewards in developing mobile solutions.

Chapter 2, "Mobile Devices," overviews the mobile device market, with an emphasis on devices that are most appropriate for corporate solutions. For each device category, we will look at the leading features, such as screen size, data input mechanisms, wireless support, and storage space.

This chapter also covers key criteria for selecting mobile devices, with a focus on data input mechanisms and wireless connectivity options. The chapter concludes with information on mobile device manufacturers and the classes of devices they provide.

Chapter 3, "Wireless Networks," addresses all aspects of wireless network coverage, including wireless personal area networks (WPANs), wireless local area networks (WLANs), wireless wide area networks (WWANs), and satellite networks. For each category, the prevalent technologies are examined, followed by a discussion about what the future holds. From this chapter you will gain an understanding of the wireless network protocols that are being used today and for what types of applications. This knowledge will be valuable as you continue through the book to learn more about the design and development of mobile and wireless applications.

Chapter 4, "Mobile Application Architectures," introduces you to the leading application architectures available for mobile computing: thin client (wireless Internet), smart client, and messaging. The chapter starts with a list of key criteria that should be considered when determining which application architecture is most suitable for a given application. It then proceeds with an overview of each application model, which includes a discussion of the advantages and disadvantages they present.

Chapter 5, "Mobile and Wireless Messaging," takes a look at the key messaging technologies currently available. It begins with an overview of the common messaging systems, such as email and paging, then moves on to SMS, EMS, and MMS, and finishes with push and application-to-application messaging. After explaining the various messaging systems, it covers the messaging value chain, from device manufacturers to messaging middleware providers.

Chapter 6, "Mobile and Wireless Security," starts with a security primer on the key aspects of creating a secure environment; it then provides information on each of the technologies involved in building secure applications. Next the chapter offers insight into issues surrounding WAP security, such as the WAP gap, before addressing issues related to securing smart client applications. The goal of this chapter is to provide developers with enough information to make educated decisions when implementing security in their mobile solutions.

Chapter 7, "Smart Client Overview," highlights the main components of a successful smart client solution. It then takes an in-depth look at the major mobile operating systems that are available for smart client solutions. The combination of the mobile operating system and the device hardware often dictate whether a smart client solution is possible to implement.

Chapter 8, "Smart Client Development," prepares you for some of the technical challenges you will encounter while developing smart client applications. It also gives you some pointers on how to get started with development. The chapter steps through each part of the development process, taking a look at technology that is available to help you build your mobile solutions. It also discusses the pros and cons of developing native versus Java applications. Information regarding device emulators, SDKs, and development tools is also provided, to help you get started with your mobile solutions.

Chapter 9, "Persistent Data on the Client," explores one of the fundamental components of smart client applications: persistent data storage. This is the technology that allows you to maintain data on the device, removing the requirement for wireless network coverage. When it comes to how the data is stored, you have a variety of options from which to choose. You can use the device's file system to store data, build your own data storage mechanism, or purchase a commercial solution. This chapter evaluates all of these options, in addition to taking a closer look at the reasons why databases are an important component of smart client applications.

Chapter 10, "Enterprise Integration through Synchronization," provides information on the primary way in which smart client applications access enterprise data, using synchronization. This chapter covers the fundamental concepts involved in enterprise synchronization, including synchronization architectures and techniques. It also covers some of the synchronization technologies available commercially, and provides an overview of SyncML and where it fits into data synchronization.

Chapter 11, "Thin Client Overview," defines the thin client application architecture by highlighting the main components that comprise a successful solution. The overview is followed by a comparison of J2EE and .NET for server-side development. The chapter concludes with information on the leading wireless Internet protocol, Wireless Application Protocol (WAP), and the steps involved in processing a wireless Internet request.

Chapter 12, "Thin Client Development," explains how to start developing wireless Internet applications. It steps you through the various stages of the development process, starting with needs analysis phase and finishing with deployment options. As you move through this process, helpful hints are provided for avoiding common pitfalls of developing thin client applications. At the end of the chapter is a section on the common thin client application models, which outlines the target audiences, technical challenges, and types of solutions available for each application type.

Chapter 13, "Wireless Languages and Content-Generation Technologies," investigates the range of markup languages being used for wireless Internet applications, including HDML, WML, HTML, cHTML, and XHTML. For each markup language, sample code is given to demonstrate its syntax. The second part of the chapter delves into the various techniques that can be used to generate dynamic wireless content. This includes server-specific technology, such as CGI and ASP, as well as cross-platform technologies such as Java servlets, JSPs, and XML with XSL style sheets.

Chapter 14, "Wireless Internet Technology and Vendors," looks at the technologies commonly used when implementing wireless Internet applications. These technologies have been divided into four categories: microbrowsers, wireless application servers, development tools, and wireless service providers. For each category the key technology features are investigated, and a summary of related vendor solutions is provided. The goal is provide you with enough information to make educated decisions as to which technology and vendors you will want to evaluate further.

Chapter 15, "Voice Applications with VoiceXML" explains how voice applications are built using the Voice eXtensible Markup Language, VoiceXML. Unlike other applications discussed in this book, VoiceXML provides a voice interface into enterprise systems, rather than a visual one. The VoiceXML architecture is very similar to that of Internet applications, but the Web browser is replaced by a voice browser, and the handheld device is replaced by a telephone. Voice interfaces give true universal access to your applications. After the history of VoiceXML is examined, the VoiceXML architecture is discussed, followed by information on building VoiceXML applications.

Chapter 16, "Mobile Information Management," covers two separate but related technologies: personal information management (PIM) and mobile device management. PIM applications include email, calendars, task lists, address books, and memo pads. Access to these applications is often the reason why consumers purchase mobile devices. Mobile device management software can provide substantial benefits for both the deployment and management of software and devices. This chapter divides these topics into separate sections to focus on the capabilities that each solution provides.

Chapter 17, "Location-Based Services," examines location positioning technology, specifically addressing how location information can be used in both consumer and corporate applications and previewing the standardization efforts that are underway for location information. It also covers what, why, and when location-based solutions are relevant to mobile and wireless computing.

Chapter 18, "Other Useful Technologies," focuses on technologies that are just beginning to be adopted by mobile application developers. The four technologies covered include Mobile Web Services, BREW, SALT, and M-Services. All have been developed to improve upon previous technologies in the same market space and to become the standard in their respective fields. Of these technologies, Web services are clearly the leader in market acceptance and standardization. The others, BREW, SALT, and M-Services, are still working to achieve meaningful vendor and developer acceptance. The goal of this chapter is to introduce these technologies and explain how they relate to mobile computing.

Who Should Read This Book

If you or your organization is planning to build mobile and/or wireless applications, this book is for you. The content is appropriate for anyone who wants or needs to learn about mobile and wireless technology and how it applies to building successful applications. Though the book is technical in nature, it does not necessarily require you to have a strong technical background. If you are comfortable with general computing architectures and have some understanding of how applications are developed, then you will be able to gain valuable information from reading this book. More specifically, if you have in-depth development experience, and are already knowledgeable about mobile computing, the design and development chapters will prove to be beneficial, as will the chapters on related technologies such as location-based services and mobile information management. Many of my colleagues who are very proficient in mobile computing were able to garner new knowledge and ideas by reading this book.

Depending on your background, you can read this book in a number of different ways. Those of you who are new to mobile and wireless computing or who do not have a technical background will benefit most by reading the first four chapters in order. These will give you a solid understanding of the business aspects of mobile and wireless computing and of mobile devices, wireless networks, and mobile application architectures. You may then want to continue reading through each chapter sequentially, or you may want to jump to another section that is of interest to you. An entire part of the book is dedicated to each of the leading mobile application architectures; and the final part delves into technologies that are useful, but not necessarily required to build mobile applications.

For those of you who have experience with mobile devices and wireless networks, and who want to learn more about mobile application architecture design and development techniques, you may want to jump straight to Chapter 4, "Mobile Application Architectures," and go from there. After reading Chapter 4, you will have a solid understanding of your development options, enabling you to move on to the chapters that are relevant to the applications you are planning to develop. As you are reading, you can always reference the chapters on key technologies, such as wireless networks, mobile devices, and security, to refresh your knowledge of those topics.

However you read this book, it will prove to be a useful reference tool. Whether you read it from cover to cover or only read the chapters most pertinent to your project, you will always be able to go back and review content as required. Each chapter also contains a list of helpful Web links where you can go to get more in-depth information, updates on technology specifications, or another perspective on how the technology relates to your project. This book is intended as an all-in-one resource containing concise, yet insightful, information on all aspects of the mobile and wireless industry.

Where to Go from Here?

The mobile and wireless industry is in a very exciting phase in its evolution. It has moved beyond the marketing hype into the design and development of concrete enterprise solutions. This book provides an in-depth look at the technologies that are available to extend enterprise systems to mobile users.

Those of you who want an overview of mobile and wireless application terminology and business issues should start reading at Chapter 1. Those who want to dive right into the technology can start with an overview of mobile devices in Chapter 2 or wireless networks in Chapter 3. Regardless of what else you decide to read in this book, it is recommended you read Chapter 4, as it provides an overview of the leading mobile application architectures, with the advantages and disadvantages of each.

By reading this book, you will gain a solid understanding of mobile and wireless technologies that will help you to develop many successful mobile applications. Ultimately, it is these applications, not the hardware or software infrastructure itself, that will lead to the overall success of the mobile and wireless industry.

Introduction to the Mobile and Wireless Landscape

Mobile and wireless computing has the power to change the way business is conducted. It allows employees, partners, and customers to access corporate data from almost anywhere. Universal data access, combined with increased worker productivity and effectiveness, is driving the demand for enterprise mobile applications. As the demand continues to increase, the mobile infrastructure that makes creating sophisticated mobile applications possible is maturing. We have moved past the irrational exuberance that surrounded consumer wireless applications into the reality of creating advanced, integrated enterprise solutions that bring true value to enterprises that are adopting mobile and wireless technology as part of their core infrastructure.

Creating successful mobile and wireless applications requires a profound knowledge of various technologies, including network protocols, portable devices, application design and development, and security. In addition, enterprise managers need to understand the risks and rewards of introducing mobility into their organizations. The combination of business rationale and technical expertise can lead to the successful development and deployment of mobile and wireless solutions today!

The first part of this book is aimed at introducing the reader to the mobile and wireless landscape. It lays the foundation on which the rest of the book is based, covering mobile devices, wireless networks, mobile application architectures, mobile and wireless security, and messaging technology. Each of these topics is covered in its own chapter, as follows, to provide a concise introduction to mobile and wireless computing:

- Chapter 1, "Welcome to Mobile and Wireless"
- Chapter 2, "Mobile Devices"
- Chapter 3, "Wireless Networks"
- Chapter 4, "Mobile Application Architectures"
- Chapter 5, "Mobile and Wireless Messaging"
- Chapter 6, "Mobile and Wireless Security"

Welcome to Mobile and Wireless

The mobile and wireless industry is entering an exciting time. Demand for mobile technology is growing at a tremendous rate. Corporations are deploying mobile applications that provide substantial business benefits, and consumers are readily adopting mobile data applications. Exciting new mobile devices are constantly being introduced, and wireless networks are providing access to data from almost anywhere. In short, mobile and wireless technology has matured to the point where it is ready for wide-scale adoption.

Actually, mobile applications have been successfully deployed for many years, but these have been largely the effort of early adopters, who have gained a clear competitive advantage by implementing a mobile solution. Wide-scale deployment of mobile and wireless applications is just starting to take place. Corporations are just beginning to realize the benefits that mobility provides. Oddly enough, many of the advancements are taking place in the enterprise market, contrary to early industry hype, which was largely focused on consumers. There are both business and technical reasons why this is taking place.

Return on investment (ROI) has become an important figure for determining whether to implement a mobile solution. It is pretty simple really. If a company can foresee a reasonable return on its investment, it can justify building the application. If it cannot, then the application will have to wait. It is much easier to determine the ROI for internal business applications than it is for consumer applications, so they get implemented first. On the technical side, corporations have much more control over the technology being used, so they can build applications that require specific hardware, software, or network connectivity. This allows them to deploy a variety of application types that can run in either online or offline modes. There is no such control over the consumer market, often limiting the type of applications that can be deployed.

Before we get into the technical aspects of building mobile applications and discussing the various application architectures, we are going to look at some of the business issues surrounding the mobile industry. This chapter provides information on mobile and wireless from a less technical perspective. It focuses on the issues around implementing a mobile solution, including the business benefits and expected challenges, as well as the main mobility enablers.

Definition of Mobile and Wireless

The definition of mobile and wireless varies from person to person and organization to organization. In many cases, the terms mobile and wireless are used interchangeably, even though they are two different things. Let's start with the term mobile. Mobile is the ability to be on the move. A mobile device is anything that can be used on the move, ranging from laptops to mobile phones. As long as location is not fixed, it is considered mobile. Areas that are not included in our definition of mobile include remote offices, home offices, or home appliances. While these are definitely remote, they are not considered mobile.

Wireless refers to the transmission of voice and data over radio waves. It allows workers to communicate with enterprise data without requiring a physical connection to the network. Wireless devices include anything that uses a wireless network to either send or receive data. The wireless network itself can be accessed from mobile workers, as well as in fixed locations. Figure 1.1 depicts the relationship between mobile and wireless. As you can see, in most cases, wireless is a subset of mobile; but in many cases, an application can be mobile without being wireless.

For an application to be considered mobile or wireless, it must be tailored to the characteristics of the device that it runs on. Limited resources, low network bandwidth, and intermittent connectivity all factor into the proper design of these applications.

Wireless applications that are not mobile use fixed wireless networks. These are wireless networks that provide network access in a fixed environment. An example is a wireless local area network (WLAN) that is used to give desktops network access. Many businesses as well as home users are installing WLAN technology to avoid having to install network cables throughout their buildings. Another example is network access via satellites in remote locations where there are no other connectivity options.

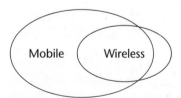

Figure 1.1   Relationship between mobile and wireless.

On the other side, we have mobile applications that are not wireless. There are many examples where this is the case. Any application that can be used on the move and that does not have wireless connectivity fits into this category. This includes many laptop and personal digital assistant (PDA) applications. Until only a few years ago, it was actually rare to have wireless data access for mobile devices. For these mobile applications, data is often synchronized using a fixed connection and stored on the device for use at a later time. It is worthwhile to note that even though these applications do not require wireless connectivity, they can often benefit from it when it is available. A sizeable portion of this book is dedicated to looking at these types of applications, which are referred to as smart client applications.

Now that we have defined mobile and wireless, it is time to look at some of the areas in which mobile applications are being deployed. Similar to the terms mobile and wireless, there is often confusion around the terms m-commerce and m-business.

m-Commerce

Mobile commerce, or m-commerce, refers to the purchase of products or services from a mobile terminal. Most of the early hype surrounding wireless technology was related to m-commerce. Companies envisioned consumers everywhere using mobile devices to purchase products. With the success of e-commerce for many types of purchases, it just made sense that m-commerce would be at least as successful. At the time, new wireless Internet technology was just coming to market, bringing with it enormous expectations. How could m-commerce not succeed? It is essentially the same as e-commerce without the constraints of a wired connection.

To meet these wild expectations, wireless operators paid billions of dollars for wireless spectrum licenses so they could upgrade their networks to meet the upcoming demand. For example, in Britain, the auctions for wireless spectrum raised $35 billion (U.S.). This equates to several hundred dollars from each person in the country. Other countries followed suit, selling their wireless spectrum for billions of dollars as well. But there is one problem: The wireless operators cannot cover these costs based on current revenue, and voice and wireless Internet access do not even come close to compensating for the expenditures. The idea was that m-commerce would help grow revenue for wireless carriers as well as the service providers. Unfortunately, as of early 2003, this has not yet come to fruition, and many operators are in serious financial trouble. The idea of making purchases from a constrained device is not that attractive when there are so many other avenues available to make the same transaction. Making the purchase at a physical store location, or even on a desktop Web browser is much more productive and enjoyable than using a wireless device, especially when the device is a mobile phone.

m-Commerce is best suited where the consumer has a sense of urgency, when they are required to have their goods or services immediately for upcoming functions and events. For product purchases, the sense of urgency is overcome by the fact that the consumer will still have to wait for the product to be delivered. In the near future, this limits m-commerce to products and services that can be obtained instantly, such as movie tickets or information services. We are still far from the time when consumers will be using their mobile devices to purchase appliances or to apply for a mortgage. Thankfully, there are some m-commerce offerings that show potential. The following solutions should help to drive the m-commerce market in the coming years:

Digital purchases. The most suitable purchase for a mobile user is for a product that can be downloaded and used immediately. The two biggest markets for digital applications are ringtones and games. Many carriers allow users to download new ringtones for their devices for a nominal fee. This provides users with a way to personalize their device. Another surefire success is mobile and wireless games. The advances made in mobile devices make them great vehicles for game playing.

Mobile banking. There are two benefits to mobile banking that a wireless device can provide. The first is providing access to personal bank accounts to view account history and execute transactions. This is an extension to Internet banking that has been very successful. The second is using a mobile device for payments, essentially acting as digital cash. A great deal of interest has been expressed about this area, so expect it to become a reality in the near future.

Information services. Although mobility has many advantages, mobile users often feel out of touch with their daily routines. Information services help address this by providing information that the user is accustomed to having, such as stock quotes, weather information, and sport scores. With the mobile messaging technology growing in popularity, many forms of information can actually be pushed to the user in the form of an alert or notification.

Location-based services. The ability for merchants to capture and react to a user's current location and requirements can be a powerful tool for selling services. Location-based services allow consumers to find the precise information they need at the exact time they want to use it. This will be an important enabler for m-commerce solutions, although privacy concerns will have to be addressed before location services are widely utilized.

Mobile shopping. Most forms of shopping are not going to be popular from mobile devices anytime soon. It is impractical to surf for items using constrained devices, making other methods of shopping much more productive and enjoyable. At the same time, there are some forms of purchases that lend themselves well to m-commerce. For example, having the ability to purchase movie tickets for a show playing the same evening can quite valuable. Mobile devices can also be used for comparison-shopping. Before making a purchase, a shopper in a retail store may want to first see what the current price of a product is from an Internet vendor to ensure they are getting a good price.

Mobile advertising. As mobile users start to take advantage of m-commerce solutions, mobile advertising is sure to follow. The mobile operator has access to several types of information that is attractive to advertisers, such as where users are located and what they use their mobile phone for. With this type of information, advertisers can send out personalized messages. The biggest obstacle to mobile advertising is customer backlash. If users start to get unsolicited messages and advertisements on their devices, they are likely to switch service providers, or worse, stop using their device. For this reason, in the near future, we will most likely only see requested advertisements, such as the nearest gas station or restaurant.

The lack of m-commerce acceptance can be blamed on both technical and business issues. On the technical side, the devices and networks do not have compelling features for consumers to use them for purchases. The interface on most phones is inadequate for any real data entry, and the wireless networks only recently have the capacity for the content that is associated with making a purchase. Grayscale images do not do a great job of selling a product. There are also security concerns that need to be addressed. Consumers are not convinced that e-commerce is secure, so confidence in m-commerce security still has a long way to go. On the business side, there are two major issues. The first is the lack of compelling applications. There are no killer m-commerce applications that are driving adoption. The second revolves around billing. The wireless industry has to address billing and pricing policies between the consumer and the company providing the service. Most consumers are not looking to establish billing relationships with every service provider they do business with. All of the billing has to be managed through a central source (most likely the wireless carrier) for adoption of new services to increase.

m-Commerce will eventually succeed. It is just taking longer than most people expected. Devices are steadily improving in both performance and usability, and wireless networks now offer communication speeds surpassing traditional desktop modems. At the same time, wireless operators continue to offer new products and services that are suitable for mobile users. Finally, as mobile devices start to be used as digital wallets, a new form of m-commerce will emerge. Devices have the potential to replace credit cards as the primary means of making purchases. So rather than using the wireless Internet to purchase a product that will be delivered at some point in the future, it can be used to pay for a product at a physical store location, just as you would currently use cash or debit cards.

m-Business

m-Business relates to e-business as m-commerce relates to e-commerce. Typically, m-business solutions are used by corporations to provide secure mobile access to enterprise data from any location. They allow employees to be more effective at their jobs. By having critical data at their fingertips, employees can respond more quickly to inquiries, remove inefficiencies brought about by manual data entry and retrieval, and make decisions based on current information. Most of the current demand for mobile and wireless technology surrounds the m-business market.

While consumers continue to purchase inexpensive mobile phones, business users are adopting high-end PDAs and laptops to conduct business. They require devices that have adequate power and functionality for advanced business applications. When wireless data access is required, business users are not deterred by the associated costs because they obtain justifiable benefits from implementing m-business applications. Corporations are able to readily see the benefits in terms of return on investment and employee productivity. In contrast to the hype around m-commerce, there is true growth and opportunity for revenue around m-business.

m-Business solutions are being deployed in several industries, using a wide variety of mobile and wireless technology ranging from public wireless networks to local area wireless networks. Devices range from mobile phones to PDAs to ruggedized tablets. The users of these applications include mobile business users, sales agents, and field service workers. Some of the leading m-business solutions being deployed today include:

Mobile office. Mobile office solutions integrate with enterprise groupware systems, such as Microsoft Exchange and Lotus Notes, to provide timely access to enterprise information such as email, calendars, task lists, and address books. Access to these applications has become a core requirement for many mobile professionals. (This is not to be confused with Microsoft Office productivity tools such as Word and Excel.)

Field sales. Current information is one of the most important assets for a sales force to have. The difference between making a sale or not often depends on having immediate access to information while you are still meeting with the customer. The best way to ensure the sales staff has this information is to deploy a mobile Sales Force Automation (SFA) application.

Field service. The nature of field service makes it an ideal candidate for mobile solutions. Service workers can improve their efficiency by having job-related information on hand, such as customer service history, inventory information, technical specifications, and data on repair procedures. In addition, information on customer issues and schedule changes can be sent to technicians while they are traveling.

Transportation and logistics. Providing delivery drivers with a mobile solution allows transportation and logistics businesses to have up-to-the minute information on deliveries, completed checkpoints, package statuses, and vehicle locations. In addition, delivery drivers can receive wireless dispatches and respond instantly, improving productivity and increasing customer satisfaction.

The applications just listed are only a sample of the numerous m-business solutions being developed and deployed today. There are literally hundreds of other mobile line-of-business solutions being used to help corporations become more successful. Mobile business applications are now starting to move beyond the early adopter stage to become a core component of a successful enterprise solution.

NOTE The information provided in this section is just an overview of what m-business solutions can provide. The focus of this book is primarily on the design and development of m-business solutions. The topics discussed do lend themselves well to m-commerce as well, but the intent is to provide information that is relevant to m-business.

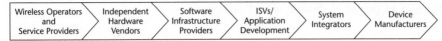

Figure 1.2   Wireless solution value chain.

Components of a Wireless Environment

Putting together a wireless solution requires products and services from several parties. The combination of all of these parties constructs the mobile value chain. Each "link" in the chain corresponds to one aspect of an end-to-end enterprise solution. Figure 1.2 shows how the components of a complete enterprise mobile solution relate to one another. No single vendor can provide all of these offerings on its own, but with the formation of strategic partnerships, many vendors are working toward providing an integrated end-to-end solution. For many development managers, this is preferable to having to piece together a solution on their own.

Wireless Operators and Service Providers

For any wireless solution, some form of wireless operator or service provider will be required. These are companies that provide the communications infrastructure for wireless wide area networks and/or wireless local area networks. Most of these providers are the same companies who provide wireless voice access for mobile phones. In some cases, other companies package the wireless communication component with other services such as wireless gateway software and client device software. Companies that provide such packaged offerings are referred to as mobile virtual network operators (MVNOs). In either case, it is important to select a wireless operator that can provide the required services in the geographic area in which your solution will be deployed.

Independent Hardware Vendors

In addition to the mobile devices, other hardware is often required for wireless solutions. This may include wireless modems, network cards, and wireless access points. Most of the laptops and PDAs available require additional hardware for wireless connectivity. This is usually in the form of a Compact Flash (CF) or PCMCIA card. This is starting to change with integrated wireless devices such as the Palm i705, Handspring Treo, RIM BlackBerry, and wireless-enabled laptops, but it will still be some time before this is standard on all devices.

Software Infrastructure Providers

In many aspects, the wireless software industry is just starting to mature. Many of the niche players have either fallen due to poor technology or business planning, or have been acquired by larger organizations. Leaders are emerging in all fields, including wireless application servers, mobile data storage and synchronization, wireless Web browsers, mobile office software, and device management. When implementing an m-business solution, you will want to select a software provider that has a proven track record and can demonstrate its offering in the environment you have selected. More details of software vendors are provided throughout the book as we cover the technology related to their specific offerings.

Independent Software Vendors (ISVs)

In addition to software infrastructure providers, independent software vendors, or ISVs, can provide industry solutions that require less work to deploy than building a solution from the ground up. Many of these solutions are either aimed at horizontal markets, such as field sales or field service, or at vertical industries, such as healthcare, finance, and transportation and logistics. If you are deploying a solution that is common to many other companies, it is likely that a turnkey solution is available from an ISV. Many of the software infrastructure providers will have lists of ISVs for various industry solutions if you are looking for a place to get started.

System Integrators (SIs)

System integrators, or SIs, are the companies that can help bring all of these components together. Many of the large SIs have practices focused on mobile enterprise solutions. If you are uncertain where to start, or can foresee that there will be difficulty implementing a mobile solution, using a system integrator can provide valuable benefits.

Device Manufacturers

Mobile devices are a requirement for a mobile solution. These devices range in size, functionality, and cost. It is important to choose the right device for the application being deployed. For m-business solutions, PDAs and laptops are more commonly used than mobile phones. This is because they offer more effective user interfaces, and can run more sophisticated applications than what is offered on mobile phones. This situation may start to change as smart phones (a combination of phone and PDA) become more popular in the business world. Chapter 2, "Mobile Devices," provides an overview of the range of mobile devices currently being used.

Keep in mind that not all of these components are required for every solution. It is entirely possible to utilize only a subset of the value chain and still produce the application that is right for the task at hand. For example, a mobile field service application can be created using only a software infrastructure provider, application developers, and mobile devices. If the solution is mobile, but not wireless, then the wireless operator is not required. Many solutions do not require additional hardware other than the device itself, and system integrators are required only when development and integration is not performed by in-house resources. As you start to plan your particular solution, you will quickly determine which components your solution will require.

As we continue through the remaining chapters of Part I, more information on these components will be provided. From there, the other parts of the book will provide everything that you need to know to design, develop, and deploy effective mobile and wireless enterprise solutions.

The Mobile Market Evolution

Every technology goes through many phases as it evolves. It starts with market hype, where the media and leading-edge companies set lofty expectations for the benefits that this new technology will provide. If the market hype is successful, early adopters will start to use the technology to attract new customers to their business, with the expectation of increased profits. These early adopters usually help to bring on the next phase—reality. This is when companies realize either that the market is not ready for the technology or that the technology is not ready for the market. In either case, the result is a period of apprehension and reevaluation. This leads to a lull in the market, giving both the technology providers and purchasers an opportunity to mature. The result is technology that is now ready for mass adoption, that is finally able to meet the potential that was widely discussed several years prior. Mobile and wireless technology fits this evolution to a tee:

Irrational exuberance. In the late 1990s, companies could not say enough about the benefits that mobility would offer. Industry analysts predicted that within a few years hundreds of millions of people would be using wireless phones for all kinds of tasks such as Internet banking, purchasing goods, or sending text messages to friends and colleagues. Wireless networks would be pervasive, allowing for instant data access from anywhere, at speeds approaching that of wired networks, and devices would be powerful enough for advanced multimedia applications. This phase for mobile and wireless did not last too long.

Market reality. By the year 2001, the reality of wireless applications became apparent. Technology such as Wireless Application Protocol (WAP) that promised to provide data access anywhere did not meet expectations. Wireless networks introduced new challenges such as limited coverage, low bandwidth, and high latency that proved to be difficult to overcome. Mobile devices did advance, but many of them were targeted at the consumer market, which turned out not to be quite ready for wireless data on a large scale. Improvements were required if the predictions were going to come true.

Mass adoption. After two years of technical advances and modest growth, we are finally reaching the stage of mass acceptance of mobile and wireless applications. The wireless networks have improved with increased bandwidth and always-on data access; mobile devices have evolved to be enterprise-ready; and corporations have had an opportunity to make concrete plans for the introduction of mobility into their organizations. Early adopters showed that tremendous benefits could be achieved by implementing mobile applications, and it is now the time to realize those benefits.

Why Go Mobile?

For most organizations, a time comes when a decision about mobile applications needs to be made. Why go mobile? What benefits will it provide to the organization, to the application users, and to their customers? Does it make financial sense to deploy mobile applications? These are just a few of the questions that organizations need to answer. Unfortunately, there is no single answer that relates to everyone. The answers to these questions will vary depending on the industry, application type, and the current infrastructure in place. For this reason, we will not attempt to provide concrete numbers on going mobile, but rather cover some of the general benefits that both businesses and end users can hope to achieve with a mobile solution.

We are at a time where mobile solutions can provide tremendous benefits to a business. The advances in devices, software infrastructure and wireless networks have made mobile solutions more practical than they ever have been. The cost of these solutions is going down, while the performance is going up, allowing the price/performance ratio to reach a level that is acceptable for most businesses.

Business Benefits

From a business perspective, the main reason to implement a mobile solution is to increase profit. There may be other side benefits, but when it comes right down to it, any expenditure has to have a positive impact on the bottom line to make it justifiable. The question then becomes, How can a corporation increase its profit by implementing a mobile solution? Profits can be increased in two ways: by increasing revenue or by reducing costs. This section explains how mobile solutions can aid in accomplishing each of these goals.

Increasing Revenue

A mobile solution can help to increase revenue in several ways. Most are realized by empowering employees to be more effective at their jobs. Examples include providing access to accurate information at the time it is needed and reducing inefficiencies brought about by dated business procedures. Some of the main advantages that mobile solutions provide to increase revenue include the following:

- Increased employee productivity
- Faster response times to business changes
- Streamlined business processes
- Improved customer satisfaction
- Increased competitive advantage

In positions such as field sales and field service, employees require access to many forms of enterprise data. A mobile solution allows them to access this data when it is required, without having to print out work orders, inventory lists, price charts, customer history records, and other similar pieces of information. For many workers, being able to access corporate information from a mobile device helps them avoid unnecessary visits to the office so they can spend more time working with customers.

Having accurate data on hand also makes them more effective at their given position. They can respond to change in the business environment much more quickly, allowing them to capitalize on new opportunities. With the ability to add, alter, or remove jobs throughout the day, field service workers can quickly and easily fill in any gaps in their work schedule, preventing costly downtime. In addition, once a job is complete, the required billing information will automatically be updated in the enterprise system without having to manually enter the data. This allows companies not only to respond to a customer's needs faster, but also to collect for services rendered in a timely fashion. Customer satisfaction will increase, leading to repeat business.

Having a way to measure the benefits the solution provides is very important. For a sales representative, this may include looking at the average revenue per salesperson, measuring the time it takes to close an opportunity, or counting the number of upsells that were executed. For field service workers, it may include counting the number of visits required to fix a problem, measuring customer downtime, and looking at the increase in the number of service calls that are made with the new solution in place. Once these metrics are calculated, they need to be compared with the costs associated with the solution to determine the overall benefit.

Reducing Costs

In many cases, the cost reductions obtained by implementing a mobile solution are more substantial than the increased revenue. Most paper-based systems are very inefficient, with much of the time being spent manually transferring data from printed forms into electronic systems. This introduces an opportunity for errors and adds an extra layer to the process that can easily be avoided. Mobile solutions allow for data to be captured electronically where it is first acquired. This leads to more detailed data with higher accuracy than paper forms. The data is then directly integrated into the appropriate system for processing. The result is quicker turnaround times on orders, shorter billing cycles, more accurate time and expense reporting, reduced reject rates due to errors, and overall more efficient systems that do not require redundant data entry.

Other areas for cost reduction include reducing the number of printed materials. For many businesses, the cost of printed materials such as manuals or schematic diagrams alone is quite high, especially if they have to be reprinted often. Being able to replace these materials with an electronic system can provide tremendous cost savings with little effort. There are also opportunities to manage the supply chain more efficiently, allowing companies to reduce required inventory levels while providing more timely service to customers.

End-User Benefits

In addition to the business benefits, there are also advantages for the end user. Rather than being confined to the office, mobile solutions provide flexibility to work in whatever surroundings are convenient. This is particularly important to mobile professionals who are frequently traveling for business purposes. While traveling, they often have extended periods of downtime waiting in airports, between meetings, or when in a hotel. During these periods, remote access to enterprise applications can provide a tremendous benefit. Having the ability to do time and expense tracking, enter sales orders, or check corporate email allows them to get back to their families and friends on their return home without having to visit the office. For employees who travel a lot, this is a great way to help them balance both work and family priorities.

Challenges

As companies look to reap the benefits of providing mobile enterprise applications, they are sure to come across some of the many challenges that are commonplace in the current state of mobile computing. These challenges do not reflect any particular vendor solution, but rather are inherent in mobile computing today. By understanding these challenges, companies will be able to make better decisions as to which devices, networks, and mobile architectures are best suited for their particular solution.

Oddly enough, many of the challenges are raised by the very technologies that enable mobile solutions: mobile devices, wireless networks, and software infrastructure. In a way this does make sense, since these technologies represent new methods of computing, introducing complexities that development teams may not have experience with. Actually, as the challenges are investigated in more depth, it is clear that many of them represent the relative immaturity of the wireless industry, and will be minimized as the market continues to evolve.

Wireless Network Issues

Even though wireless access is not a requirement for a mobile solution, it is often considered one of the primary challenges of implementing mobile applications. This is because it affects all aspects of the solution, including design, development, and deployment. Even when companies are building an application that will not initially take advantage of wireless connectivity, they usually want to plan for its introduction in a later phase of the application's life cycle. Wireless network issues have to be incorporated into the preliminary planning stages so future adoption is possible. The following are some of the issues with wireless networks that have to be considered when a solution is being developed:

Coverage and penetration. Wireless coverage is not pervasive. Wireless operators typically provide coverage in areas of higher population density, surrounding a particular geography. Even when the physical geography has network coverage, there is no guarantee that the signal can penetrate physical barriers such as buildings or tunnels. These interruptions can wreak havoc on applications that are not designed to handle such circumstances.

Bandwidth. The speed of wireless connections can differ dramatically depending on the wireless protocol being used. Some of the early wireless networks have very slow data transfer rates, around 9.6 kilobits per second (Kbps), while more modern networks may surpass 56 Kbps.

Latency. Sometimes even more important than bandwidth is network latency. Latency is defined as the time it takes for a request to reach the target server. Many users blame poor performance on bandwidth, when it is really the latency in the network that is taking the majority of the connection time. For every round-trip to the enterprise server, network latency can add between 1 and 30 seconds to the overall connection time. The amount of traffic affects latency, so during peak periods of usage, latency will be the highest.

Reliability. Dropped connections are common among wireless networks. This can lead to usability issues for certain types of applications. Many of the reliability issues are related to coverage and penetration of the network.

Cost. The cost of wireless connectivity can sometimes be prohibitive. Some networks charge by connection times, while others charge based on the amount of data transferred.

Lack of standards. Each network provider maintains different network protocols for connecting to wireless devices. Some of these are IP-based, while others are not. This issue is more profound in North America and Asia-Pacific where the leading network providers are implementing several competing protocols. In Europe, there is more of a consensus on network protocols, leading to less complexity.

These wireless network issues have to be addressed on all levels of application design. The mobile device will work only with specific wireless networks, so you have to consider the wireless network during device selection phase. Network issues also affect the type of application that should be deployed. It is important to design the mobile application with wireless challenges in mind. Some mobile architectures are better suited for today's networks than others. (Thorough information on wireless networks is provided in Chapter 3, "Wireless Networks," and a summary of the mobile application architectures is provided in Chapter 4, "Mobile Application Architectures.")

Mobile Device Diversity

The type of device selected is paramount to the success of the mobile application. It is critical that the devices match the types of application being developed. So before selecting which device is most appropriate, you have to ask yourself the following questions:

- What type of data access is required? Is wireless required? If so, is an integrated modem required, or is an add-on component suitable?

- What type of application is going to be developed? Is it a smart client business application, email and messaging, or a wireless Internet application?

- How does the device connect to the enterprise? Does it use wireless connections or is it cradle-based?

- Which mobile operating systems do we want to use? On which operating systems do we have experience developing?

- Who is going to use this device? Does it have to be ruggedized to withstand a little punishment or is regular protection adequate?

- How long does the device have to last without recharging? Does the user have the ability to charge it periodically during the day, once daily, or only every few days?

- What form factor does this device need to have? What type of data input is required? Is a mobile phone suitable or is a PDA or laptop required?

- What cost is acceptable?

These are just a few of the questions that need to be addressed before being able to choose the correct device(s). Once a device (or set of devices) has been selected, the development team will have to address some challenges. These include selecting the type of application that works with the mobile operating system employed by the devices, being able to integrate with wireless gateways, learning the APIs for the messaging technology being used, and designing the application to take advantage of the limited screen size and input mechanisms on the device. Very often this requires some degree of knowledge beyond what many developers currently possess.

Sometimes, however, it may not seem like you are choosing a device as much as it is chosen for you. If your company already has equipped your sales force with a certain device, or if the geographic area you are targeting has an extremely limited wireless characteristic, you may feel as if you have no choice. With m-commerce solutions, when a company identifies and targets an end-user group, it often will not have the ability to choose which device that group will use. For instance, if a bank wants to target a certain clientele to provide special mobile services to that group, it will have to determine an application that best fits within the broad range of devices that its customer base is already using.

One final challenge with the device diversity is the management of all of these devices and the applications on the devices. Organizations need to define a device management strategy early in the deployment of mobile solutions before the mobile environment gets too chaotic. If multiple mobile operating systems are supported, a solution needs to be found that provides a single environment for managing all platforms you are deploying.

Software Infrastructure Choices

With the vast number of mobile devices and network connectivity options, it is important to select a software platform that will work effectively with other parts of your solution. This may involve having the ability to create wireless Internet applications using a variety of markup languages, having a persistent datastore with synchronization capabilities, or having mobile messaging services.

Based on the application being developed and which devices it is going to be deployed to, there may be a preference for the server platform that is being used. Is a Java 2 Enterprise Edition (J2EE) platform best suited for your environment? Or is a .NET solution from Microsoft a better choice? For the application on the device, does it have to have offline capabilities or is a WAP online solution acceptable? Each platform has its strengths and weaknesses. You need to find one whose strengths meet your core requirements. This task can prove to be difficult, as mobile and wireless software is constantly changing to meet the growing demands for mobile solutions and to adhere to emerging mobile standards.

Some other concerns that have to be addressed include the following:

- Mobile and wireless security
- Introduction of new services, such as location-based services
- Support of the communication protocols being used
- Integration with existing enterprise applications and data
- Support for current, and possibly future, device selections that may be made
- Development team expertise

All of these aspects of a mobile platform have to be considered when implementing a mobile solution. Additionally, it is a recommended that you anticipate future enhancements you may want in your solutions and adopt a platform that is moving in a similar direction. This may initially seem to be a daunting task, but it is not as difficult as it appears. Mobile and wireless technologies are currently converging on a defined set of standards that are being adopted by network operators, device manufacturers, and software infrastructure vendors alike. This will not happen overnight, but as the mobile and wireless market matures, it will continue to move in the right direction.

Mobility Enablers

Four main market enablers are driving the adoption of mobile and wireless solutions: wireless networks, mobile devices, software infrastructure, and standardization. Without advancements in all of these categories, building mobile enterprise solutions would not be practical. It is no coincidence that mobile and wireless is growing in acceptance just as these technologies and standards are starting to mature.

The previous section covered some of the challenges introduced by these technologies. This section looks at the features these technologies offer and the trends that are emerging to enable the development of robust mobile solutions.

Wireless Networks

As with other mobile and wireless technologies, the initial marketing focus for wireless carriers was aimed at the consumer market. The emphasis was on inexpensive handsets and packaged calling plans to meet the general consumer's needs. This focus was successful in many regards, as mobile phone usage and device sales have skyrocketed. The problem was that the average consumer does not generate much revenue for wireless carriers, and the revenue it does generate is not high-margin. Wireless data usage was considered to be the answer carriers were looking for, but once again, the consumer market did not provide the revenue or profit that was required. Since then, much of the focus of wireless data services has moved to the corporate market.

In an effort to provide attractive functionality for enterprise users, wireless network operators have spent significant resources to upgrade their networks. The first phase of the upgrade was adding data services in conjunction with their voice offerings. The second phase was making these data services faster and more reliable. Actually, network operators are implementing networks with features that will help to overcome many of the problems that were discussed previously in this chapter. The upgrade is not yet complete, but early indications show that the wireless network improvements will have a meaningful impact on mobile and wireless solution deployments. The following are some of the most significant features:

Increased bandwidth. Many of the voice-oriented wireless networks provide very limited wireless data capabilities. This is no longer the case with the next-generation wireless networks being deployed today. These networks are designed for wireless data, and provide communication speeds between 56 Kbps and 384 Kbps. This speed is very adequate for the limited amount of data that is typically required from a wireless device. That said, some third-generation wireless networks are boasting speeds approaching 2 Mbps, allowing for advanced multimedia applications.

Always-on capabilities. The term always-on refers to the users' ability to access data at any time, without having to establish a connection to the wireless network for each session. This ability has been enabled by packet-switched networks. Unlike circuit-switched networks that bill users on the amount of time they are connected, packet-switched networks charge based on the amount of data transferred. This allows users to stay connected to the network at all times without having to be worried about huge wireless charges. (Chapter 3, "Wireless Networks," has more information on circuit-switched and packet-switched networks.)

Lower costs. The introduction of packet-based networks allows wireless operators to offer new types of wireless packages that are based on data usage instead of call times. Many flat-rate packages are now available. These come in various sizes, some offering unlimited data access for a fixed fee. This has become very popular for many corporations as it allows them to know exactly how much the service will cost, making budgeting much easier. Another factor is roaming charges. Many carriers have drastically reduced, if not totally eliminated, roaming charges for using other carriers' services. Historically, these charges have been quite substantial, often limiting the practicality of some mobile solutions.

Enhanced services. New services are being offered that can add value to mobile solutions. Many carriers now allow users to download additional applications for their devices, providing a wide range of new capabilities. In addition, some integrated services, such as location-based services, are providing companies with valuable features that can impact the success of custom mobile solutions.

Interoperability between carriers. Wireless operators are starting to work together to help promote the benefits of mobility. This cooperation has resulted in a new level of interoperability for both data and voice communication. An example is the ability for users in North America to send text messages to users on other networks, a feature made available in 2002.

Another trend is the introduction of public wireless LANs (PWLANs) to targeted high-traffic areas such as airports, hotels, and conference centers. PWLANs provide high-speed speed data access at a very low cost, and can either augment a wireless wide area network (WAN), or be deployed on their own. These features, along with additional network coverage and new handsets, are an important piece of building wireless enterprise solutions. (Chapter 3 provides complete details on the range of wireless networks and protocols being used.)

Mobile Devices

The advancements being made in mobile devices are truly incredible, and the selection is astounding. A few years ago the choice was between a wireless phone and a simple PDA. Now there is a long list of options ranging from high-end PDAs with integrated wireless modems down to small phones with wireless Web-browsing capabilities. Even the simplest devices provide enough computing power to run small applications, play games, and of course make voice calls.

The proliferation of devices in the enterprise is a key driver for the growth of mobile solutions. As more personal devices find their way into the enterprise, corporations are realizing the benefits that can be achieved with mobile solutions. The trend is for smaller devices and more processing power. A device that fits in your hand has as much computing power as desktops did less than 10 years ago. The same device is also able to communicate over a wireless network and view office documents at the same time. This combination of size, power, and flexibility is a key enabling technology for enterprise mobile solutions. Other key advancements include increased storage space, high-resolution displays, built-in wireless support, reduced energy consumption and increased battery capacity, improved ergonomics, and support for peripherals such as wireless modems, digital cameras, Global Positioning System (GPS) units, portable keyboards, and large-capacity storage.

With the device landscape changing so rapidly, it is recommended that organizations select devices based on immediate business need, not for future intent. Features that are attractive today may not be deemed important for future applications. The device market is currently fragmented, with many specialty-purpose devices targeted at specific sets of users. It is expected that this will continue for the foreseeable future as integrated voice and data devices are not yet experiencing widespread acceptance in the enterprise. (For more details on the range of devices on the market, see Chapter 2, "Mobile Devices.")

Software Infrastructure

Along with wireless networks and mobile devices, a strong software platform is also required to develop and deploy robust mobile solutions. These platforms offer support for the leading mobile computing models, including wireless Internet, smart client, and mobile messaging architectures. In many cases, these platform vendors have partnered with other mobile infrastructure providers to supply an end-to-end solution. This helps ease the process of integrating disparate technologies.

Software vendors have made significant advances in several key areas that enable mobile application development. For wireless Internet applications, the introduction of wireless application server frameworks, device emulators, WAP, and development tools has had an impact. For smart client applications, the advent of small mobile databases, advanced synchronization technology, and comprehensive mobile operating systems has allowed many m-business solutions to be possible. In addition, new technologies from Microsoft in the form of .NET Compact Framework, and Java community in the form of the Java 2 Micro Edition (J2ME), have provided developers with the tools they need to implement these solutions.

We do not want to forget about the advances being made in the areas of messaging, security, enterprise integration, mobile device management, and location services. All of these technologies are key in successful development and deployment of mobile solutions.

NOTE This section is intentionally brief, since most of the remainder of this book deals with the design and development of mobile applications using mobile software infrastructure. The first introduction to the mobile application architectures is provided in Chapter 4, "Mobile Application Architectures." These architectures are then explained in more depth in Chapter 5, "Mobile and Wireless Messaging," and in Parts II and III of the book.

Standardization

Many companies have cited the lack of standards as a major obstacle in developing mobile solutions. They are worried that the technology they use will become obsolete before their application is even deployed. This is a valid concern. Having an established standard provides some comfort that their applications will work now and in the future. As with any new industry, it takes time for standards to develop. Fortunately, we are now at the point where standards are emerging at all levels of the mobile environment.

The leading wireless standard is WAP, which defines technologies for developing and deploying wireless Internet applications. The early incarnations of WAP received mixed reactions, partially due to the standard itself, and partly due to the state of the wireless industry. WAP 2.0 has addressed many of these concerns by working much better with common Internet standards.

For markup languages, the eXtensible Hypertext Markup Language (XHTML) is emerging as the language of choice. Many of the existing markup languages such as Wireless Markup Language (WML), Compact HTML (cHTML), and HTML itself are all converging on XHTML. To accommodate differences in client browser capabilities, different profiles of XHTML are targeted at wireless devices. Another markup language that has reached widespread adoption is VoiceXML. VoiceXML is a markup language for the creation of interactive voice applications.

For the development of client-side applications, two technologies are competing for the developers' attention: J2ME and Binary Runtime Environment for Wireless (BREW). BREW is targeted at smart phones, while J2ME spans from smart phones to high-end PDAs. Application developers can feel confident that these standards will provide a stable application environment for some time to come. Although not officially a standard, Microsoft's .NET Compact Framework (.NET CF) will help to drive mobile development for Windows CE-based devices. Similar to J2ME, .NET CF is a viable development platform that is unlikely to fade away. It is also worthwhile to note that the other common development tools based on C/C++ and Visual Basic are good choices for smart client application development.

For synchronization, SyncML is a leading candidate for standardization. The SyncML Initiative now part of the Open Mobile Alliance (OMA) is actively working on synchronization specifications for a variety of data types, including personal information such as calendar and address books, corporate business data, and mobile device management. (More information on SyncML can be found in Chapter 10, "Enterprise Integration through Synchronization," and Chapter 16, "Mobile Information Management.") Finally we have the wireless networks. While there is no agreed-upon standard for wide area wireless, underlying support for the Internet Protocol (IP) is an emerging trend. This is a dramatic improvement over the many proprietary wireless protocols that have been used in the past. For wireless LANs, 802.11b and 802.11a are the leading standards being deployed. Both of these allow for interoperability between both hardware and software products. Bluetooth provides similar support and interoperability for personal area networks (PANs).

Although standards help to promote interoperability and stability, they are not mandatory for creating robust mobile solutions. Many products on the market do not adhere to the standards mentioned previously; nevertheless, they still provide highly functional, stable platforms for building advanced m-business applications.

Summary

Corporations often evaluate mobile solutions by their return on investment (ROI). ROI measures the gains that can be expected in return for the expenditures. This is a very effective approach to determining whether a mobile solution is worthwhile. Many vendors are starting to realize that ROI is an important measurement for success and are providing this information to perspective customers along with the technical merits of their solution. This is helping m-business applications grow in adoption, which helps other mobile and wireless solutions succeed as well. Wireless carriers will continue to increase the capabilities of their networks and provide new services, and device manufacturers will continue to deliver innovative offerings that will be used by both consumers and business users alike.

Many challenges need to be overcome when implementing a mobile solution. Many of these challenges have been introduced by mobility-enabling technology such as mobile devices, wireless networks, and software infrastructure providers. As the industry continues to mature, these challenges will become less significant, and the opportunities mobility provides will become more meaningful. As an industry, we are well on our way to accomplishing the three requirements for substantial industry growth to occur:

- The availability of compelling applications
- The availability of suitable mobile devices
- The introduction of always-on, easy-to-use wireless networks

In the next chapter we will take an in-depth look at mobile devices and their impact on implementing a successful mobile solution.

Helpful Links

Since most of the technical content from this chapter is expanded in upcoming chapters, the links provided in this chapter are for industry analyst Web sites. These sites provide technical and business research and analysis on the mobile and wireless marketplace.

Gartner Group	www.gartner.com
Giga Information Group	www.gigaweb.com
IDC	www.idc.com
META Group	www.metagroup.com
Summit Strategies	www.summitstrat.com
Yankee Group	www.yankeegroup.com

CHAPTER 2

Mobile Devices

A wide variety of mobile devices are available to address a broad range of applications and users. They range from very inexpensive Web-enabled devices to high-end customized tablets, with laptops, a variety of PDAs, and smart phones in between. Along with size differences come variations in the features and performance that these devices provide. No matter which type of mobile application you are looking to deploy, a device is available that will meet your needs.

This chapter provides an overview of the mobile device market, with an emphasis on the devices that are most appropriate for m-business solutions. For each device category we will look at the leading features, such as screen size, data input mechanisms, wireless support, and storage space. A table summarizing the device manufacturers and the class(es) of devices they sell is also provided for quick reference. This information should be helpful when deciding which device classes are most appropriate for a particular application and which manufacturers produce these devices.

Device Overview

Of all the components in a wireless solution, the mobile device gets the most attention. It is the only part of a mobile solution that the end user is in contact with. When a problem arises, some aspect of the mobile device is often held accountable. This may be the wireless connectivity, mobile operating system, input mechanisms, or performance characteristics. In any case, the application developer absolutely must be intimately familiar with the device or devices being used in order to take advantage of their unique capabilities.

Mobile devices come in all sizes with a wide range of features. Choosing the correct device involves evaluating a variety of criteria other than just cost. Two of the major factors that need to be considered are the data input mechanism and wireless connectivity options. Each of these is described in detail later in this section. Some of the other major areas that should be given special attention include the following:

- Device size and weight.

- Available memory for applications and data.

- Processor speed: This will affect the types of applications that will be able to run on the device.

- Screen characteristics such as size, color depth, indoor and outdoor suitability.

- Mobile operating system support: Can it be upgraded when a new version of the operating system is released?

- Expansion slots for adding peripherals such as more memory, wireless modems, GPS receivers, or digital cameras. Looking into what peripherals are available is also a good idea.

- Battery life (this is a big one).

- Integrated features such as digital cameras, keyboards, infrared, and Bluetooth.

- Software support, including third-party applications, development tools, mobile browsers, hardware drivers, as well as others that may be applicable to your particular situation.

These cover some of the major factors that should be considered, but there are bound to be others specific to the solution being implemented. When you are researching mobile devices, the vendor Web sites are a great place to look first, but keep in mind the information provided will not be very objective. Many other information sites are aimed at developers and business managers alike that can help make the device selection process more effective. The overview of the device categories later in this chapter will also help you become familiar with which devices are available for the enterprise market.

Input Mechanisms

Along with device size and power, the data input mechanism is one of the most important aspects of selecting a mobile device. This does depend on the application, and the levels of user interaction required, but for most m-business applications where a substantial amount of data is entered into the application, data input options have to be given strong consideration.

Mobile data devices receive their properties from two areas: mobile phones and desktop PCs. Both of these have specific design goals and constraints. Mobile phones are designed to be efficient and simple, allowing for single-handed operation by inexperienced users. PCs are designed for power and richness. Mobile data devices have to compromise between the two to provide various levels of functionality and richness. When it comes to methods of data input, several options are available.

Keypad Input

Mobile phones typically implement a 12-digit keypad. This is very intuitive and effective for numerical input but cumbersome and awkward for entering text. This is because each key on a keypad represents three characters. Users have to press a key multiple times to enter a single character. Since mobile phones are primarily designed for voice usage, it is unlikely that the keypad will change anytime soon. That said, there are technologies available that make entering text easier. One of these is called T9, which stands for Text on 9 Keys. This is a predictive text input technology that allows a user to input words by pressing only one button per letter. T9 translates the button press sequences into words. It does not work in all cases, but when it does, it is about twice as fast as using traditional multi-tap systems. America Online, the owner of T9, estimates that it is accurate 95 percent of the time.

Device manufacturers are constantly looking for ways to make keypad text entry more effective. Toggle sticks, rollers, and touch screens will help to alleviate this problem in some areas, but do not expect fast data entry to be a reality on a keypad-driven device anytime soon.

Pen-Based Input

One of the breakthroughs for handheld devices was the introduction of touch screens for pen-based input. This allows a user to enter data using a stylus, without requiring any form of physical keyboard or keypad. Several pen-based input mechanisms are available. Deciding which ones to implement will depend on the mobile operating system and device being used. The most common types of pen-based input are:

Soft keyboards. The most approachable method of entering text on mobile devices is by using a soft keyboard, a keyboard displayed on the screen. Users enter data by pressing "keys," just as they would on a physical keyboard. This method is very easy to learn and use, but it is also takes up screen space and can be limiting in terms of input speed. It is often used as a backup mechanism when other forms of data input are not effective.

Character recognition. Character recognition works by interpreting which letters a user is entering on the screen. There is usually a specific area on the screen for users to construct the characters. The Windows CE operating system from Microsoft provides this capability. Software runs locally on the device to do the interpretation. The letters have to be constructed fairly accurately for them to be understood. This is usually not a problem for small amounts of data, but can be a challenge for users who need to enter data quickly. They often find the characters difficult to write, leading to mistakes.

Graffiti. Palm OS introduced a new form of character recognition based up on simple character-set called Graffiti. The Graffiti characters allow for quicker data input since the letters are easy to input and recognize. There is a learning curve involved, but once Graffiti is mastered, data can be input quickly with few errors.

Handwriting recognition. Some forms of handwriting recognition work by taking screen captures of the inputted text, while others actually interpret written characters. The interpretation solutions take character recognition to a new level. Instead of being able to interpret only a defined set of characters with predetermined strokes, handwriting recognition attempts to interpret a user's personal style of text entry. This requires sophisticated software and powerful hardware to do the analysis. Microsoft is implementing forms of handwriting recognition on its Tablet PC platform.

Keyboard Input

Even with the advances being made in other forms of data input, the keyboard remains the most efficient and easy to use (especially for Western languages such as English). Mobile devices have many ways to take advantage of keyboard data entry. Laptops and handheld PCs come equipped with physical keyboards, while the smaller palm-sized PDAs often support keyboards as a peripheral. These keyboards are usually attached as a clip-on unit or via a cable, but wireless connections are also an option. In addition, most PDAs have PC companion software, allowing the user to enter data such as contact information on the PC. Some handheld devices have taken the step of incorporating a keyboard into the unit itself. This is accomplished with small thumb-based keyboards at the bottom of the unit. Research in Motion (RIM) was the first to incorporate this concept into its BlackBerry devices. With its revolutionary design, RIM has attracted a wide variety of users to its devices, prompting other companies, including Handspring and Sharp, to adopt the thumb-sized keyboard as the means for data input.

Voice Input

Voice input is both effective and easy to use. Simple voice commands, such as initiating a call or looking up a contact, can be executed directly on the mobile device. This works by prerecording voice commands into the device. When a voice command is entered, it is compared to the recorded message and executes the corresponding action. Many mobile phones come equipped with this feature.

For more advanced voice applications, such as looking up product information or executing a bank transaction, the voice recognition software resides on a server called a voice gateway. These gateways are run on powerful machines that are capable of running automatic speech recognition and text-to-speech software. This software is capable of understanding complex commands—such as "What is the weather going to be like on Friday in Waterloo?"—and returning the appropriate spoken response. Many of these applications use proprietary Interactive Voice Response (IVR) systems or VoiceXML, an emerging voice standard. (If you are interested in learning more about VoiceXML, see Chapter 15, "Voice Applications with VoiceXML.")

Wireless Communication

Three basic connection options are available for obtaining wireless communication from a wireless device: two-unit, detachable, and integrated.

Two-Unit Configuration

Two-unit connections require two pieces of equipment to work together, such as a PDA and a cell phone, as shown in Figure 2.1. One unit provides the wireless connectivity for the other unit to use. These devices can communicate with each other in a variety of ways:

Cable connection. This involves having a physical cable connecting the cell phone and mobile device. The phone must have wireless data support as well as an interface cable to the device of choice.

Infrared connection. This involves lining up the Infrared ports on the cell phone and mobile device for communication. A direct line of sight is required for this solution to work. In addition, only selected phones have infrared support available.

Bluetooth connection. Bluetooth may be the best option for two-unit connectivity. It allows a mobile phone to provide connectivity through a personal area network (PAN) up to a range of about 10 meters. A direct line of sight is not required, so it is possible to have the cell phone in a different location from the mobile device. This will become an increasingly popular option as more Bluetooth-enabled devices are released.

The advantage of the two-unit configuration is that you can choose each device based on its own functionality, and you do not have to sacrifice features for wireless capabilities. The configuration also provides a great degree of flexibility. If one device becomes outdated or malfunctions, it can be replaced without having to replace the entire system. This is especially appealing when the data device is a laptop or PDA that may not need to be replaced as frequently. In addition, the cell phone can potentially provide the wireless communication for several other devices.

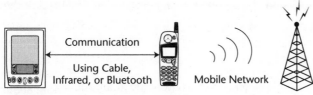

Figure 2.1 Two-unit configuration.

The downside to this configuration is its complexity. Both the cell phone and the mobile device have to be configured properly for this to work. When Bluetooth is not available, the cable or Infrared connections can become cumbersome.

This setup is recommended only where occasional wireless connectivity is required. When frequent access is necessary, it is worthwhile to investigate some of the other options available.

Detachable Configuration

A detachable configuration involves using a plug-in module or clip-on attachment to a mobile device to provide wireless connectivity. These additional modules can provide connectivity to a variety of wireless networks with little configuration. For the plug-in modules, either CompactFlash or PCMCIA cards can be used. This obviously requires a device that has a slot for this type of peripheral. For devices that do not have integrated CompactFlash or PCMCIA slots, external jackets are often available. These jackets (often called sleds) provide the support for CompactFlash or PCMCIA cards.

NOTE Some of the wireless modems on the market also include support for voice calls. An example of this is the Sierra AirCard 555, which provides CDMA2000 1x wireless data as well as voice capabilities. The Sierra Wireless Web site can be found at www.sierrawireless.com.

The benefit of using these interfaces is that multiple wireless modems are available, providing connectivity to several wireless networks including WLANs, WANs, and PANs. You can then switch interface cards as you move into different wireless environments. Cards are also available that support connecting to multiple network protocols in one unit.

In some cases, the only available option is a clip-on modem. This is similar to the external jacket, except that the modem is incorporated directly to the unit. These modules are often purchased in conjunction with a wireless service plan since they only provide access to a single wireless protocol, most commonly a wireless WAN.

The majority of the PDAs available now use a detachable configuration for wireless connectivity. It allows users to select the PDA and wireless component separately, but at the same time have them integrate well together. The disadvantage is that the wireless modem uses the open expansion slot so other peripherals cannot be used at the same time.

Integrated Configuration

An emerging trend is to enclose wireless connectivity within the mobile device. This has always been the case for voice-oriented devices such as cell phones and smart phones, but is a new concept for many handheld devices. This configuration has many advantages, as it alleviates complexity and provides tight integration between the mobile OS and wireless modem. Applications for these devices can be designed to take advantage of the wireless modem, knowing that it is always going to be present. Troubleshooting is also simplified, since there is only one manufacturer involved in the solution.

On the downside, some flexibility is lost. The user is now limited to the wireless network type that has been integrated into the device. In addition, it may become more difficult for developers to calculate the amount of power an application requires, as they will have a hard time differentiating between the power that the application requires and that of the wireless modem.

For users who will be using wireless communication frequently or who are looking for a very intuitive, user-friendly experience, integrated configurations are ideal. Devices are available for a variety of wireless networks. Many ruggedized devices with integrated wireless LAN access have been on the market for some time, while integrated public WAN devices are just now being deployed on a large scale. As devices continue to evolve, it may become practical to carry a single device for both data and voice communication.

Mobile Device Classifications

Much has changed since 1996 when the PalmPilot was first released. Even though previously there were other palm-sized devices, such as the Apple Newton, the PalmPilot changed the way we look at mobility. Users now had the option of using a small, palm-sized device to store their schedules, calendars, to-do lists, as well as perform other simple applications. This was clearly an option that users liked, signaled by the phenomenal pace of adoption of Palm devices. By 2000, the vast majority of all palm-sized devices were based on Palm operating system (Palm OS).

Because of the success of the Palm devices, many other companies released mobile device offerings in an attempt to get a slice of this burgeoning market. As these new companies entered the market, they came out with new devices, with new features. In this section we will look at each of the major device categories. As depicted in Figure 2.2, a distinct relationship exists between number of units sold and price: As the cost rises, fewer devices are sold. Devices costs vary anywhere from under $100 to several thousand, depending on the features required.

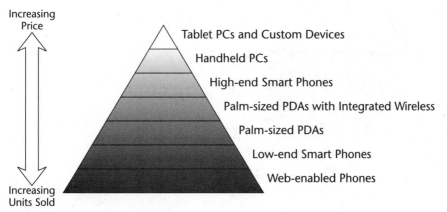

Figure 2.2   Device classifications.

Web-Enabled Phones

Cellular phones are by far the most widely used wireless devices on the market. In some European countries, over 75 percent of the population owns a cellular phone. In most cases, the primary use is voice, but with the introduction of WAP, text messaging, and other wireless Internet technologies, data applications have become much more common. With their widespread popularity, Web-enabled phones become an immediate target for wireless Internet applications.

Figure 2.3 shows a typical Web-enabled phone. As you can see, it provides a very limited display (typically between 4 to 12 lines of text), with the regular 12-button phone keypad for data entry. These limitations make cellular phones a poor choice for surfing the Web, since the amount of data that can be received or input is rather limited. The strength of cellular phones is their widespread use, making them very suitable for targeted applications, often aimed at the consumer market. Some examples include stock quotes, traffic reports, flight updates, ticket purchases, and news headlines. In all of these applications, limited amounts of data are required to get the expected response. The amount of data being retrieved is not overwhelming, so the device can display it in a meaningful format. If, however, you are considering implementing corporate applications, such as sales force and field service automation, a more capable device might be a better choice.

Figure 2.3   Nokia 8390 Web-enabled phone.
Image courtesy of Nokia.

As you would expect, cellular phones have built-in wireless modems, making it very easy for users to connect to wireless networks. Once connected, the user can use the modem for either voice calls or data applications; or, using the Wireless Telephony Application (WTA) support in WAP, both voice and data can be used simultaneously. In addition, with the always-on capabilities, cellular phones are ideally suited for text messaging applications. Since these messages are typically limited to 160 characters in length, the input capabilities of these devices is suitable. One other side benefit is the long battery life. With their limited processing power, cellular phones are able to conserve battery power, allowing them to last longer than the more sophisticated devices such as smart phones and PDAs. Keep in mind, however, that the more the device is used for wireless communication, the faster the battery power will be consumed.

The primary use of these devices is for voice access, so quality of voice communication, network coverage, and calling packages usually takes precedence over data services when choosing the best device.

Two-Way Pagers

Two-way pagers are designed for mobile users to exchange short, text-based messages. Very often the applications are loaded by the manufacturer and are not upgradeable. These devices do a great job of paging, but often are not suitable for most sophisticated applications. This is true for most two-way pagers, but there are some exceptions.

RIM and Motorola both offer more advanced paging devices, which provide PDA-like capabilities. They both started out using proprietary operating systems, but this has changed with the advent of J2ME, which both have moved to since. The RIM Blackberry devices and Motorola Timeport pagers have expanded upon the basic paging capabilities by offering support for personal information management (PIM) applications and wireless Internet applications. In addition, they offer advanced user input capabilities using a standard—QWERTY—keyboard that is operated using your thumbs. You can see an example of this in Figure 2.4, which shows the RIM 957 (BlackBerry) and RIM 950.

Figure 2.4 RIM two-way pagers.
Image courtesy of Research In Motion.

Several microbrowsers are available for these devices, often provided by the company who provides your wireless connectivity. Very often these browsers are WAP-based, using WML as their markup language, although some HTML browsers are available. Since these devices do have a decent means of data input, they are capable of running wireless applications geared at data capture.

One of the most convenient features of two-way pagers is that they are always connected to a wireless network. Users do not have to explicitly connect to a wireless network when they want to access an application. They simply start to use the application and it will transmit the data as required. This is possible because the devices use networks that are packet-switched; users are only charged for the data being transmitted, not per minute as on circuit-switched networks. This feature is required for receiving pages, email, and other types of messages, and also comes in handy for using wireless Internet applications. Also, considering that these devices are always connected to the network, the battery life is quite exceptional, often lasting for weeks without recharging.

Low-End Smart Phones

Smart phones are called such because they provide the ability to run local applications as well as make voice calls. Like Web-enabled phones, these are primarily voice devices, so selection should be based largely on their voice capabilities. The support for applications is somewhat limited due to the constrained storage, processing power, and display sizes. They are similar to Web-enabled devices in form factor and display size, but they provide the ability for applications to run locally without requiring network coverage.

One of the major attractions to smart phones is their simplicity. The average user is able to have her or his device working in minutes without having to worry about extensive configuration. Also, since the devices have very small processors and little memory, they are able to last for days on a single battery charge. Of course, the fact that smart phones offer cell phone capabilities, as well as the support for applications, makes them ideal for those who want to limit the number of devices they have to carry around.

Users can extend the features of the device by downloading new applications over the air. This is called over-the-air (OTA) provisioning. Both handset manufacturers and wireless carriers are creating electronic storefronts where developers can submit their applications for download. Users download applications of interest to them for a nominal fee [usually between $1 and $5 (U.S.)]. The profit is then divided between the developer and the storefront provider.

Much of the interest surrounding the Java 2 Micro Edition (J2ME) is for building applications for these devices. Due to the limitations imposed by the device manufacturers, most m-business applications are not suitable for these devices. At the time this book was written, games were by far the most common applications being downloaded.

Nokia has predicted that it will sell tens of millions of J2ME-based smart phones by the end of 2003. This is not a small number, and presents a very large application development opportunity for J2ME programmers. It also represents a very large number of the new devices that will be coming to market. (Note: This number is only for Nokia devices. Motorola has also released J2ME devices to the market, and a handful of other manufacturers are adopting J2ME as well.)

Palm-Sized PDAs

The palm-sized PDA device is currently in the sweet spot of the mobile device market. This device has touchscreen capabilities for user input. The form factor is in between handheld PCs and smart phones, typically with a one-quarter video graphics array (VGA) screen (320×240), most often in color. The palm-sized PDA has instant on/off capabilities, meaning the device does not have to boot up before being used. The more recent devices on the market also incorporate built-in wireless modems for wireless connectivity. The devices that do not offer a built-in modem usually have an expansion slot into which a wireless modem can be plugged.

The most common devices in this market are the Palm and Pocket PC. Palm devices are currently the market leaders in this category, although their market share lead is shrinking as Pocket PC devices add more capabilities that are well suited for corporate users. Both the Palm and Pocket PC devices have touch-sensitive screens with built-in character recognition support. Palm devices are available in both monochrome and color, while Pocket PCs typically offer color screens. Palm devices are typically somewhat smaller than Pocket PCs, making them very easy to carry around as they can be attached to a belt or easily slipped into a briefcase or purse. They typically weigh around half a pound, or 225 grams.

While the form factor of Palm devices may be attractive, the performance and storage capabilities are often subpar. Many Palm OS-based devices have 8 to 32 megabytes (MB) of total memory and processors running at 33 megahertz (MHz) or less. Historically, this has been acceptable for the applications running on these devices, but as companies roll out enterprise line-of-business applications, this has started to become a problem. As these applications become more advanced, they require more data storage and processing power to meet the application requirements. In addition, as Java applications are deployed on these devices, they require more power to handle the requirements of the Java Virtual Machines (JVMs) and related applications. The new breed of Palm devices running Palm OS 5.0 have overcome this problem by moving to ARM-based processors running at 206 MHz. This puts them on more even ground with the Pocket PC devices on the market.

Most of the Pocket PC-based devices do have slightly better performance characteristics. The latest devices on the market come with between 32 MB and 64 MB of memory along with X-Scale processors running at 400 MHz. The extra computing power is required to run the Windows CE operating system, which is heavier than its Palm OS counterpart. One downside of the increased power is its effect on battery consumption. Palm devices can often last several days on a single battery charge, while Pocket PC devices typically have to be recharged daily.

The original market for many of these devices was as a personal digital assistant, hence the name PDA. Palm grabbed a huge amount of the PDA market by offering an intuitive user interface to access common applications such as calendars, contact lists, and to-do lists. In addition, it has a well-regarded development program to promote the growth of Palm OS-based applications. This broad range of applications, along with the relatively low cost of Palm devices, continues to make Palm OS devices an attractive purchase for many PDA users.

Pocket PC devices came to market a few years after Palm devices, with a different target audience in mind: the high-end PDA audience looking for a rich user interface with multimedia capabilities. These devices of course also target those that wish to be able to use their Microsoft Office products, such as Microsoft Word and Excel, on their PDAs. To accomplish these goals, Pocket PC devices are required to have more hardware capabilities, which also affects the price. In general, most Pocket PC devices are 30 to 50 percent more expensive than Palm OS devices. Although this has started to change with the entry of Dell into the Pocket PC with less expensive devices.

To more clearly distinguish their consumer and enterprise offerings, Palm Inc. have created two subbrands of devices: Tungsten and Zire. The Palm Tungsten family of products is geared for mobile professionals and enterprise work forces. These solutions combine high-powered hardware and software solutions to solve the complex problems IT professionals are faced with. The Palm Zire family of products is aimed at the consumer market, offering low-price devices with easy to use designs. Palm introduced these product families at the end of 2002 with the goal of maintaining their handheld marketshare lead.

When it comes to wireless connectivity, the majority of PDA devices require the modem to be an add-on component. This is usually accomplished by using a sled, or through an expansion slot. In either case, the modem adds extra cost and complexity to the PDA wireless solution. In most cases, users are required to establish a connection to the wireless network when data needs to be transferred and disconnect when they are finished. This means that they are not always connected to the network, so pushing data to the device is more complex.

The manual connection to wireless networks is changing with the introduction of integrated wireless devices such as the Palm i705, shown in Figure 2.5. These devices come with built-in modems for quick and easy data access. These integrated devices are growing in popularity in both the consumer and business markets. The ease of use and always-on network connections make them very attractive choices for a variety of mobile applications.

The years to come are going to be very interesting as the PDA device market continues to evolve. Manufacturers are making devices aimed at corporate users who require high-performance devices to deploy more advanced applications. It is at this market that many of the smart client application vendors are targeting their solutions. With the introduction of Windows CE .NET and Palm OS 5.0, it is very clear that this market is going to remain competitive for some time to come. This will only benefit the consumer as exciting new features are constantly being added to attract their attention.

NOTE The market is currently seeing a convergence between PDA devices and voice devices. At the end of 2001 and during 2002, many vendors released combined voice/PDA devices based on the Palm OS. In addition, Pocket PC Phone Edition and Microsoft Smartphone 2002 use Windows CE as their operating system, providing many capabilities present in Pocket PC PDAs with additional voice capabilities.

Figure 2.5 Palm i705 with Integrated Wireless.

Image courtesy of Palm Inc.

High-End Smart Phones

As the wireless application market matures, we are seeing a move toward more capable devices in the form of high-end smart phones. The leading cellular phone manufacturers, including Nokia, Sony Ericsson, and Motorola, among others, are producing these devices, which are targeted at the corporate market. These devices provide the ability for voice communication as well as both thin and smart client applications. This makes them a good choice for people who do not want to carry around multiple devices, but still want support for a variety of applications. Figure 2.6 shows the Sony Ericsson P800 Smartphone in voice and data modes.

Figure 2.6   Sony Ericsson P800 Smartphone.

Image courtesy of Sony Ericsson.

The form factor of a smart phone falls in between that of a cell phone and a PDA. They typically have a flip-top mechanism for displaying the full screen and keyboard. When they are closed, they look like a large cellular phone, with the regular 12-digit keypad, and small amount of screen showing. When they are opened, they have screen sizes ranging from 640×200 to 320×240. They also often provide a keyboard for data input. The processors in these devices are powerful enough to run sophisticated local applications as well as advanced wireless Internet applications. The browsers for these devices often have support for color graphics and multimedia, using either WML or HTML as the markup language.

Similar to the two-way pagers and cellular phones, smart phones can often be used for days on a single battery charge. Until recently, the services provided on smart phones were not very customizable. The user was not able to add applications or change the content available on the device. This is changing with some of the latest smart phones, which offer full operating systems with memory available for third-party applications. The most common operating systems in smart phones are Symbian OS, Palm OS, Pocket PC Phone Edition, and Microsoft Smartphone 2002. Support for J2ME is also common.

In North America, smart phones have not been accepted as readily as they have been in Europe and Asia. There are many reasons for this, including wireless network support, the operating systems used, and the manufacturers' marketing and distribution plans. Typically, smart phone manufacturers have been rolling out their new devices in Europe before introducing them to North America. A good example of this is the Nokia Communicator. The Nokia 9210 was released in Europe at the end of 2001, while its counterpart, the Nokia 9290 wasn't released in North America until the middle of 2002.

One thing that may kick-start the smart phone market in North America is Microsoft's entry into the market with its Windows CE-based solution called Microsoft Smartphone 2002. Since it uses the same base operating system as Pocket PC devices, we may start seeing application developers include Microsoft Smartphone 2002 in their supported platforms list.

Another category of device that gets included in the high-end smart phone category is the PDA with integrated voice support. Kyocera and Handspring both have Palm OS devices with integrated voice. RIM has released the BlackBerry 6710/6750 with integrated support for voice communication, and a variety of Pocket PC Phone Edition devices by vendors such as AudioVox and Samsung are now available. These hybrid product offerings are blurring the line between the wireless PDA and the high-end smart phone classes of devices.

Handheld PCs

The concept of the handheld PC is not new. The design can be similar to a laptop, where the screen folds over the keyboard creating a compact shell around the internal device. For this reason, handheld PCs were commonly known as clamshell computers. The original clamshell device came out long before current PDAs were available.

In the mid-1980s, Psion introduced an organizer product that offered the unique capability to run applications. It allowed users to run financial, scientific, and data-driven applications locally on the device. This was in addition to the calculator functions for which it was mainly used. Even though this device was not hugely popular, it laid the foundation for the current line of handheld PCs.

By the early 1990s, Psion released a more functional clamshell computer that had a keyboard as well as a graphical user interface (GUI) screen for running more sophisticated applications. Other companies such as Casio released similar offerings, most often based on a proprietary operating system specific to the device. These operating systems did not support third-party applications, a major limitation. Not long afterward, Microsoft, looking to get a foothold in the PDA operating system market, came out with Windows CE. Most of the companies put aside their proprietary systems and adopted Windows CE as the operating system of choice for their devices, Psion being a notable exception. The specific implementation of Windows CE for devices with this form factor is called the Handheld PC.

Jump to the present day. Now, most Handheld PCs have one-half VGA (480×320) color screens with full keyboards. They have the ability to run a variety of smart client and Web-based applications. In general, Handheld PCs are not used to replace laptops, but rather complement them. The common use of the Handheld PC is not as a general computing device, but rather as an information-gathering device. The one-half VGA screen and integrated keyboard promote rapid data capture, allowing corporations to increase productivity for processes that previously required manual data capture.

Figure 2.7 Samsung NEXiO Handheld PC.

There are many benefits of running line-of-business data collection applications to replace paper-based processes: faster, more accurate data collection, increased employee productivity, faster business process transactions, and reduced operational costs. Even though these benefits could be realized using a laptop computer, a handheld PC is ideal for the job for many reasons. The handheld PCs deliver instant on/off capabilities, making data access immediate. They have a long battery life due to the low power consumption chips used in their design. A single battery charge can often last for an entire day's usage. Additionally, since there are no moving parts, they can handle being knocked around, so can be used in many field environments. These characteristics, along with a smaller form factor than a laptop, make the Handheld PC an ideal device for users who are interested in immediate data access and capture. These devices can host a variety of applications that provide clear benefits for users working away from the office environment. The fact that they often weigh around one pound, or 500 grams, is a nice feature for people who spend a lot of time out of the office.

The handheld PC market has been under pressure from increasingly powerful PDAs and smaller and more efficient laptops. HP is one of the only manufacturers that still sells the clamshell Handheld PC. Other manufacturers have gone with a tablet form factor as illustrated by the Samsung NEXiO in Figure 2.7. The Samsung NEXiO runs Windows .NET 4.1 and has integrated 802.11 capabilities. It is a good example of how handheld PCs have evolved, breathing new life into the handheld PC market.

Tablet PCs

The Tablet PC market is very interesting. Over the past several years, devices in this category have not been terribly successful, and yet Microsoft and its partners are promoting this type of device quite heavily. These devices are aimed at horizontal markets such as field sales and field service, hoping to provide a compelling alternative to a full-blown laptop. Size-wise, they are slightly smaller than a laptop, with some of them having the ability to morph between laptop and tablet layouts. In the laptop position they have a keyboard for data entry; and when in tablet format, they have touch screen, pen-based input. Figure 2.8 shows the Acer TravelMate 100 Tablet PC as it morphs from laptop to tablet mode.

Figure 2.8 Acer TravelMate100 Tablet PC.
Image courtesy of Acer.

Microsoft has developed many new applications (such as the Tablet PC Journal) to take advantage of the touch screen features. The journal software allows users to jot down notes and diagrams in their own handwriting, which can then be converted into standard text at the touch of a button. It is like a form of electronic paper.

Tablet PCs run Windows XP as their operating system and have the usual set of laptop add-on peripherals. They also have similar power and storage capacities to laptops. Microsoft is marketing them as an evolution of the laptop, having all of the features of a laptop, with added tablet features and longer battery life.

It is yet to be seen whether the reincarnation of the Tablet PC is here to stay. They do have some attractive features, but until businesses can justify the additional cost and support, laptops will continue to dominate this market.

Notebook/Laptop

In this day and age, laptops are commonplace, and often are not grouped with the other classes of mobile devices. Companies rarely show the benefits of their wireless solution on a full-sized laptop device; they usually prefer to demonstrate the technology using the latest handheld or smart phone to emphasize the small form factors on which the solution is capable of running. In reality, while mobile applications are capable of running on handheld devices, they are often deployed to laptop computers. For this reason, laptop computers are included in the mobile device classification in this chapter. Laptop computers offer many benefits for mobile applications: They run the same operating systems that users are accustomed to on their desktop computer; they can run the same applications as the desktop computers; and they have the same computing power that users experience on their desktop computers. So the laptop is really just a portable desktop computer, with few restrictions as to what types of applications can be executed. This makes it very easy for IT departments to build laptop applications for the company's mobile users.

Many of the Customer Relationship Management (CRM) and mobile Enterprise Resource Planning (ERP) applications that are currently deployed are laptop-based. Yes, there are now versions coming out targeted at the sub-laptop market, but the current deployments are mainly on laptops. As discussed previously, this is because the application does not have to be altered for deployment on the laptop. Implementing a mobile application on a laptop is as straightforward as implementing a desktop client/server-based application. The developer does not have to be concerned with limited screen sizes, memory, or processing power.

Most laptops on the market today have equivalent capabilities when compared to desktop computers. They have processors running around 2 gigahertz (GHz); they have 60-gigabyte (GB) hard drives; and they have 256 to 512 MB of RAM. In addition, they have full keyboards, 14-inch display screens, and CD-ROM or DVD drives. They include everything that a user requires when away from the office.

One of the major downsides to laptops is their weight and size. Even the lightest machines weigh 2 to 4 pounds, or 1 to 2 kilograms. They also require the user to carry around a separate bag or case, whereas many PDAs can be stored in a pocket or purse, or even clipped to a belt.

When making your device selection, do not limit yourself to the handheld mobile device market when a laptop may meet your requirements for size at the same time as providing excellent performance. At the same time, laptop computers do cost significantly more than the smaller devices, so if the smaller device does meet the needs of the end user, it may be more cost-effective and convenient to use.

Device Manufacturers

Many device manufacturers are leading the way with innovative new capabilities in the mobile arena. Table 2.1 summarizes the leading device manufacturers along with the class(es) of device they provide.

Table 2.1   Device Manufacturers*

DEVICE MANUFACTURER	DEVICE CLASSES	MANUFACTURER URL
Acer	PDA, Tablet PC	www.acer.com
AudioVox	Web-phone, PDA/smart phone	www.audiovox.com
Casio	Tablet PC, Handheld PC, PDA	www.casio.com
Dell	PDA	www.dell.com
Good Technology	PDA	www.good.com
Handspring	PDA, smart phone	www.handspring

Table 2.1 *(Continued)*

DEVICE MANUFACTURER	DEVICE CLASSES	MANUFACTURER URL
Intermec	Handheld PC, PDA, Tablet PC	www.intermec.com
Motorola	Web-phone, two-way pager, smart phone	www.motorola.com
Nokia	Web-phone, smart phone	www.nokia.com
Palm	PDA	www.palm.com
Research in Motion (RIM)	Two-way pager, PDA/smart phone	www.rim.com
Samsung	Web-phone, smart phone, Handheld PC	www.samsung.com
Siemens	Web-phone, PDA/smart phone, and Handheld PC	www.siemens.com
Sony	PDA	www.sony.com
Sony Ericsson	Web-phone, smart phone	www.sonyericsson.com
Symbol	PDA, custom**	www.symbol.com

*Because laptop computers have been around for some time, and most organizations are aware of the leading vendors in this device class, they are not included in this table.

**"Custom" refers to devices that do not fit into the horizontal categories defined in this chapter. They are typically aimed at vertical markets and have features specific to those markets.

Summary

Many factors must be considered when selecting a mobile device: size, cost, performance, support for peripherals, data input mechanism, and wireless connectivity. It is important to select the device that is appropriate for the job at hand. This means paying special attention to data input mechanisms and wireless connectivity options. The decision is further complicated by the number of mobile device classes available. You will have to select from Web-enabled phones, smart phones, PDAs, Handheld PCs, Tablet PCs, and laptops. Each of these devices has its strengths and weaknesses.

In the next chapter, we are going to cover the wide range of wireless networks that are available for personal area, local area, and wide area networks. This information should help with the device selection process, as wireless is often a key criterion.

Helpful Links

Here are some links that may prove to be useful if you want to learn more about some of the topics covered in this chapter.

AOL Mobile T9 site	www.t9.com
J2ME device list	http://wireless.java.sun.com /device
Microsoft Mobile Devices	www.microsoft.com/mobile
Palm OS Web site	www.palmsource.com
Symbian OS Web site	www.symbian.com
Wireless Developer Network PDA Channel	www.wirelessdevnet.com /channels/pda/

Wireless Networks

Wireless network technology is one of the hottest topics in mobile computing. Everyone has an opinion on the state of the third-generation (3G) wireless networks, the effect of Bluetooth for personal networks, and which wireless local area network technology will dominate the market. Even though not all mobile applications require wireless connectivity, there is no doubt that wireless technology is one of the main driving forces behind mobile computing.

This chapter provides an overview of the four main categories of wireless networks: wireless personal area networks (WPANs), wireless local area networks (WLANs), wireless wide area networks (WWANs), and satellite networks. For each category we summarize the prevalent technologies and discuss what the future holds. By reading this chapter you will gain an understanding of which wireless network protocols are being used and for what types of applications. This knowledge will be valuable as you continue through the book to learn more about the design and development of mobile and wireless applications.

Overview of Wireless Networks

Wireless networks serve many purposes. In some cases they are used as cable replacements, while in other cases they are used to provide access to corporate data from remote locations. Much of the industry hype surrounds third-generation wide area networks that provide broadband wireless connectivity to users on a national basis. These networks are now commercially available (in larger urban centers) in most first-world regions. At the same time, breakthroughs in short-range networks are also generating excitement. As users carry around multiple devices, a need arises for an easy, effective way for them to communicate; and what is easier than wireless?

For the purpose of our discussion, wireless networks will be divided into two broad segments: short-range and long-range. Short-range wireless pertains to networks that are confined to a limited area. This applies to local area networks (LANs), such as corporate buildings, school campuses, manufacturing plants or homes, as well as to personal area networks (PANs) where portable computers within close proximity to one another need to communicate. These networks typically operate over unlicensed spectrum reserved for industrial, scientific, medical (ISM) usage. The available frequencies differ from country to country. The most common frequency band is at 2.4 GHz, which is available across most of the globe. Other bands at 5 GHz and 40 GHz are also often used. The availability of these frequencies allows users to operate wireless networks without obtaining a license, and without charge.

Long-range networks continue where LANs end. Connectivity is typically provided by companies that sell the wireless connectivity as a service. These networks span large areas such as a metropolitan area, a state or province, or an entire country. The goal of long-range networks is to provide wireless coverage globally. The most common long-range network is wireless wide area network (WWAN). When true global coverage is required, satellite networks are also available.

Four Categories of Wireless Networks

Table 3.1 provides more detail about the four wireless networks categories. Information such as coverage area, function, relative cost, and throughput are some of the main areas where these networks differ. As mentioned earlier, short-range networks operate on unlicensed frequency bands, therefore, there are no airtime fees associated with their usage. The same is not true of WWANs and satellite networks, which charge either by the minute or by the amount of data transferred.

Table 3.1 High-Level Differences between WPANs, WLANs, WWANs, and Satellite

TYPE OF NETWORK	COVERAGE AREA	FUNCTION	ASSOCIATED COST	TYPICAL THROUGHPUT	STANDARDS
Wireless personal area network (WPAN)	Personal operating space; typically 10 meters	Cable replacement technology, personal networks	Very low	0.1–4 Mbps	IrDA, Bluetooth, 802.15
Wireless local area network (WLAN)	In buildings or campuses; typically 100 meters	Extension or alternative to wired LAN	Low–medium	1–54 Mbps	802.11a, b, g, HIPERLAN/2

Table 3.1 *(Continued)*

TYPE OF NETWORK	COVERAGE AREA	FUNCTION	ASSOCIATED COST	TYPICAL THROUGH-PUT	STANDARDS
Wireless wide area network (WWAN)	Coverage provided on national basis from multiple carriers	Extension of LAN	Medium–high	8 Kbps–2 Mbps	GSM, TDMA, CDMA, GPRS, EDGE, WCDMA
Satellite networks	Global coverage	Extension of LAN	Very high	2 Kbps–19.2 Kbps	TDMA, CDMA, FDMA

Application developers do not have to know the internal workings of wireless networks to be successful, but having knowledge of how they work helps them understand why certain wireless technologies behave the way they do. This knowledge also lays a foundation on which application design architectures can be based. For example, if an application architect knows that the wireless network will only provide 9.6 Kbps of throughput, he or she will want to limit the frequency and amount of data transfer. Similar rationale applies to coverage and cost issues.

For each category of wireless networks, we will provide a summary of the leading standards as well as some insight into where they are being used.

Frequency Fundamentals

Before we can delve too deeply into wireless technologies, a brief primer on radio frequencies is required. Many of the wireless technologies in the WPAN, WLAN, and WWAN categories transmit information using radio waves. For this to take place, the data is superimposed onto the radio wave, which is also known as the carrier wave, since it carries the data. This process is called modulation. There are many modulation techniques available, all with certain advantages and disadvantages in terms of efficiency and power requirements. This summary of the various mechanisms will be helpful as you read this chapter. The modulation techniques are as follows:

Narrowband technology. Narrowband radio systems transmit and receive data on a specific radio frequency. The frequency band is kept as narrow as possible to allow the information to be passed. Interference is avoided by coordinating different users on different frequencies. The radio receiver filters out all signals except those on the designated frequency. For a company to use narrowband technology, it requires a license issued by the government. Examples of such companies include many of the wide area network providers discussed later in this chapter.

Spread spectrum. By design, spread spectrum trades off bandwidth efficiency for reliability, integrity, and security. It consumes more bandwidth than narrow-band technology, but produces a signal that is louder and easier to detect by receivers that know the parameters of the signal being broadcast. To everyone else, the spread-spectrum signal looks like background noise. Two variations of spread-spectrum radio exist: frequency-hopping and direct-sequence.

Frequency-hopping spread spectrum (FHSS). FHSS uses a narrowband carrier that rapidly cycles through frequencies. Both the sender and receiver know the frequency pattern being used. The idea is that even if one frequency is blocked, another should be available. If this is not the case, then the data is re-sent. When properly synchronized, the result is a single logical channel over which the information is transmitted. To everyone else, it appears as short bursts of noise. The maximum data rate using FHSS is typically around 1 Mbps.

Direct-sequence spread spectrum (DSSS). DSSS spreads the signal across a broad band of radio frequencies simultaneously. Each bit transmitted has a redundant bit pattern called a chip. The longer the chip, the more likely the original data can be recovered. Longer bits also require more bandwidth. To receivers not expecting the signal, DSSS appears as low-power broadband noise and is rejected. DSSS requires more power than FHSS, but data rates can be increased to a maximum of 2 Mbps.

Orthogonal Frequency Division Multiplexing (OFDM). OFDM transmits data in a parallel method, as opposed to the hopping technique used by FHSS and the spreading technique used by DSSS. This protects it from interference since the signal is being sent over parallel frequencies. OFDM has ultrahigh spectrum efficiency, meaning that more data can travel over a smaller amount of band-width. This makes it effective for high-data-rate transmissions. The drawbacks of OFDM are that it is more difficult to implement than either FHSS or DSSS, and consumes greater amounts of power.

Wireless Personal Area Networks (WPANs)

The market for wireless personal area networks is expanding rapidly. As people use more electronic devices at home and in the office, and with the proliferation of periph-erals, a clear need for wireless connectivity between these devices has emerged. Exam-ples of the devices that need to be networked are desktop computers, handheld computers, printers, microphones, speakers, pagers, mobile phones, bar code readers, and sensors. Using cables to connect these devices with a PC and with each other can be a difficult task in a stationary location. When you add mobility into the mix, the challenge becomes daunting. If the setup and administration of a WPAN becomes sim-ple and intuitive in the future for the end user, then the most concrete scenario for WPAN technology is cable replacement. This provides a compelling reason to use WPAN technology, and will open the door for more advanced applications in the future. Here are the main characteristics of a WPAN:

■ Short-range communication

■ Low power consumption

■ Low cost

■ Small personal networks

■ Communication of devices within a personal space

While providing these features, a WPAN has to achieve two main goals: broad market applicability and device interoperability. It is important that the WPAN specification addresses the leading device categories that require wireless connectivity in a way that is both easy to implement and affordable. The price point to make a technology attractive is $5 (U.S.) or less. At this level, device manufacturers are willing to incorporate a technology into a broad range of devices for both the consumer and business markets. Interoperability is also imperative. Wireless capabilities are not very useful if they do not allow a device to communicate with other devices and peripherals.

Three wireless standards are leading the way for WPANs: IrDA, Bluetooth, and IEEE 802.15. Each of these standards enables users to connect a variety of devices without having to buy, carry, or connect cables. They also provide a way to establish ad hoc networks among the abundance of mobile devices on the market. Each of these standards is discussed in the following subsections.

WPAN Standards

Many standards are available for personal area networks. Each standard has strengths and weaknesses, making it suitable for specific application scenarios. In some cases, more than one technology will be able to perform a required task, hence nontechnical factors such as cost and availability will factor into the decision as to which technology is more appropriate. Here we take a look at the leading standards in this space. The information provided will give you a solid understanding about where each standard is being used and for what purposes.

IrDA

IrDA, the acronym for Infrared Data Association, is an international organization that creates and promotes interoperable, low-cost infrared data connection standards. IrDA has a set of protocols to support a broad range of appliances, computing, and communication devices. These protocols are typically aimed at providing high-speed, short-range, line-of-sight, and point-to-point wireless data transfer. IrDA protocols use IrDA DATA as the data delivery mechanism, and IrDA CONTROL as the controlling mechanism.

Chances are that you currently own a device that has support for infrared communication. The Infrared Data Association estimates that more than 300 million IrDA enabled devices have been shipped, making it one of the most pervasive wireless technologies in existence. The original goal of IrDA was to provide a cable replacement technology, much like the other PAN standards. The idea was that two computers could communicate simply by pointing them at each other. For example, to print a document, you would simply point the infrared (IR) port at the printer and be able to send the data. No cables would be required.

Technically, infrared technology is well suited for such tasks. The following are some of infrared's features:

- Communication range of up to 1 meter, although a distance of 2 meters can often be reached.

- A low-power option for communication up to 20 centimeters. This requires 10 times less power than the full-power implementation.

- Bidirectional communication.

- Data transmission from 9600 bps to a maximum speed of 4 Mbps.

In theory, using IR for data transfer is a great idea. Unfortunately, even with such ubiquity it is rarely used for its original intent. This may be due to technical challenges in many early implementations, or more plausibly, to the line-of-sight restriction. For IR to work, the communicating devices have to maintain line of sight. This means that they have to situated within the operating range (typically up to 2 meters apart), point at each other, and have no physical impediments. In most office environments, this limitation is not practical for many peripherals such as printers or scanners. Using infrared to transfer data between two devices is more realistic. Two device users can use infrared to transfer information, such as electronic business cards, between one another. Users with Palm devices call this type of transfer *beaming*, as in, "Can you beam me your contact information?" Beyond user-to-user data transfer, infrared is not commonly used for information transfer, since most users do not use two devices with IR ports. While nearly all portable devices have one, the majority of desktops do not. Once again, this limits the effectiveness of IR as a mass-market data transfer protocol.

That said, there are some areas where infrared is frequently used. The IrDA CONTROL standard allows wireless peripherals such as keyboards, mice, game pads, joysticks, and pointing units to interact wirelessly with a host device, very often a desktop PC or gaming unit. A host device can communicate with up to eight peripherals simultaneously. The data transmission rate for IrDA CONTROL typically reaches a maximum at 75 Kbps, which is easily fast enough for the type of data being transferred by these types of devices.

One of the major advantages of IrDA from a device manufacturer's perspective is cost. IR ports can be incorporated into a device for as low as $1 (U.S.). This is a very low cost for implementing wireless communication into a device compared to other WPAN standards.

Bluetooth

Bluetooth is a standard for enabling wireless communication between mobile computers, mobile phones, and portable handheld devices. Unlike IR, Bluetooth does not require a line of sight between devices to be effective. It is able to communicate through physical barriers, typically with a range of 10 meters, although with power amplifiers,

100 meters is possible. Bluetooth uses the unlicensed 2.4-GHz spectrum for communication, with a peak throughput of 720 Kbps. It is expected that this throughput will increase to around 10 Mbps with future Bluetooth specifications.

The origins of Bluetooth date back to 1994 when Ericsson was researching ways to enable mobile phones to communicate with peripherals. Four years later, in 1998, Ericsson, along with Nokia, Intel, Toshiba, and IBM, formed the Bluetooth Special Interest Group (SIG) to define a specification for small form-factor, low-cost wireless communication. Since then, 3COM, Lucent, Microsoft, and Motorola have joined the Bluetooth SIG as Bluetooth promoters. In addition, well over 2,000 companies have joined the SIG as Bluetooth Adopter/Associate members. This all happened before a single Bluetooth product was commercially available, leading to unprecedented market excitement.

People were excited about the futuristic products that would soon be available, expecting that every device, from portable computers to home appliances, would soon incorporate Bluetooth technology. These devices would then interact with one another, transferring data files, contact information, security credentials, and even perform financial transactions. All of this would happen seamlessly without any technical knowledge required from the user.

Needless to say, the hype once again surpassed the technology. While Bluetooth will indeed enable those scenarios someday, right now it is most effective as a cable replacement technology. Since a line of sight is not required for communication, getting Bluetooth devices to interact with one another is trivial. Bluetooth provides an autodiscovery mode, whereby Bluetooth devices will automatically discover other devices that are within range. Once they are detected, they can start communicating. There is some concern that this will overload the 2.4-GHz spectrum as more Bluetooth devices become available. To address this issue, the Bluetooth specification defines three device modes:

Generally discoverable mode. This allows a Bluetooth device to be detected by any other Bluetooth device within its proximity.

Limited discoverable mode. Only well-defined devices will be able to detect a device in this mode. This mode will be used when a user has many Bluetooth devices and wants them to discover each other automatically.

Nondiscoverable mode. This makes the device invisible to other devices so it cannot be detected.

When two or more devices connect, they form a piconet, an ad hoc network that can consist of a maximum of eight devices. Every device in a piconet can communicate directly with the other devices. It is also possible to have networks with more than eight devices. In this case, several piconets can be combined together into a scatternet. In a scatternet configuration, not all devices can see each other; only the devices within each piconet are able to communicate. Figure 3.1 helps to illustrate how this works. In this figure there is one scatternet consisting of five piconets; the hands-free mobile phone is a member of three different piconets and is able to communicate directly with the headset, the Bluetooth pen, and the access point, but is not able to communicate directly with the laptops, printer, or fax machine.

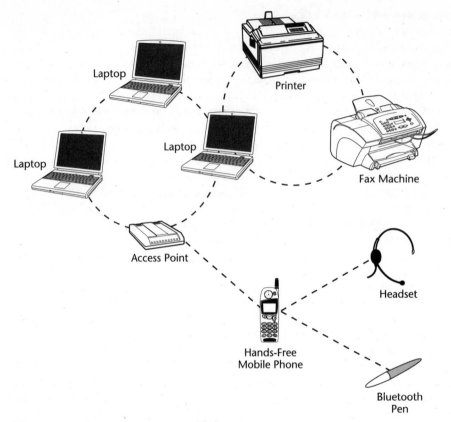

Laptop

Printer

Laptop

Fax Machine

Laptop

Access Point

Headset

Hands-Free
Mobile Phone

Bluetooth
Pen

Figure 3.1   Bluetooth scatternet with five piconets.

The number of Bluetooth devices on the market is growing every day. It is common for mobile phones, PDAs, laptops, and peripherals to come equipped with Bluetooth chips. This has been made possible by the lowered cost of Bluetooth chipsets [currently around $20 (U.S.), with a targeted range of $5 to $10 (U.S.)] in conjunction with increased market demand. Users are now aware of the many compelling features that Bluetooth offers. The leading ones include:

- Cable replacement
- Mobile device networking
- Global ad hoc networking
- Support for both voice and data communication
- Worldwide vendor and product support

Bluetooth Profiles

In order for Bluetooth to realize true ubiquity, interoperability is a key. Bluetooth devices from different vendors have to be able to communicate seamlessly. In order to promote this level of interoperability, the Bluetooth SIG has defined 13 profiles that device manufacturers can use when implementing their products. These profiles help ensure that Bluetooth products are built on a single foundation, allowing for true interoperability.

The entire second volume of the Bluetooth v1.1 specification is dedicated to profile definitions. Each profile is designed for a specific task. Four profiles are foundation profiles, providing the building blocks upon which other profiles are constructed. The other nine profiles are usage profiles. These describe actual usage cases where Bluetooth technology excels. Bluetooth profiles are not meant to be the definitive way to use Bluetooth technology but rather are aimed at providing standards for implementers to build upon. Device manufacturers will base their Bluetooth offerings on these profiles, ensuring that all Bluetooth devices will be able to communicate with one another.

Detailed summaries of each profile are provided in the Bluetooth profile definition book, which is over 450 pages long. If you are interested in obtaining this information, visit the Bluetooth Web site at www.bluetooth.com.

Bluetooth Security

Because cable replacement is one of Bluetooth's primary uses, the overall goal of Bluetooth security is to make the wireless connection at least as secure as cables would be. The Bluetooth specification defines security at the link level. Application-level security is not specified, leaving the developer to choose the security mechanism that is most appropriate for each particular application.

The Bluetooth specification defines several security measures that can be employed in various situations. Additionally, each profile definition outlines when security should be implemented for particular usage scenarios. Bluetooth communication can be encrypted for over-the-air communication and has built-in device authentication. The level of encryption is user-defined and can have a key size between 8 and 128 bits. This allows the user to determine what level of security is required. Note that a trade-off exists between speed and security: Greater key lengths lead to slow communication. For authentication, each Bluetooth device has a unique address so the user can have some faith in the device with which they are communicating. (For an overview of mobile and wireless security, see Chapter 6, "Mobile and Wireless Security.")

HELPFUL REFERENCE The Bluetooth SIG is aware of the importance of security, and has put together a concise Bluetooth security white paper that can be downloaded from the whitepapers section of the official Bluetooth Web site: www.bluetooth.com/dev/wpapers.asp.

802.15

802.15 is a specification driven by the Institute of Electrical and Electronics Engineers (IEEE) to develop consensus standards for short-range wireless networks or wireless personal area networks. It has similar goals to Bluetooth in that it looks to address wireless networking of portable and mobile computing devices such as PCs, PDAs, mobile phones, peripherals, and consumer electronics. The 802.15 WPAN Working Group was established in 1999 as part of the Local and Metropolitan Area Networks Standards Committee of the IEEE.

At the time of establishment, the 802.15 WPAN Working Group was aware of the Bluetooth specification and used parts of it as the foundation for the 802.15 standard. The 802.15 WPAN specification is aimed at standardizing the Media Access Control (MAC) and Physical (PHY) layers of Bluetooth, in the attempt to accommodate wider adoption of short-range wireless technology. 802.15 also deals with issues such as coexistence and interoperability within the networks. To accomplish this goal, four task groups have been established, each working on specific components of the 802.15 specification. They are:

802.15 WPAN Task Group 1: WPAN/Bluetooth. The WPAN Task Group 1 (TG1) has created the WPAN 802.15.1 standard based on the Bluetooth v1.1 specification. To accomplish this, the IEEE licensed technology from the Bluetooth SIG. Specifically, 802.15.1 defines the MAC and PHY specifications for wireless connectivity of devices that are either fixed or portable within the personal computing space. The spec also takes into consideration coexistence requirements with 802.11 wireless local area network (WLAN) devices.

802.15 WPAN Task Group 2: Coexistence Mechanisms. The 802.15 WPAN Task Group 2 (TG2) is developing the recommended practices to facilitate the coexistence of WPAN (802.15) and WLAN (802.11) technologies. Part of this task involves developing a coexistence model to quantify the mutual interference of a WPAN and a WLAN. Once approved, this outcome of TG2's work will become the IEEE 802.15.2 specification.

802.15 WPAN Task Group 3: High Rate WPAN. The 802.15 WPAN Task Group 3 (TG3) is chartered to publish a new standard for high-rate (20 Mbps or higher) WPANs. In addition to high data rates, 802.15.3 also has to provide a means for low-power and low-cost solutions to address the needs of portable consumer electronics, digital imaging, and multimedia applications.

802.15 WPAN Task Group 4: Low Rate-Long Battery Life. The 802.15 WPAN Task Group 4 (TG4) is chartered to establish a low-data-rate (200 Kbps maximum) solution with long battery life (many months to many years) and low complexity. It is intended to operate in an unlicensed international frequency band and is targeted at sensors, interactive toys, smart badges, home automation, and remote controls.

The 802.15 specification is still a work in progress as each of the task groups is at different stages in the specification process. TG1 has completed the 802.15.1 specification and has gotten approval from the IEEE Standards Association (IEEE-SA), while the other groups are still working toward that level. Once completed, the 802.15 WPAN specification will cover all of the current issues surrounding WPAN technology, including Bluetooth compatibility, coexistence with 802.11, high-data transfer rates, and low-power consumption solutions. The combination of all of these will make the IEEE 802.15 specification very attractive for WPAN infrastructure providers.

WPAN Comparison

Of the three WPAN standards, IrDA, Bluetooth, and 802.15, IrDA has been around the longest, and has the highest market penetration, with more than 300 million enabled devices shipped. At the same time, infrared also is the most limiting, as the range is up to 2 meters, and it requires a line of site between communicating devices. The Bluetooth specification addresses these issues by using unlicensed 2.4-GHz spectrum for communication. This allows for communication through physical barriers, as well as larger ranges, typically up to about 10 meters. Bluetooth has also garnered a lot of industry attention, with more than 2,000 companies joining the Bluetooth SIG. In order to provide further standardization for WPAN technology, the IEEE 802.15 specification was developed. The 802.15 specification uses Bluetooth v1.1 as a foundation for providing standardized short-range wireless communication between portable and mobile computing devices. Table 3.2 provides a summary of the leading WPAN technologies.

There is no clear leader, as we are still in the early stages of WPAN technology development. Bluetooth has generated the most industry attention so far, but 802.15 is just as exciting. Since 802.15 is interoperable with both Bluetooth and 802.11, it will have a solid future in the WPAN space. In many ways, IR is not a competing technology to either Bluetooth or 802.15 since it addresses a separate market need. IrDA is included in nearly all mobile devices, providing a quick and easy way for reliable short-range data transfer. With its low implementation costs, many low-end devices will continue to support IrDA, while more advanced devices with more robust wireless needs will implement Bluetooth or 802.15.

Table 3.2 Comparison of WPAN Technologies

STANDARD	FREQUENCY	BANDWIDTH	OPTIMUM OPERATING RANGE	POINTS OF INTEREST
IrDA	875 nm wavelength	9600 bps to 4 Mbps. Future of 15 Mbps	1–2 meters (3–6 feet)	Requires line of site for communication.
Bluetooth	2.4 GHz	v1.1: 720 Kbps; v2.0: 10 Mbps	10 meters (30 feet) to 100 meters (300 feet)	Automatic device discovery; communicates through physical barriers.

(continues)

Table 3.2   Comparison of WPAN Technologies *(Continued)*

STANDARD	FREQUENCY	BANDWIDTH	OPTIMUM OPERATING RANGE	POINTS OF INTEREST
IEEE 802.15	2.4 GHz	802.15.1: 1 Mbps 802.15.3: 20-plus Mbps	10 meters (30 feet) to 100 meters (300 feet)	Uses Bluetooth as the foundation; coexistence with 802.11 devices.

Another area of interest is the increasing range that these technologies can address. Initially, Bluetooth was aimed at a personal operating space of 10 meters. Now, with power-amplified Bluetooth access points, the range has extended to 100 meters. 802.15 is in the same situation. The increased range for these technologies blurs the line between wireless personal area networks and wireless local area networks.

Wireless Local Area Networks (WLANs)

Wireless local area network solutions comprise one of the fastest growing segments of the telecommunications industry. The finalization of industry standards, and the corresponding release of WLAN products by leading manufacturers, has sparked the implementation of WLAN solutions in many market segments, including small office/home office (SOHO), large corporations, manufacturing plants, and public hotspots such as airports, convention centers, hotels, and even coffee shops.

In some instances WLAN technology is used to save costs and avoid laying cable, while in other cases it is the only option for providing high-speed Internet access to the public. Whatever the reason, WLAN solutions are popping up everywhere.

To address this growing demand, traditional networking companies, as well as new players to the market, have released a variety of WLAN products. These products typically implement one of the many WLAN standards, although dual-mode products that support multiple standards are starting to emerge as well. When evaluating these products, some key areas should be considered, including:

Range/coverage. The range for WLAN products is anywhere from 50 meters to 150 meters.

Throughput. The data transfer rate ranges from 1 Mbps to 54 Mbps.

Interference. Some standards will experience interference from standard household electronics and other wireless networking technologies.

Power consumption. The amount of power consumed by the wireless adapter differs between product offerings, often depending on standards they implement.

Cost. The cost of a solution can vary significantly depending on the requirements of the deployment and which standard is being implemented.

In this section we provide some insight into typical WLAN configurations, as well as the leading WLAN standards.

WLAN Configurations

Wireless LAN configurations range from extremely simple to very complex. The simplest WLAN is an independent, peer-to-peer configuration where two or more devices with wireless adapters connect to each other, as depicted in Figure 3.2. Peer-to-peer configurations are often called ad hoc networks since they do not require any administration or preconfiguration. They also do not require the use of an access point, as each adapter communicates directly to another adapter without going through a central location.

Peer-to-peer networks are very useful when a group of users need to communicate with one another in an unstructured way. These networks can be extended by adding a wireless access point (AP) to the configuration. The AP can act as a repeater between the devices, essentially doubling the range of operation. In addition, access points can provide connectivity to a wired network allowing wireless users to share the wired network resources. Figure 3.3 illustrates this configuration.

In a SOHO environment, access points can be used to provide multiple users access to a single high-speed connection without having to run Ethernet wires to each computer. In a corporate environment, many access points can work together to provide wireless coverage for an entire building or campus. The coverage area from each access point is called a microcell. To ensure coverage over a large area, the microcells will overlap at their boundaries, allowing users to freely move between cells without losing connectivity. This movement between a cluster of access points in a wireless network is called roaming. Roaming is made possible by a handoff mechanism whereby one access point passes the client information to another access point. This entire process is invisible to the client.

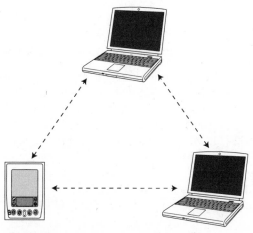

Figure 3.2 Peer-to-peer WLAN configuration.

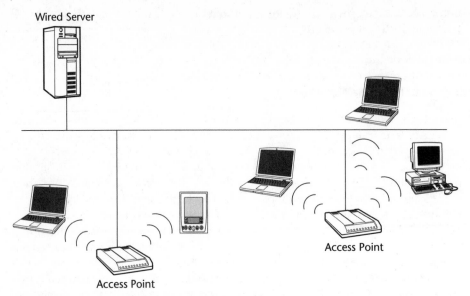

Figure 3.3   WLAN configuration with access point.

In more advanced configurations, extension points (EP) may be used in conjunction with access points. These EPs extend the range of the network by relaying signals to client devices, other EPs, or to an access point. They do not have to be tethered to the wired network, making it possible to service far-reaching clients. One other piece of WLAN equipment is a directional antenna. It allows a signal to be extended to locations many kilometers away. At the second location, the antenna is then connected to an access point, which provides wireless LAN connectivity for the rest of the facility.

WLAN Standards

Two standards bodies, IEEE and European Telecommunications Standards Institute (ETSI) and one technology alliance (HomeRF) promote WLAN standards. In the IEEE 802.11 family of WLANs, three standards deserve individual attention, and a handful of others are worth a quick mention. The leading standard is 802.11b, or Wi-Fi, short for Wireless Fidelity. The clear challenger is 802.11a, which provides increased throughput at a higher, less cluttered frequency; the outside contender is 802.11g, which just completed the final stage of IEEE approval at the time of writing. Other WLAN standards that are worth consideration are HIPERLAN/1 and HIPERLAN/2. We will provide information on each of these standards to give you a firm understanding of their technical and business advantages.

802.11

The IEEE 802.11 specification was approved in July 1997, making it the first wireless LAN standard to be defined. It uses the same switching protocols as wired Ethernet, but allows communication to happen without wires, instead using unlicensed 2.4-GHz frequency radio communication. Two frequency modulation techniques are supported in 802.11: FHSS and DSSS. 802.11 products are not commonly sold anymore, as updated versions (802.11a and 802.11b) have taken its place, providing higher bandwidths at a lower cost.

NOTE We are starting with 802.11b before 802.11a because it has achieved a higher level of commercial adoption. The letter after the name represents the time at which the specification was first proposed, but not necessarily which one was first adopted.

802.11b/Wi-Fi

802.11b is the most popular standard in the 802.11x family. The specification was approved at the same time as 802.11a in 1999, but since then has achieved broad market acceptance for wireless networking. 802.11b is based on the DSSS version of 802.11, using the 2.4-GHz spectrum. Since DSSS is easier to implement than orthogonal frequency division multiplexing (OFDM), as used in 802.11a, 802.11b products came to market much sooner than their 802.11a counterparts. In addition, the 2.4-GHz spectrum is available globally for WLAN configurations, while the 5-GHz spectrum that 802.11a uses is for limited uses in many countries.

802.11b is able to reach a maximum capacity of 11 Mbps. This surpassed the 10 Mbps speed that is part of the original Ethernet standard, making 802.11b a practical alternative to, or extension of, a wired LAN. To help foster interoperability between 802.11b products, the Wi-Fi Alliance [formerly the Wireless Ethernet Compatibility Alliance (WECA)] has set up certification the aforementioned Wireless Fidelity, or Wi-Fi. Obtaining Wi-Fi certification ensures that 802.11b products will be able to interoperate with other Wi-Fi products globally. This certification, combined with the release of 802.11b products by leading networking companies such as Cisco, Lucent, Agere Systems, Proxim, and 3Com, has made 802.11b the leading WLAN standard.

The use of the 2.4-GHz band for communication has advantages and disadvantages. On the plus side, the 2.4-GHz spectrum is almost universally available for WLAN configurations. Initially, a few countries did not allow for its usage, but this has changed thanks to lobbying by industry groups. Additionally, 2.4-GHz signals are able to penetrate physical barriers such as walls and ceilings more effectively than higher frequencies can. The downside of using the 2.4-GHz spectrum is congestion. Since it is unlicensed, meaning anyone can use it without obtaining a special license, other electronic products also use this frequency for communication. Two common examples are cordless phones and microwave ovens. With the widespread use of this spectrum, there is a possibility that it will become overcrowded, resulting in too much interference for effective data communication. Hopefully, this will not be the case since any manufacturer of any product that uses the 2.4-GHz band is required to take interference into account in its product design.

One interesting point about the 802.11b specification is how it handles roaming between access points. The specification requires a method for roaming, but leaves the implementation up to the AP manufacturer. This will make roaming between different vendors' access points difficult, as it is unlikely that manufacturers will employ the same handoff routines.

In typical indoor office configurations, an 802.11b access point can communicate with devices up to 100 meters (around 300 feet) away. The further away a terminal is from the access point, the slower the communication will be. Devices within about 30 meters can usually achieve a raw data transfer rate of 11 Mbps; beyond 30 meters, the rate drops to 5.5 Mbps, and then to 2 Mbps around 65 meters away, and finally, to 1 Mbps around the outer edge. These numbers represent the anticipated coverage area and transmission speeds, but the products from each vendor will differ in performance. If you are looking to implement an 802.11b WLAN, it is recommended that you do a site survey to obtain the actual operating range and associated bandwidth for your location.

802.11b Security

When the IEEE created the 802.11 specification, it implemented a feature called Wired Equivalent Privacy (WEP) with the intent of providing basic levels of authentication and data encryption. As the name suggests, the goal of WEP is to provide an equivalent level of security as normally present in an unsecured wired LAN. This is clearly important, as wireless networks do not have the physical protection that wired environments do. Both 802.11a and 802.11b specifications use WEP.

For authentication, an access point that has WEP enabled will send a text request to the client to verify the client's identity. The client uses RC4 encryption with a secret key to encrypt the text, then returns the encrypted text back to the access point. Once received, the access point decrypts the text using the same key. If the text matches the text that was sent, then the client is authenticated and granted access. For encryption, WEP provides a 24-bit initialization vector that augments the WEP key. This vector changes with each packet, thereby providing a basic level of data encryption.

Unfortunately, both forms of WEP security present some concerns. For authentication, WEP supports no more than four keys and provides no mechanism for refreshing those keys on a regular basis. The result is that the same keys are used by multiple clients and access points and are never changed. This means that malicious users can "listen" to the communication stream and, by using freely available software, very quickly authenticate themselves to the access point. For the encryption layer, WEP uses RC4 in what is called a one-time-pad manner. The security of a one-time-pad is only as secure as the pad being used, which for WEP is the 24 bits. This means that the one-time-pad is repeated at least every 2^{24} packets. For access points with moderate amounts of traffic, this is a matter of hours, hence, attackers monitoring the datastream could detect two messages encrypted with the same 24-bit initialization vector and be able to determine the keys and decipher the plain text.

Companies should realize that WEP was never designed to provide end-to-end security. It is intended for usage in conjunction with existing security mechanisms such as firewalls, virtual private networks (VPNs), and application-level security. The following are some suggestions for corporations that are using, or planning on using, WEP security as part of their WLAN:

- Use a firewall to separate the wireless network from the wired network.
- Have the wireless users authenticate with a VPN to access the corporate network.
- Incorporate security at the application level for highly confidential information.
- Implement dynamic key refreshing for the WEP keys.
- Do not assume that WEP guarantees absolute data privacy.

Not all of the 802.11 WLAN security issues can be attributed to problems with WEP. Many of these issues have resulted from companies not using the WEP for its original purpose or from not using it at all. By implementing additional security mechanisms as listed, corporations can ensure secure wireless communication. In addition, the 802.11i Task Group is working additional levels of security for 802.11 WLANs. The first component of the 802.11i draft is currently being implemented in the form of Wi-Fi Protected Access. Wi-Fi Protected Access offers increased security over WEP. (More information on mobile and wireless security can be found in Chapter 6 "Mobile and Wireless Security.")

802.11a

802.11a is a high-speed alternative to 802.11b, transmitting at 5 GHz and speeds up to 54 Mbps. Unlike 802.11 and 802.11b, 802.11a uses OFDM modulation technology. This, along with the difference in frequency, makes 802.11a networks incompatible with 802.11b networks. Due to the increased complexity of 802.11a, the first products did not reach the market until early 2002, with all the chipsets being provided by a single vendor, Atheros Communications. Since then other vendors have released 802.11a chipsets, helping 802.11a gain broader market acceptance and interoperability certification.

The Wi-Fi Alliance has included certification for 802.11a products within the Wi-Fi certification program. They are using the same name for both 802.11b and 802.11a to help reduce confusion in the market and to foster growth of the emerging 802.11a products. The Wi-Fi Alliance are hoping that Wi-Fi certification will have the same effect on the 802.11a market as it did on the 802.11b market. Certification gives consumers confidence that the products they are purchasing will work with other products based on the same specification.

The move to the 5-GHz band and OFDM modulation provides two important benefits over 802.11b. First, it increases the maximum speed per channel from 11 Mbps to 54 Mbps. This is a tremendous boost, especially considering that the bandwidth is shared among all the users on an access point. The increased speed is especially useful for wireless multimedia, large file transfers, and fast Internet access. Second, the bandwidth available in the 5-GHz range is larger than available at 2.4 GHz, allowing for more simultaneous users without potential conflicts. Additionally, the 5-GHz band is not as congested at the 2.4-GHz band, resulting in less interference.

These advantages come with some downsides. The higher operating frequency equates to a shorter range. This means that to maintain the high data rates, a larger number of 802.11a access points are required to cover the same area, versus 802.11b. While 802.11b access points have a typical range of 100 meters, 802.11a access points are often limited to between 25 and 50 meters. In addition, OFDM requires more power than DSSS, leading to higher power consumption by 802.11a products. This is definitely a disadvantage for mobile devices that have limited battery power. Another downside is that 802.11a and 802.11b products are not compatible. With the large number of 802.11b products on the market, this will have a negative effect on the adoption of 802.11a products. That said, both standards can coexist, and products are now on the market that support both 802.11a and 802.11b in a single chipset. This dual-mode approach is very attractive for users who want the advantages of 802.11a, with the backward compatibility and market penetration of 802.11b.

One final item to note about 802.11a is that the 5-GHz frequency is not universally available for WLAN products. Many European countries, as well as Japan, are resisting the adoption of 802.11a as a standard, leaving some doubt as to whether it will become a global standard as 802.11b has.

802.11g

IEEE 802.11g brings high-speed wireless communication to the 2.4-GHz band, while maintaining backward compatibility with 802.11b. This is accomplished on two layers. First, 802.11g operates on the same 2.4-GHz frequency band as 802.11b, with the same DSSS modulation types for speeds up to 11 Mbps. For 54 Mbps, 802.11g uses the more efficient OFDM modulation types, still within the 2.4-GHz band. In practice, an 802.11g network card will be able to work with an 802.11b access point, and 802.11b cards will work with an 802.11g access point. In both of these scenarios, the 802.11b component is the limiting factor, so the maximum speed is 11 Mbps. To obtain the 54-Mbps speeds, both the network cards and access point have to be 802.11g compliant. In all other aspects, such as network capacity and range, 802.11b and 802.11g are the same.

Since 802.11g offers the same speed as 802.11a, comparisons between them are inevitable. And because they both use OFDM modulation, the main differences result from their frequency ranges and corresponding bandwidth. The total available bandwidth at 2.4 GHz remains the same as with 802.11b. This results in lower capacity for 802.11g WLANs when compared to 802.11a. In addition, fewer channels are available, leading to a higher potential of conflicts. When we take into consideration the backward compatibility that 802.11g has with 802.11b, 802.11g becomes an attractive option for companies that have 802.11b installations.

Other 802.11 Standards

Just as 802.11g improved upon 802.11b, other 802.11 task groups are in place to improve upon the exiting 802.11x standards. The areas of concentration are security, quality of service, compliance, and interoperability. All of these are still in the task group stage of the specification process:

IEEE 802.11e. Aimed at providing quality of service (QoS) capabilities to enable reliable voice communication to complement 802.11b systems. 802.11e will also provide enhanced security and authentication mechanisms. It is expected to receive final IEEE approval in 2003.

IEEE 802.11f. Aimed at developing the recommended practices for an Inter-Access Point Protocol (IAPP) to achieve multivendor access point interoperability.

IEEE 802.11h. Aimed at enhancing the 802.11a High-Speed Physical layer in the 5-GHz band to make IEED 802.11a products compliant with European regulatory requirements.

IEEE 802.11i. Aimed at enhancing the 802.11 MAC layer to increase security and authentication mechanisms.

HomeRF

As the name suggests, HomeRF is a wireless LAN technology aimed at home wireless networking. It is based on the 802.11 FHSS standard, but enhancements have been made to meet the unique needs of the average consumer. HomeRF uses the Shared Wireless Access Protocol (SWAP). One of the major enhancements of SWAP is its support for high-quality voice communication. Additionally, the HomeRF specification incorporates the Digital Enhanced Cordless Telephony (DECT) standard. This allows cordless phones to use the same home networking infrastructure as PCs and appliances while providing advanced telephony features, including call waiting, caller ID, call forwarding, and personal ringtones.

The HomeRF specification has been designed for ease of use and price rather than bandwidth and performance. HomeRF networks provide a range of up to 50 meters (around 150 feet) with maximum speeds at 10 Mbps. The first generation of HomeRF products provided throughput of 1.6 Mbps. HomeRF uses the 2.4-GHz frequency band so it will experience similar interference as 802.11b from household appliances such as microwaves.

With wide industry adoption of 802.11b, HomeRF products have not been able to reach critical mass. The major advantages of HomeRF over 802.11b are ease of use, cost, and telephony support, all areas being addressed by 802.11b products and upcoming 802.11x specifications. For this reason among others, the HomeRF Consortium disbanded in early 2003, making HomeRF a defunct standard.

HIPERLAN/1 and HIPERLAN/2

The European Telecommunications Standards Institute (ETSI) proposed the High-Performance Radio Local Area Network (HIPERLAN) standard in 1992 to address the need for high-speed short-range wireless communication. This first version is commonly referred to as HIPERLAN/1. It is based on Ethernet standards, with its radio transmission taken from GSM. It uses the 5-GHz frequency band. The operating range and bandwidth is difficult to determine since HIPERLAN/1 did not experience commercial success. According to the specification, HIPERLAN/1 has data rates approaching 23.5 Mbps.

HIPERLAN/2 is the next-generation WLAN specification from ETSI Broadband Radio Access Networks (BRAN). It continues to use the 5-GHz frequency band, but with OFDM technology. It is able to achieve peak speeds of 54 Mbps with an approximate range of 150 meters (450 feet).

HIPERLAN/2 has been designed to address the various market segments where WLANs are used: enterprise networking, SOHO, and 3G wireless hotspots. To this end, it has incorporated QoS for real-time multimedia communication, efficient power consumption for portable devices, strong security and interoperability with Ethernet, IEEE 1394 (Firewire), and 3G mobile systems. The specification also permits roaming between HIPERLAN/2 access points, making it suitable for corporate environments. As of late 2002, HIPERLAN/2 still has not seen any meaningful adoption in either the consumer or corporate space.

WLAN Summary

The market demand for WLAN solutions is growing at a rapid pace across several market segments, including the home, office, and public hotspots. Deciding which technology is appropriate to use is not a trivial task, as all three WLAN specifications have strengths and weaknesses and because each WLAN deployment is unique. When evaluating a WLAN solution, make sure you take into consideration future needs, since the technology upgrade paths depend upon the original choice. The following are some key criteria for selecting the right high-speed WLAN solution for you:

Capacity requirements. If you are installing a WLAN for a large number of users, and population density is a concern, then 802.11a may be a good choice since it provides larger bandwidth to accommodate more users per access point. If not, 802.11b/g might be more appropriate.

Interoperability of wireless devices. Wireless LAN solutions from different vendors may not be interoperable, perhaps because of the frequency band used, the frequency modulation technology (FHSS, DSSS, or OFDM), or just due to the implementation of a particular vendor. Wi-Fi certification helps to ensure that 802.11b and 802.11a products will work with products using the same standard. (The first 802.11g products became available in January 2003.)

Timing of high-speed requirement. If high-speed access is needed immediately, a WLAN technology such as 802.11a or HIPERLAN/2 is probably the right choice. If it can wait, then 802.11b with an upgrade to 802.11g might be suitable.

Migration plan. If a WLAN solution is already in place, or if you are looking to take advantage of proven technology such as 802.11b, keep in mind the migration plans for incorporating higher speeds or, possibly, other frequencies. A range of dual-mode WLAN products are available that support both 802.11a and 802.11b.

Interference concerns. If interference is expected on the 2.4-GHz frequency band from products such as Bluetooth, cordless phones, or even microwave ovens, it might make sense to select a product that is using the less-crowded 5-GHz frequency band.

Range/penetration. Higher-frequency signals have shorter range and worse penetration than lower-frequency signals. In some ways, these effects are mitigated by the system manufacturers, but it is still a worthwhile consideration. In some cases, you may prefer a solution that cannot penetrate walls, to prevent eavesdropping from outside parties. For longer range and better penetration, the 2.4-GHz standards such as 802.11b and 802.11g are better choices than those using the 5-GHz frequency band.

Power requirements. Does the device using the WLAN technology have a limited power source? If so, power requirements for each standard must be a factor. The rule of thumb is that higher frequencies require more power to transmit the signal the same distance as lower frequencies. This may not apply in all cases, but it is a safe guideline to go by.

Regulatory factors. Are there limitations imposed by your geographic location that you have to consider when choosing a technology? How about the availability of products in your region? These should be taken into account before making any decision.

Table 3.3 provides an overview of the characteristics that you need to evaluate when determining which solution is best for your situation.

Table 3.3 Comparison of WLAN Technologies

STANDARD	FREQUENCY	BANDWIDTH	RANGE	POINTS OF INTEREST
802.11	2.4 GHz	1–2 Mbps	100 meters (300 feet)	The first approved specification in the 802.11 family.
802.11a	5 GHz	54 Mbps	50 meters (150 feet)	Uses OFDM modulation to achieve high data rates; first commercial products became available in 2002.
802.11b	2.4 GHz	11 Mbps	100 meters (300 feet)	Largest market penetration of any WLAN standard, with commercial products available since 1999.
802.11g	2.4 GHz	54 Mbps	100 meters (300 feet)	Approved by the IEEE-SA in the fall of 2002. Backward-compatible with 802.11b.
HomeRF	2.4 GHz	10 Mbps	50 meters (150 feet)	HomeRF did not achieve commercial success.

(continues)

Table 3.3 Comparison of WLAN Technologies *(Continued)*

STANDARD	FREQUENCY	BANDWIDTH	RANGE	POINTS OF INTEREST
HIPERLAN/1	5 GHz	Theoretically 20 Mbps	–	HIPERLAN/1 did not achieve commercial success.
HIPERLAN/2	5 GHz	54 Mbps	150 meters (450 feet)	Designed for integration with other networks, including wired LANs, IEEE 1394 (Firewire), and 3G mobile networks. Unlikely to achieve commercial success.

One of the most exciting uses of WLAN technology is for providing high-speed Internet access to public hotspots such as hotels, airports, school campuses, and coffee shops. In this scenario, WLAN technologies are being incorporated to wireless wide area network (WWAN) deployments to provide more reliable connectivity at a lower cost. As we discuss WWAN technologies in the next section, we will take a closer look at how WLAN technology is playing a role in the third-generation (3G) wireless deployments.

Wireless Wide Area Networks (WWANs)

WWANs have been in place since the early 1980s for voice communication, and since the early 1990s for data communication. Access to these networks requires users to sign an agreement with the company that operates the network they are interested in. This agreement will allow them to use the wireless network for a fee, which is often calculated by the number of minutes the user is connected to the network, or more recently, by the amount of data transferred over the network. This fee helps the service provider cover the cost of building and maintaining the wide area network, as well as the cost required to purchase the spectrum used for communication. Unlike WPANs and WLANs, wireless wide area networks do not operate over unlicensed frequencies, meaning they have to pay to purchase (or license) the spectrum being used. In some cases the amount of money paid for this spectrum is incredibly large. For example, the sale of spectrum for next-generation wireless networks cost more than $35 billion (U.S.). In order to recover some of those expenses, a usage fee has to be charged.

Do not let the prospect of a fee turn you away from using WWANs, however, as they provide an important component of many wireless solutions. When coverage beyond the range of a WLAN solution is required, WWANs are the place to look. They can provide national, and often international, wireless coverage for both voice and data communication. The communication quality and speed depends on the technology being used. Other properties that are network-dependent include area of wireless coverage, operating frequencies, handset availability, and, of course, cost.

In the remainder of this section we will take a look at the generations of wireless networks and the protocols currently being used. We will start with an overview of the terminology and concepts for wide area wireless networks.

Communication Fundamentals

Many technical concepts are involved with wireless voice and data transmissions. We are going to cover the most pertinent ones in this section before we discuss the evolution of wide area wireless networks. You will find these terms to be very relevant to WWANs.

Analog versus Digital Signals

The first wireless networks used analog signals to transmit sound. Analog signals constantly change as the voice is transmitted, similar to fluctuations in voice itself. Due to the fluctuating nature of analog waves, they are often represented by a sine wave. The first generation of wireless networks used analog transmission for voice communication.

As wireless networks evolved and started to be used for more data as well as voice traffic, a need for digital communication arose. Digital transmissions are a stream of 1s and 0s. Since the data stored on computers is inherently digital, digital networks proved to be more efficient in terms of both spectrum and power consumption. All second-generation wireless networks with data capabilities use digital technology.

The following are some of the many benefits attained by moving to digital networks:

Efficiency. Digital networks can transfer more data over the same amount of spectrum; they also allow for compression for even higher efficiency. Additionally, digital signals consume less power than analog.

Security. Analog signals can be easily listened to by eavesdroppers with a radio tuner; even encrypted analog signals can be cracked quite easily. Eavesdroppers have a much more difficult time with digital signals as they can be encrypted to various strengths, depending on the level of privacy required. In addition, the distribution techniques of digital signals make them more difficult to decipher.

Quality. Digital signals result in better-quality sound with less interference. Advanced filters can be used to remove any noise.

Features. Digital technology allows for voice features such as call answer and caller ID, as well as provide the basis for all data traffic as required for any wireless m-business application.

Circuit-Switching versus Packet-Switching

Two switching mechanisms are used for data transfer: circuit-switched and packet-switched. Circuit-switched networks establish a physical connection between the two communicating parties. This connection is maintained for the duration of the connection and cannot be used by any other parties. Landline telephone networks are circuit-switched. When you make a telephone call, you are granted a connection to the person you are talking to. The line becomes available again only after the conversation is finished and one party hangs up. Thus, for every conversation, a dedicated line is required.

Circuit-switching works well for voice communication where there is a constant stream of data being transferred, but is very inefficient for many forms of data communication where information is requested in bursts, such as when browsing Web pages, because even when no data is being transferred, the connection has to be maintained. This adds additional cost to consumers, as they pay according to the time connected to the network, not by the amount of data downloaded. From the wireless operators' perspective it is also wasteful, as they cannot use that connection for any other purpose. Additionally, if the connection is lost, the user has to establish a new connection, a process that takes anywhere from 5 to 40 seconds.

Packet-switched networks solve these problems. They do not require dedicated connections, but rather allow several users to share a single connection to maximize spectrum. Internet traffic uses packet-based networks for data transmission. This works by dividing data into small units called packets. These packets are then assigned a destination address that they carry around as they are being routed through the network. Multiple users can share the same data path since the data packets do not have to follow any specific path to get to their destination. This allows them to take the optimal path for bandwidth efficiency. At the receiving end, the data packets are then reassembled into their original format.

In terms of the user experience, packet-switched networks are far superior to circuit-switched when it comes to transferring data. Since a dedicated connection does not have to be established, and since no bandwidth is being wasted if data is not being transferred, devices are able to maintain a constant connection to the network without incurring additional costs or network burden. This alleviates users from having to establish a network connection each time they need to transfer data. Additionally, users are charged only for the data they transmit rather than connection times. This can result in significant savings. Network operators also prefer packet networks since they can utilize all of their network bandwidth. Packet switching is the basis of always-on connectivity provided by third-generation wireless networks.

Cells, Handoffs, and Roaming

A cell is the geographical area that obtains wireless coverage from a single cell site (base station). The coverage area for a cell depends on the network protocol, signal power, and obstructions that may impede its progress. As you move further away from the base station, the signal strength is weaker. Typical cell sizes range from 1 to 40 kilometers in radius. For wireless wide area networks, multiple cells are coordinated into a cell system. These systems provide coverage over a larger geographic area. Figure 3.4a illustrates the coverage area of nine cells; Figure 3.4b illustrates the same cells as hexagons as they are often depicted in theoretical drawings.

a)
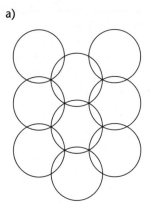

b)
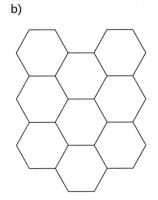

Figure 3.4 Cell coverage.

In highly populated areas, it is frequently necessary to create microcells, where base stations are positioned closer to one another to handle higher numbers of users. Microcells are often deployed in individual buildings, such as conference centers or airports, to handle the large volume of users. The coverage area of microcells is much less than a regular cell, typically around 100 meters in diameter.

When users move from one cell to another while a call is in progress, their connection has to be passed from cell to cell. This process is called a handoff (or in some cases, handover). The handoff is one of the most important aspects of mobile computing since it allows for the uninterrupted movement of mobile users. An unsuccessful handoff results in a "dropped" call, which can be very frustrating for users. This is especially true on circuit-switched networks where users have to reestablish their connection before continuing.

When users move to a cell owned by a different wireless operator, they are roaming. Historically, wireless operators have charged high premiums for using services from other carriers, especially when roaming between countries. This situation is starting to change as carrier consolidation and strategic partnerships increasingly come into being. In fact, many wireless plans allow for unlimited roaming within the country or sometimes even the continent.

Multiplexing Techniques

Multiplexing is a term used to describe how a signal can be divided among multiple users. This spectrum sharing allows wireless operators to maximize the use of their spectrum to accommodate a large number of users over fewer channels. For digital systems, three main multiplexing techniques are being used for wide area networks: frequency division, time division, and code division. A fourth method, called orthogonal frequency division, is the most complex of all of these methods. It is commonly used in high-speed local area networks, as discussed earlier in this chapter, but is starting to grow in popularity for wide area networks as well.

Frequency-division multiplexing (FDM). Numerous signals are combined on a single channel. Each signal on the channel is assigned unique frequency for communication. The caller and the receiver tune to the same frequency to communicate. This is similar to how radio stations work. Each has its own frequency band over which it broadcasts. To listen to a particular channel, you tune the receiver to that particular frequency. For person-to-person communication, this is a very inefficient use of spectrum, hence is only used by analog wireless networks.

Time-division multiplexing (TDM). As with FDM, numerous signals are combined on a single channel, but with TDM they are divided into separate time slots. The time segments are assigned to an individual user and are rotated at regular intervals. The receiver interprets the appropriate time slot (channel) to receive the information. This technique allows for variation in the number of signals sent along the line, and constantly adjusts the time intervals to maximize bandwidth. Many of the current second-generation wireless systems are based on time-division multiplexing as it provides efficient use of spectrum with minimal interference.

Code-division multiplexing (CDM). Rather than dividing the signal using frequency or time, CDM attaches a code to each signal, and sends them all over the same broad spectrum. This results in very high spectrum efficiency and low levels of interference by other signals. Even though all of the signals are being broadcast at once, a receiver will only accept the signals with the right code. This technique is used in several second-generation wireless networks and is the basis for nearly all third-generation networks.

First-Generation Networks (1G)

First-generation (1G) wireless networks were first constructed in the late 1970s in the United States and in the early 1980s in Europe. These analog networks were used only for voice communication and they suffered from high levels of interference, which led to unpredictable call quality. Early on, first-generation networks also suffered from poor handoffs, often resulting in dropped connections, low capacity, and almost no security. The devices also had to be quite large to incorporate the radio receivers necessary to capture the analog signal.

Despite these difficulties, first-generation networks were deployed commercially in many countries. In the early 1980s, the United States, along with Japan, Mexico, and Saudi Arabia, deployed networks using the Advanced Mobile Phone Service (AMPS). In Europe, several analog standards were introduced, including the Total Access Communications System (TACS) and Nordic Mobile Telephony (NMT).

NOTE Interestingly, the AMPS acronym later came to refer also to Analog Mobile Phone System, and then to American Mobile Phone System. All three terms are still in use.

Due to the limitations of the first-generation networks, European countries were quick to move to a new standard based on completely different technology. The case is different in the United States where the AMPS network was widely deployed across all geographies. In many regions, the AMPS network is still used, often as a backup when digital coverage is unavailable.

First-generation networks do not play much of a role in the current m-business environment. They are of interest only when looking at backward compatibility of second-generation networks or for broader coverage in rural areas for voice coverage.

Second-Generation Networks (2G)

Second-generation networks introduced digital capabilities to wireless in the early 1990s. This resulted in higher-quality voice as well as basic data services. In addition, other features such as voicemail, call waiting, caller ID, and three-way calling were introduced to the wireless voice market. Another incredibly important benefit of moving to digital technology was increased capacity. Digital technology allows for more users to communicate on the same amount of spectrum, thereby increasing efficiency. Additionally, digital networks also provide a means for strong security, which was missing in the first-generation analog networks.

All of the second-generation networks provide support for data communications, with the exception of TDMA, which uses Cellular Digital Packet Data (CDPD) for its data services. The rates of data transfer range between 9.6 and 19.2 Kbps. This is fast enough for simple Internet applications based on the Wireless Application Protocol (WAP) or text messaging with Short Message Service (SMS), but it is not adequate for more data-intensive applications.

Four major second-generation systems are currently in use:

- Digital AMPS (DAMPS, now known as TDMA), which can take advantage of first-generation AMPS service.

- Code Division Multiple Access (CDMA IS-95), introduced by QUALCOMM in 1995.

- Global System for Mobile Communications (GSM), which is the most popular of the 2G networks.

- Personal Digital Cellular (PDC), largely used in Japan.

These four networks are all based on different standards, making them incompatible with one another. Moreover, even the same network protocol may not be compatible due to frequency regulations. For example, in Europe, GSM networks operate on the 900-MHz and 1800-MHz frequency bands, while in North America GSM networks use the 1900-MHz frequency band.

To accommodate these differences, dual-band and tri-band handsets are available. These handsets can switch between the different GSM frequencies so a user can roam between countries, or even continents, and still use the same mobile phone. In North America there are also dual-mode phones, which allow for roaming between the digital

DAMPS network and the analog AMPS network. Other features in second-generation handsets include WAP browsers, text messaging, contact lists, calendars, games, and changeable ringtones. Kits are also available that allow for PDAs to gain wireless access using a mobile phone as the modem. The handheld devices are either attached to the phone using a cable or via short-range wireless connectivity such as infrared or Bluetooth.

When it comes to deploying data-driven applications, second-generation networks are often inadequate. They are circuit-switched, which is inefficient and expensive, and provide very limited data transfer rates. The need for higher capacity, higher data transfer rates, and global roaming were the driving forces behind the introduction of third-generation wireless networks.

Second-and-a-Half-Generation Networks (2.5G)

Just as the name suggests, 2.5G networks are a step toward third-generation networks, but they are not quite there. The good news is that they provide the main feature that users require to be successful on the mobile Internet: packet data. The move from circuit-switched systems to packet-switched systems is the major difference between 2G and 2.5G networks. This move brings along many other positive features, the leading one being high-speed data transfer at rates up to 144 Kbps, nearly 10 times that of 2G networks. For wireless operators, upgrading to 2.5G networks often requires only a software upgrade, not hardware modifications. This is a very attractive option for carriers who want to test the market for enhanced data services without incurring large capital expenditures.

NOTE Although 115 to 144 Kbps is commonly used as the theoretical throughput of 2.5G systems, in practice, a more realistic data transfer rate is approximately 40 to 56 Kbps. This is still fast enough for a broad range of new applications.

Two leading 2.5G network protocols are in use: General Packet Radio Services (GPRS) Code Division Multiple Access 2000 1x (CDMA2000 1x). Both offer many compelling enhancements over existing 2G networks. The following are five important characteristics of 2.5G networks:

Efficiency. More efficient use of spectrum by sharing connections among several users for both data and voice communication. This allows wireless operators to accommodate more users on the same network.

Speed. By implementing more efficient modulation algorithms and by being able to use multiple channels simultaneously for data transfer, 2.5G networks can provide transfer rates up to 115 Kbps. This is a vast improvement over 2G systems.

Always-on capability. Users can remain connected to the 2.5G network without having to pay by-the-minute charges, as they would on a circuit-switched network. This allows users to access data services at their convenience without incurring fees for resources they are not using. It also allows for a new range of applications that can push data to the user, rather than have them request it.

Upgrade to 2G systems. Both 2.5G technologies, GPRS and CDMA 1x, are upgrades to existing cellular networks. Users still have the same voice capabilities as before, but now have high-speed data access with the same network coverage. In most cases, moving to 2.5G from 2G involves a software upgrade for the wireless carrier, as opposed to building new infrastructure from the ground up.

Foundation of 3G infrastructure. The implementation of 2.5G technology lays the groundwork for future upgrades to 3G systems. This is true from both a business and technical standpoint. It is expected that as users experience the benefits of high-speed packet data, they will require even more speed and capacity, forcing carriers to upgrade to 3G. At the technical level, 2.5G provides the base packet network on which 3G networks can be built.

GPRS Handsets

Three classes of handsets have been defined to take advantage of the new GPRS services. Terminal is the term used to describe the GPRS unit, since it may not actually be a mobile handset, but instead a wireless network card or other module. The following are the three classes:

Class A. These terminals handle both voice and packet data at the same time. This requires two transceivers, making Class A devices the most expensive of the three.

Class B. These terminals can handle both voice and packet data, but not at the same time. This requires one transceiver that can be used for either voice or data. This keeps the cost of the handset down.

Class C. These terminals can handle either voice or data. This class of terminal may be a low-end handset or, more likely, a wireless modem.

Often it is the device that is the limiting factor for the total data rates that 2.5G networks provide. In order to achieve the theoretical 115 Kbps, the device has to utilize all available time slots for both uploading and downloading data. Most devices do not provide this capability, as it increases the overall cost of the device for speed improvements that are not required by the average user.

2.5G Applications

Along with enhanced networks and new handsets come a full range of new wireless data applications. In some cases, the 2.5G applications are simply extensions to the applications used on 2G networks; in other cases, they are applications that previously were not practical or even possible on 2G networks. A good example of this is wireless Internet applications. On 2G networks, WAP applications are typically text-based with limited graphics. With 2.5G networks, wireless Internet applications can contain full graphics with some multimedia capabilities. The applications will also run more efficiently since the networks are packet-switched.

Other applications that lend themselves to 2.5G networks include:

- Corporate email, calendar, contact lists
- Instant messaging
- Still and moving images
- Job dispatch
- Vehicle positioning using location-based services
- Remote LAN access
- File-sharing applications

All of these applications take advantage of at least one of the characteristics mentioned earlier in this section. The increased data transfer rates apply to nearly all applications, while others, such as instant messaging and job dispatch, make particular use of the always-on functionality. In addition, since GPRS networks behave similarly to a typical LAN, they allow developers to build applications the same way they would for a typical IP network (of course, keeping the device and network limitations in mind).

Third-Generation Networks (3G)

Third-generation (3G) networks started with the vision to develop a single global standard for high-speed data and high-quality voice services. The goal was to have all users worldwide use a single standard that would allow for true global roaming. Unfortunately, as companies and standards bodies from Europe, North America, and Japan met, they could not agree on a single protocol for 3G systems. To help things progress, the leading standards bodies from these countries created a new group called the Third-Generation Partnership Project (3GPP). This group became the driving force behind the development of 3G standards.

After much negotiation, it was realized that backward compatibility with 2G networks and frequency differences among countries were too much to overcome, so agreement on a single 3G implementation could not be reached. Instead, three branches of 3G systems were created: Wideband CDMA (WCDMA), CDMA2000, and Enhanced Data Rates for Global Evolution (EDGE). In 1999, The International Telecommunication Union (ITU) approved an industry standard for third-generation wireless systems. This standard is called International Mobile Telecommunication-2000 (IMT-2000). Each of the 3G network protocols is explained in more depth in the section entitled "Network Protocols" later in this chapter.

> **NOTE** Actually, there are more than three branches of 3G, since WCDMA varies in implementation between European and Japanese operators. Also note that, in Europe, 3G systems are often referred to as UMTS (Universal Mobile Telecommunications System) rather than WCDMA.

Interestingly, 3G systems are based on CDMA technology, whereas the majority of 2G implementations use TDMA. This fact was also an item of contention among the standards bodies. Several proposals were made to the 3GPP, but the eventual winner was the CDMA-based solutions mentioned previously. The first 3G systems were implemented on a trial basis in Japan and Europe in late 2001. Worldwide commercial rollouts of these networks began in late 2002, and will continue into 2005. By 2005 to 2006, 3G networks are expected to reach a critical mass in both the consumer and corporate markets. This is the point at which the high-speed wireless networks become a necessity, not a luxury. Over time, the goal of the IMT-2000 is to converge the three 3G standards into a single universally accepted standard. Only time will tell if this will occur.

Once there was agreement on the standards, details were provided on which features these networks would offer. Two main enhancements characterize third-generation networks:

Higher data rates. 3G systems provide data transfer rates ranging from 144 Kbps to 2 Mbps, depending on the level of mobility. The IMT-2000 has defined three levels of mobility. All 3G networks must support the following minimum requirements:

- *High mobility:* 144 Kbps for outdoor rural users who are traveling at speeds greater than 120 kilometers per hour (75 miles per hour).

- *Full mobility:* 384 Kbps for users traveling less than 120 kilometers per hour in urban areas.

- *Limited mobility:* 2 Mbps for users who are moving at less than 10 kilometers per hour (around 6 miles per hour).

Enhanced quality of service. QoS is supported from end to end in 3G systems, unlike their 2.5G predecessors. This allows users to establish agreements with network operators for certain network properties such as data transfer rate and network latency.

Looking at the preceding information, note that most vendors talk about 2 Mbps bit rates for 3G networks, when in reality that is only for users in stationary indoor environments. A more typical rate for the average mobile user is around 384 Kbps, which is still close to 50 times the rates achieved in most 2G systems. This is also notable, because when users move away from 3G network coverage, they will typically drop into 2.5G or 2G networks that have wider coverage areas. This may result in an unacceptable decrease in speed, as users could potentially move from 384 Kbps to 56 Kbps, or possibly even 9.6 Kbps transfer rates.

Another important point is that the data transfer rates in 3G systems are dependent on the user's distance from the base station. The further a user moves from the base station, the more difficult it becomes to achieve high speeds. The QoS features of 3G networks often mitigate this, but distance still affects network performance.

NOTE We do not go into depth about the physical structure of 3G networks, as it is not pertinent to the development of m-business applications. Since 3G networks are IP-based, the applications created for one network should work equally well on others. This is a significant advantage introduced with 2.5G and 3G networks.

3G Devices

Device manufacturers typically trail network operators when releasing new devices. This means that we have yet to see many of the devices for third-generation networks. That said, many manufacturers have already started releasing devices for 3G networks. These devices usually have a few things in common, namely:

- Support for multimedia content such as video streaming
- Large, high-resolution screens
- Form factors similar to PDAs
- Integrated voice and data as seen in high-end smart phones
- Mobile operating systems that allow for sophisticated client-side applications

Looking at the device characteristics, it becomes clear that 3G networks are targeted at wireless data as much as they are at voice communication.

3G Applications

Many of the applications aimed at 3G networks are similar to those being developed for 2.5G systems. Both generations of networks support IP applications, providing an easy migration path for application developers. That said, there are several new applications that are enabled specifically by the increased bit rates 3G networks offer, namely:

- Streaming video applications
- Downloadable audio such as MP3s
- Over-the-air download of software programs
- Workplace collaboration
- Voice-over-IP (VoIP)
- Location-based services
- Multimedia messaging services
- Support for corporate email and a full range of attachments

This is just a sampling of the new applications that 3G networks will enable. As developers start to gain access to these networks, and the programming interfaces they provide, new and innovative applications will start to enter the market. This is a very exciting time for application developers, as they are just beginning to see the capabilities that wireless has to offer.

Network Protocols

Many network protocols were mentioned in the discussion of the different generations of wireless systems. This section offers more information on these protocols, such as the frequencies over which they operate, their data transfer rates, and the locations where they are deployed. We start with paging networks and data-only networks, and progress to third-generation networks.

Paging Networks

Paging networks provide an alternative to the networks discussed earlier in this chapter. While many people consider pagers to be behind the times, these devices continue to provide some advantages over the current wireless networks. Paging networks operate at a very low frequency, which allows them to have good in-building coverage and wider coverage areas per base station. This has enabled nationwide coverage for paging networks, even in very remote areas. Paging networks are packet-based, allowing for information to be pushed to the user. This is clearly an important characteristic as push data is fundamental to receiving a page.

Three leading paging networks are currently in use:

Flex/reflex. A popular paging network in North America, invented by Motorola.

ERMES. The European Radio Message System, developed for use in Europe, has been adopted by the ITU as the recommended paging system.

POCSAG. This older paging system, named after the Post Office Code Standardization Group, has been replaced by more advanced systems, and is not widely used.

Even though paging networks are often overlooked when evaluating wireless network options, they still provide a service that is essential to many consumers and corporate workers: They are a low-cost alternative for users who do not require the additional features found in mobile phones or wireless PDAs.

Data-Only Networks

Several data-only networks are worth discussing. These networks are either used in conjunction with a voice network or on their own for data-driven applications.

CDPD

Cellular Digital Packet Data (CDPD) was introduced as a packet-switched option for AMPS and DAMPS. This allowed carriers to offer both voice and data services with their existing network infrastructure. It is the only way to send data over the analog AMPS network. Additionally, it is used as the data layer for TDMA networks.

CDPD has a maximum download speed of 19.2 Kbps and upload speed of 9.2 Kbps. In practice, these rates are lower, because they are shared among multiple users in a cell. Even with slow data rates however, CDPD became a popular option for wireless data using laptops and PDAs. In this scenario, wireless cards with embedded CDPD terminals are used for connectivity. For application developers, CDPD is a familiar protocol since it can run IP applications without modification. Every user connected to a CDPD network has a unique IP address. CDPD network coverage is found in most major cities in North America.

Mobitex

Mobitex is a narrowband, packet-switched protocol that provides data transfer rates up to 8 Kbps. It was first developed by Ericsson in the 1980s and has since been deployed in 23 countries worldwide. Mobitex networks provide low data rates at a low cost. The most substantial Mobitex network in North America is offered by Cingular Wireless; consequently, it is used by both Palm.net and BlackBerry.net networks in the United States.

Mobitex operates on three different spectrum bands: 400 MHz, 800 MHz, and 900 MHz. Any application developed for the Mobitex specification can operate equally well on any of these frequencies.

DataTAC

DataTAC stands for Data Total Access Communications. Like Mobitex, DataTAC offers packet-switched narrowband wireless data access, in this case with data transfer rates up to 19.2 Kbps. Three types of DataTAC networks have been deployed:

- DataTAC 4000, which operates in North America.
- DataTAC 5000, which operates in Asia and the Pacific.
- DataTAC 6000, which operates in Europe.

The DataTAC network is often referred to as ARDIS, which is the name brand it operates under in North America. According to ARDIS, its network covers more than 90 percent of the North American business population. Notably, the RIM BlackBerry 850 and 857 devices use the DataTAC network for wireless connectivity.

TDMA (2G)

In the Americas, the term Time-Division Multiple Access (TDMA) is used to describe the DAMPS network, which is used to provide digital capabilities to the AMPS network. TDMA by itself does not provide any data capabilities; these are provided by CDPD, which is commonly integrated into TDMA networks.

TDMA networks are the leading 2G technology in the Americas; in the United States, it has nearly 100 percent coverage, and it is the only national network in many South American countries. In this region, TDMA subscriber growth is outpacing all other technologies, although this is likely to change as third-generation networks become mainstream. AMPS operators are able to add a TDMA layer on their existing infrastructure. By moving to a TDMA network, AMPS operators can achieve a threefold increase in capacity.

In other locations, TDMA is viewed from more of an engineering level. It is a multiplexing technology that uses time slices to divide a signal among multiple users (as described earlier in this chapter). In this sense, TDMA, also referred to as ANSI-136, is the leading 2G technology, as it provides the base for other network protocols such as GSM.

GSM (2G)

The Global System for Mobile Communications (GSM) is the most extensively used 2G technology in the world. GSM is a circuit-switched technology based on TDMA (ANSI-136). TDMA operators in North America are increasingly seeing the benefits of GSM technology: high voice quality, international roaming capability, large number of handset manufacturers, and broad developer base. Additionally, GSM offers a clear path of migration to 3G networks, which is attractive for many TDMA carriers.

GSM operates on three distinct frequencies: 900 MHz, 1800 MHz, and 1900 MHz. The 900 MHz and 1800 MHz are used in Europe, while 1900 MHz is used in North America. In order to provide international roaming, handsets are available that support all three GSM frequencies. These devices are called world phones because they provide worldwide access to wireless networks.

The data transfer rate of GSM networks is 9.6 Kbps. This is enough for the popular text-messaging applications and WAP browsing, but inadequate for other types of applications. To address this concern, a transitional technology called High-Speed, Circuit-Switched Data (HSCSD) was introduced. HSCSD allowed for multiple time slots to be used in a single connection for data transfer up to 57.6 Kbps. This was the expected evolution of GSM networks until more attractive packet-based networks such as GPRS came along. Now HSCSD is rarely discussed as the future for GSM networks. Instead, GSM networks have an upgrade path to 3G using GPRS as the 2.5G technology.

cdmaOne (2G)

Similar to TDMA, Code-Division Multiple Access (CDMA) is both a multiplexing technique and a 2G wireless network protocol. In order to differentiate the two, the network protocol is commonly referred to as cdmaOne, while the standard itself is CDMA, or IS-95.

cdmaOne is a circuit-switched technology that uses a spread-spectrum signal propagation technique. A signal is divided using distinct codes, then spread over the entire spectrum available to it. The advantages of this approach are increased user capacity and high voice quality. CDMA is very efficient when it comes to using spectrum; cdmaOne has been shown to increase the user capacity of an AMPS network by 8 to 10 times.

cdmaOne operates over two main frequencies: 800 MHz and 1900 MHz. The 800 MHz implementation competes with TDMA (DAMPS) in the United States, while the 1900 MHz implementation competes with GSM. Each channel of a cdmaOne system is 1.25 MHz wide and offers data transfer rates of 14.4 Kbps. Like GSM, this is sufficient for text messaging and simple wireless Internet applications, but insufficient for richer applications using images and multimedia content. This has been addressed in the newer CDMA standards called CDMA2000 and WCDMA. Each of these is described later in this section.

QUALCOMM, the pioneer of CDMA, commercially introduced CDMA in 1995. The evolution and promotion of CDMA is now handled by the CDMA Development Group (CDG), which is composed of the leading CDMA service providers and manufacturers. CDMA has experienced broad commercial success in the United States and Asia Pacific where GSM is not as prevalent. It has also been selected by the ITU as the basis for third-generation networks.

PDC (2G)

Personal Digital Cellular (PDC), a system used in Japan, is based on TDMA and offers backward compatibility to the country's existing analog networks. PDC offers a 9.6 Kbps data transfer rate over circuit-switched networks.

In the 1990s, Japan's largest telecom company, NTT DoCoMo, added a packet-switched addition to PDC called Packet PDC (P-PDC) that provides packet data capabilities. It also launched a service called i-mode that provides wireless multimedia over this network. Due to the phenomenal success of i-mode, PDC is now one of the world's most popular wireless standards, even though it is used only in one country.

The evolution of PDC goes directly to third-generation networks using WCDMA. There is no intermediate transition to 2.5G networks.

GPRS (2.5G)

General Packet Radio System (GPRS) introduced packet-switched capabilities to wireless networks. It is a simple, cost-effective upgrade to GSM and TDMA networks that provides increased bit rates and an improved user experience. As discussed in the earlier section on 2.5G networks, packet data systems provide always-on capabilities that allow users to remain connected to the wireless network for extended periods of time without incurring large usage fees.

GPRS operates on the same frequencies as the 2G networks it upgrades, specifically 900 MHz, 1800 MHz, and 1900 MHz. It can provide high-speed data transfer at theoretical rates approaching 115 Kbps, although in practice the throughput has proven to be closer to 40 Kbps. Since GPRS is an upgrade to existing infrastructure, users are able to roam between the GPRS network and GSM/TDMA networks. This ability allows operators to upgrade the major urban centers first, and gradually extend the GPRS network into rural areas.

GPRS is considered a safe point between 2G and 3G networks. It allows operators to experiment with new data services without incurring high capital expenditures. As users become more familiar with the benefits that GPRS networks provide, it is expected they will want even faster data rates, therefore increasing demand for 3G networks. GPRS is the base for EDGE and WCDMA systems.

CDMA2000 1x (2.5G)

CDMA2000 1x is difficult to categorize. Officially, it is classified as a third-generation technology, as defined by the ITU. Practically, it is a 2.5G technology, as it does not meet the minimum speed requirements for 3G networks, and is very similar to GPRS, which is clearly defined at 2.5G. For this reason, we are including CDMA2000 1x as a 2.5G standard.

CDMA2000 1x supports both voice and data services over the standard 1.25 MHz CDMA channel. The 1x in the name signifies that it uses one 1.25 MHz channel. Due to improved modulation, power control, and overall design, it can achieve theoretical data transfer rates of 144 Kbps. In practice, the data throughput for CDMA2000 1x is closer to those of GPRS at 40 to 56 Kbps. Additionally, CDMA2000 1x provides nearly double the capacity of previous CDMA systems and even more substantial increases for TDMA and GSM systems. It also adds packet-data capabilities to provide always-on capabilities.

CDMA2000 1x is backward-compatible with cdmaOne, operating on the same 800-MHz and 1900-MHz frequency bands. It is a cost-effective upgrade to existing cdmaOne networks, paving the way for third-generation CDMA2000 1XEV and CDMA2000 3x networks (3x networks use three 1.25-MHz channels simultaneously, thereby providing increased data throughput).

EDGE (3G)

Enhanced Data Rates for Global Evolution (EDGE) is a technology that allows current GSM networks to offer 3G services within existing frequencies. TDMA or GSM/GPRS infrastructure can be upgraded to EDGE with minimum impact. Since EDGE is a narrowband (200-kHz channels) technology, wireless operators are able to deploy 3G EDGE services without obtaining a 3G license. These two factors make EDGE a low-cost, fast solution for providing 3G services on a nationwide level.

EDGE is able to increase the over-the-air data rates using the same spectrum as 2G/2.5G systems. The peak data rates with EDGE are typically around 384 Kbps, with actual download rates ranging between 75 and 150 Kbps, depending on the user's distance from the base station, among other factors.

EDGE provides a seamless upgrade from 2G GSM networks to 3G UMTS (WCDMA) networks. The frequency bands that EDGE operates in complement the UMTS/WCDMA technology, allowing for the deployment of networks that use GSM, EDGE, and WCDMA in a fashion that is transparent the to user.

CDMA2000 1x EV (3G)

There are two members of CDMA2000 1x EV family: CDMA2000 1x Evolution Data Optimized (CDMA 1x EV-DO) and CDMA2000 1x Evolution Data and Voice (CDMA 1x EV-DV).

CDMA2000 1x EV-DO is a data-optimized version of CDMA2000 that can provide peak data rates of up to 2.4 Mbps using a single CDMA 1.25-MHz channel. With the introduction of improved technology for modulation and dynamically assigned data rates, 1x EV-DO is able to achieve download rates near the theoretical levels. CDMA2000 1x EV-DO supports the complete range of available frequencies, including 450 MHz, 700 MHz, 800 MHz, 1800 MHz, 1900 MHz, and 2 GHz. CDMA 1x EV-DO can work on all IP networks, giving wireless operators an opportunity to gain experience with IP-based technologies before moving their voice networks to IP. It is the natural path of evolution for CDMA 2000 1x networks.

CDMA2000 1x EV-DV networks are expected to be available in 2003. They will provide similar wireless data capabilities as CDMA2000 1x EV-DO, but with the addition of integrated voice capabilities. CDMA2000 1x EV-DV will provide peak data rates topping 3 Mbps, with typical throughput around 1 Mbps. This technology will enable real-time packet services for two-way conversational communication.

WCDMA (3G)

Two types of Wideband CDMA (WCDMA) are in use: Frequency-Division Duplex (FDD) and Time-Division Duplex (TDD). TDD is a hybrid of CDMA and TDMA technologies and is not commonly used commercially for WWAN implementations. It is better suited for indoor usage. The FDD version is the focus here, as it is being deployed commercially in several countries. The "wideband" part of its name refers to the requirement for channel bandwidth of 5 MHz. This is four times larger than CDMA2000 1x (1.25-MHz channel width) and 25 times larger than the GSM (200-kHz channel width). The wider bandwidth allows for higher data transfer rates. WCDMA supports all three modes of mobility as defined in the IMT-2000. As a refresher, these modes are high mobility (144 Kbps), full mobility (384 Kbps), and limited mobility (2 Mbps).

The version of WCDMA being deployed in Europe is commonly referred to as Universal Mobile Telecommunication System (UMTS). It uses direct-sequence (DS) CDMA and therefore it is also called DS-WCDMA. For our purposes, we will continue to refer to it simply as WCDMA. WCDMA has been designed as an upgrade to existing GPRS and EDGE networks, operating on the 2-GHz frequency band, and providing peak data transfer rates of 2 Mbps in stationary environments. Unlike the upgrade to GPRS and EDGE, WCDMA requires coverage to be built from scratch. Since this requires significant capital expenditures by carriers, WCDMA coverage will initially be available only in urban centers. Fortunately, WCDMA is compatible with GSM, GPRS, and EDGE networks, so users do have the ability to roam between the various network implementations. This means that as users move out of WCDMA coverage, they will automatically be handed off to a GSM, GPRS, or EDGE network and continue communicating at a slower data rate.

In Japan, NTT DoCoMo has deployed a WCDMA-based 3G network called Freedom of Multimedia Access (FOMA). It uses a slightly different version of WCDMA than that used in Europe. It was designed as an upgrade from PDC, and therefore does not require backward compatibility with GSM-based networks. In North America, the 2-GHz frequency band is already in use, so WCDMA either has to be introduced into the existing frequencies or new spectrum has to be allotted. This is one of the reasons why a single 3G network protocol cannot be adopted worldwide.

WWAN Operators

One observable trend in the wide area wireless industry is consolidation. In order to remain viable, many network operators are purchasing or partnering with companies that offer complementary technology. This makes it very difficult to maintain an accurate list of the WWAN operators. For this reason, in Table 3.4, rather than attempting to provide an exhaustive list of the wireless operators, we list information and URLs for the organizations that oversee the network protocols. These Web sites will have the most up-to-date information on the carriers in each region for their respective protocols. Additionally, these sites serve as a good resource for information on these operators' respective technologies.

Table 3.4 Oversight Organizations for Network Protocols

ORGANIZATION	WEB SITE	DESCRIPTION
GSM Association	www.gsmworld.com	Country-by-country listing of GSM/GPRS/EDGE carriers
CDMA Group	www.cdg.org	Regional listings of carriers using CDMA technology
UMTS Forum	www.umts-forum.org	Listing of countries with UMTS (WCDMA) networks
Mobitex Operators Association	www.mobitex.org	Listings of all public and private Mobitex networks
Motorola DataTAC	www.datatac.com	Listings of public DataTAC operators

Criteria for Selecting a WWAN Operator

There are literally hundreds of WWAN operators worldwide, each of which supports at least one of the network protocols covered in this chapter. If you are investigating a wireless provider for your organization, the following are the leading characteristics that you will want to consider before making your decision:

Coverage. This is potentially the most important aspect. If your user base is outside of the carrier's coverage area, then the service will be useless to them. Inter-carrier roaming helps to alleviate this problem, but does not solve it entirely since the cost of roaming is usually higher than the rate charged by the primary carrier. When looking at coverage for 2.5G and 3G networks, find out which areas the updated networks service and which the 2G networks service.

Cost. Many areas of cost need to be investigated. The first item to remember is that cost is often based on call times rather than data usage. This usually results in higher bills than would be accrued with a packet-switched network, which only charges for the actual data transferred. If you are signing up for a call bundle, try to avoid unused minutes, as they end up costing more than you would save by getting the bundle in the first place. Many of the packet data carriers provide unlimited data transfer at a fixed rate. These packages work well for budgeting purposes.

Capacity. With the increasing number of wireless users, it is important that the operator you choose has enough capacity to handle them. Insufficient capacity results in the inability to connect to the network and dropped calls for circuit-switched networks, and slow data rates for packet-switched networks. For corporate contracts, wireless carriers should be willing to share their capacity data.

Device support. Mobile devices cannot operate on all networks. If a device has already been selected, make sure there is a modem available for the network you select. In cases where the device has a built-in modem, you may want to evaluate the network options before purchasing the devices. In many cases, wireless operators will have formed partnerships with specific device manufacturers to ensure their offerings work well together.

Upgrade plans. Consider both current and future deployments in your decision. If you expect that high-speed data will be a requirement in the future, evaluate the carrier's 3G deployments to ensure it will cover the geographies you require. You will also want to find out whether new handsets are required in order to upgrade.

Evolution to 3G

The WWAN market is a complex mesh of network protocols and standards. Second-generation network deployments are largely based on two competing technologies: TDMA and CDMA. The overwhelming majority of second-generation wireless customers use TDMA-based networks (TDMA, GSM, PDC). Interestingly, this looks to change, as the third generation of networks is based largely on CDMA technology (WCDMA, CDMA2000 1xEV). Figure 3.5 illustrates the move to third-generation networks, showing that TDMA is dominant in 2G, while CDMA is prevailing in 3G.

Without some analysis, this seems quite odd. Why are 3G networks based on CDMA when the majority of existing networks are TDMA-based? Is it because CDMA is a superior technology? If that is the case, then are network providers deploying an inferior technology?

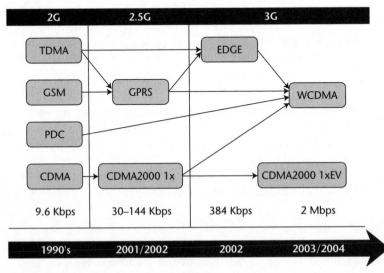

Figure 3.5   Wireless network evolution.

Let us look at why this indeed may be the case. Network providers have to evaluate more than just the technology when implementing a wireless network. Some of the factors beyond the spectral efficiency of the protocol include ease of implementation (which affects time to market), technology maturity, backward compatibility, and implementation costs. TDMA technology has been around longer than CDMA and was less risky for companies deploying large networks. Furthermore, CDMA is more complex for operators to deploy, resulting in higher implementation costs and longer time to market. When 2G networks were being deployed, these factors made TDMA a safer choice.

Things have changed. The number of wireless users is growing rapidly. These users are demanding high-speed wireless access, in addition to high-quality voice services. The existing second-generation networks are not capable of providing these services. To accommodate these demands, networks have to be upgraded to provide higher capacity and faster bit rates. Third-generation networks had to adopt a standard with greater spectral efficiency, namely CDMA. The downside is that most of the existing network infrastructure has to be rebuilt to support the 3G networks. This can be an expensive and time-consuming task, so an intermediate step was added: 2.5G. These networks provide a stepping-stone in the evolution to 3G, as 2.5G networks can be deployed on existing 2G systems with relative ease. This enables carriers to provide the increased capacity and higher data rates while still maintaining backward compatibility. GPRS is a software overlay on TDMA/GSM networks, and CDMA2000 1x is a software upgrade to cdmaOne networks. GPRS and CDMA2000 1x also introduced some of the core 3G infrastructure in the form of packet-switched networks, making the move to 3G networks a bit smoother.

The final step (for now) is the move to the third-generation networks. This is a bit tricky as well, since not all 3G networks are the same. While most third-generation networks are based on CDMA, there is one option that is not—EDGE—and it's an attractive option, since it is a technology where TDMA and GSM/GPRS can converge, although it does not offer the same throughput as WCDMA. This situation will likely have one of two outcomes: a direct upgrade from GSM/GRPS to WCDMA, or an eventual second upgrade of EDGE systems to WCDMA.

On the cdmaOne side of the equation, we have CDMA2000 1x providing the intermediate step to either CDMA2000 1x EV, or WCDMA. Since all of these networks are CDMA-based, the upgrade is not quite as difficult. Finally, PDC is moving directly to WCDMA without a 2.5G network in between. This has led to the first commercial WCDMA deployments in Japan.

WWAN Summary

Wireless wide area networks continue to evolve. It has been over 20 years since the first-generation analog networks were deployed commercially. Since then we have seen the emergence of digital technology, resulting in higher-quality voice communication and the capability to transfer data wirelessly. The latest developments are high-speed packet-switched networks that provide always-on capabilities. These networks are a leading driver in the adoption of wireless data applications.

Table 3.5 summarizes the wireless networks that are currently being used for wireless data, showing the network frequency, data transfer rates, and deployment regions.

There is an increasing threat from WLANs to 3G deployments in highly populated areas and public hotspots. While network operators continue to deploy 3G networks, WLANS are being deployed in many public locations. WLANs provide a high-speed (usually 11 Mbps), low-cost solution for wireless data access. There is no doubt that they will have some impact on the business of 3G operators, but perhaps not as much as some might think. We have to remember that the primary revenue stream (by a long shot) for wireless wide area networks is voice communication, not data. Third-generation operators are counting on wireless data for new sources of income and profits, but at the same time voice communication remains the killer application for WWANs.

Table 3.5   Comparison of Network Protocols

NETWORK PROTOCOL	FREQUENCIES	DATA TRANSFER RATE	COVERAGE
Mobitex	400 MHz, 800 MHz, 900 MHz	8 Kbps	Widely available in North America; coverage in 23 other countries.
CDPD	800–900 MHz	19.2 Kbps	Widely available in North America.
DataTAC	800 MHz	19.2 Kbps	North America, Asia Pacific, and Europe.
TDMA	800 MHz, 1900 MHz	19.2 Kbps via CDPD	Widely used in the Americas, with some deployment in Asia Pacific.
GSM	900 MHz, 1800 MHz, 1900 MHz	9.6 Kbps	Pervasive throughout Europe; deployed in nearly 170 countries.
CDMA	800 MHz, 1900 MHz	14.4 Kbps	Widely used in North America; popular in Asia Pacific.
GPRS	900 MHz, 1800 MHz, 1900 MHz	115 Kbps	Logical upgrade to GSM and TDMA networks, therefore deployed in the same regions.
CDMA2000 1x	800 MHz, 1900 MHz	144 Kbps	Logical upgrade to CDMA networks, therefore deployed in the same regions.

Table 3.5 *(Continued)*

NETWORK PROTOCOL	FREQUENCIES	DATA TRANSFER RATE	COVERAGE
EDGE	800 MHz, 900 MHz, 1800 MHz, 1900 MHz	384 Kbps	Upgrade to GPRS, GSM, and TDMA; deployed in same regions.
CDMA2000 1x EV-DO	450 MHz, 700 MHz, 800 MHz, 1800 MHz, 1900 MHz, 2 GHz	144 Kbps– 2.4 Mbps	Upgrade to cdmaOne and CDMA2000 1X; deployed in same regions.
WCDMA	2 GHz	144 Kbps– 2 Mbps	Upgrade to PDC, EDGE, and CDMA2000, resulting in potential worldwide deployments.

Satellite Systems

Using satellite systems for wireless voice and data first came into consideration in the early 1990s. The goal was to provide a wireless network that would provide coverage for the entire planet, without the need for multiple mobile phones or to roam between many carriers' networks. The state of WWANs then was quite different from today. Many of the existing WWANs were still analog and there was no clear standard for second-generation networks. Satellite phones had the potential to become the de facto standard for wide area communication.

At the same time that satellite marketers were pitching a phone that would work across the globe, wireless carriers were implementing expansive networks; and mobile phone manufacturers were working on new lightweight attractive designs. By the time the satellite networks were ready for commercial use, they had already been overtaken by cellular systems. Satellite companies had a very difficult time promoting bulky, expensive satellite phones, when attractive, inexpensive cellular phones would provide the same benefits. If that were not enough, the airtime fees for the satellite networks would scare away even the most spendthrift customer.

Many of the early mobile Internet satellite providers discovered that their business plans were not as foolproof as they thought; some went bankrupt. Fortunately, the story did not end here. Even though widespread satellite phone usage was not destined for success, many other scenarios existed where satellite technology could prove to be helpful. Here are three examples:

Fixed locations. This refers to situations when access to voice and data systems is required from locations that are not serviced by other network types. Some examples of locations that could benefit from this service are industries such as oil and gas, utilities, and manufacturing; humanitarian workers in remote areas; military establishments; remote office locations or semifixed locations such as trailers or shelters. Consumer services are also offered where the download link is via satellite and the upload is through a telephone line.

Portable communication. Field personnel can send and receive voice and data regardless of their location. This can be useful for search-and-rescue efforts, research in remote locations, news agencies, corporate travelers in remote locations, and military efforts.

Vehicle. This may be one of the more compelling cases for satellite connectivity. Field workers can be assured that no matter what their location, they can communicate it and other pertinent information back to enterprise systems. Delivery personnel, law enforcement, and news organizations can all benefit from guaranteed coverage.

The terminals used for satellite communication range from handheld units to mobile base stations to fixed satellite dish receivers. The peak data transmission speeds range from 2.4 Kbps to 2 Mbps, depending on the solution being sought. For the everyday mobile professional, satellite communication does not provide a compelling benefit, but for people requiring voice and data access from remote locations or guaranteed coverage in nonremote locations, satellite technology may be worth consideration. Also of interest, some of the satellite service providers offer roaming between existing cellular systems and satellite systems.

To get you started, Table 3.6 lists some of the leading satellite service providers with their Web sites.

Table 3.6 Satellite Service Providers

SATELLITE SERVICE PROVIDER	CORPORATE URL
Iridium	www.iridium.com
WirelessMatrix	www.wirelessmatrixcorp.com
GlobalStar	www.globalstar.com
ICO	www.ico.com
Teledesic	www.teledesic.com
Orbcomm	www.orbcomm.com
Inmarsat	www.inmarsat.com
Hughes Network Systems	www.hns.com
Thuraya Satellite	www.thuraya.com
Asia Cellular Satellite	www.acesphones.com

Summary

Wireless connectivity is available at many levels. Personal networks can be established for communication between devices in close proximity to one another. These networks are most commonly used as a cable replacement technology. The leading technologies in this area are Bluetooth and 802.15. Moving to larger wireless networks, wireless local area networks can provide high-speed wireless coverage within 100 meters of a base station. These networks can be used to either replace wired networks or to extend them. The leading WLAN technologies include 802.11a, 802.11b, and HIPERLAN. Both of these network types operate over unlicensed frequencies so there is no charge or airtime.

For wireless communication on a larger scale, wireless wide area networks are available. These networks are in their third generation, offering high-quality voice communication and high-speed data access on a national level. These networks extend enterprise systems into new locations. There are many competing standards for WWANs, including GSM, GPRS, cdmaOne, and WCDMA. If truly global wireless access is required, satellite networks are available to provide access to voice and data applications from anywhere on the planet.

In the next chapter we will take an introductory look at the leading mobile application architectures.

Helpful Links

With so much information covered in this chapter, we list several Web sites to help you find additional information on a given topic.

Bluetooth	www.bluetooth.com
CDMA Development Group (CDG)	www.cdg.org
ETSI, for HIPERLAN/2 specifications	www.etsi.org
IEEE 802.15	www.ieee802.org/15/
IEEE 802.11	www.ieee802.org/11/; www.standards.ieee.org
Infrared Data Association (IrDA)	www.irda.org
International Mobile Telecommunication-2000 (IMT-2000)	www.imt-2000.org
International Telecommunications Union	www.itu.int/home/index.html
Mobitex Operator Information	www.mobitex.org
QUALCOMM CDMA	www.qualcomm.com /cdma/index.html
Third-Generation Partnership Project (3GPP)	www.3gpp.org

Third-Generation Partnership Project 2 (3GPP2)	www.3gpp2.org
3G Americas	www.3gamericas.com
3G information	www.3g.co.uk
Universal Mobile Telecommunications System (UMTS)	www.umts-forum.org
Wi-Fi Alliance	www.wi-fi.org

Mobile Application Architectures

Many factors contribute to the success (or failure) of a mobile solution. These include the mobile device, wireless network connectivity, enterprise integration, and most important, the application architecture. Many people do not realize that several application models are available for mobile development, each with a different set of characteristics that make it appropriate for some applications and inappropriate for others.

In this chapter we introduce three mobile application architectures: wireless Internet, smart client, and messaging. A summary of each application model is provided, along with the advantages and disadvantages it offers. But before we investigate the architectures, we will look at some of the key criteria used for determining which architecture is best suited for a given application.

Choosing the Right Architecture

Many factors come into play when selecting a mobile application architecture. Evaluating the target audience, device type, network connectivity, enterprise integration, and security requirements, along with the specific criteria in the following list will enable you to select the architecture that is most suitable for your particular situation. Finding the answers to these questions, along with any others that may arise is an important step to determining which application architecture is most appropriate for your particular application.

Application Users

- Who are the end users of this application?
- Do they have technical skills?
- What are the expected usage scenarios?

Understanding the end users is important in meeting their needs and is a fundamental requirement for any mobile application. The success of many mobile applications is often determined by the adoption and usage by end users.

Device Type

- For corporate solutions, are there devices already deployed that must be used, or are new devices being provisioned for this application?
- What type of device is most appropriate? Some factors that will affect this include the data input mechanism, wireless connectivity options, and form factor.
- Is the device a complete package? Some devices come with wireless capabilities, while others need to be coupled with wireless components.
- What are the capabilities of the components? For instance, some wireless PCM-CIA cards cannot be connected to the Internet and receive SMS messages simultaneously.

For some corporate solutions and many consumer solutions, you may not be allowed to dictate the target device. In that case, the questions must be approached from a different angle.

- What functionality is available within a specified group of devices?
- Do any devices preclude certain functionality?

Enterprise Connectivity

- How will the mobile device connect to the enterprise?
- Does it require wireless access, or is wired access (for example, USB, dial-up, serial) acceptable?
- If wireless, what type of networking will it use: WPAN, WLAN, WWAN, or satellite?
- Does the type of networking affect the amount of data transferred from the mobile application to the enterprise server?

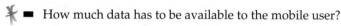

Enterprise Data

- How much data has to be available to the mobile user?
- Where does this data reside: on the client device or enterprise server?
- Is it feasible to download the data in real-time over a wireless network, or is client-side data storage required?
- What is the longevity of the data and how often must it be refreshed? For example, stock quotes are only valuable when they are current, while an inventory list may not require daily updates.
- Is it assumed that each end user will have only one device?
- Can users share a device without mixing their data?
- If a local data store is chosen, how will you reconcile local data with enterprise data?
- Do you have a conflict-resolution scheme for updates to your corporate database?

Enterprise Integration

- Does your client-side method of integration match your server-side API? For example, an application that has a local data store may choose to synchronize its changes up to the corporate database. Usually this process requires direct access to the corporate database or access to a data buffer which sends the data to the corporate data store later.
- What if the only access to the corporate data store is through an API? In that case, you will need to use business logic to call the API using the data buffer.

One of the fundamental reasons to deploy mobile solutions is to extend the reach of enterprise data to mobile workers. To accomplish this, an application architecture must integrate with enterprise data. This can range in time and complexity, from trivial to impossible, and could be considered the most important area for concern.

For these reasons, many companies that are extending existing Internet sites will choose a wireless Internet model even though other application models may be more appropriate based on the other selection criteria.

User Notification

- Do users need to be notified of new information during the day?
- If the only message needed is a "ping" to the user, can existing mobile phones or paging device be used? For example, if a field technician must be informed

that he or she must synchronize because another customer has been added to the schedule, can the message be sent via phone, pager, or to a smart client device? Does the message have to be sent at all, because the field technician will synchronize at the end of every job anyway?

- Does the notification have to communicate some specific information directly to the mobile application, allowing for a lookup value, hyperlink, or automatic login to speed up the process? Receiving a message that reads: "New customer added. Click here to view details," is certainly easier for the end user.

- What happens if the device is off or in another mode, which does not permit notifications to be received?

The ability for mobile users to be notified or updated during the day is a growing requirement for many mobile applications. This ability to push information can make mobile applications much more effective, and more manageable from the users' perspective.

Security

- Is the mobile data sensitive in nature? Sensitive data must be protected from within the corporate network, during transmission, and on the device.

- How can data be kept secure over public networks? Is Secure Sockets Layer (SSL) available for Internet content?

- Is your data store on the device protected from casual prying and/or from serious hacking?

In addition to hosting your planned application, the device can provide access to other corporate resources. For this reason, access to the device and corporate network needs to be monitored.

- How strong is your authentication method?

- Where does user authentication take place: on the device, on the server, or in both locations?

- Are viruses a concern? If so, where does virus scanning take place?

- How about the device itself? What happens if a device is lost or stolen?

Battery Life

- Is power consumption a concern?

- Does wireless connectivity consume too much power?

- Can devices be recharged during the day?

- Is it possible to provide backup batteries for the device?
- Will users abandon devices that they don't find convenient?
- If surfing the Web for one hour will drain the battery of the device, is it worth it?

The battery life of the mobile device is a major concern. Mobile phones can often last several days on a single charge, while PDAs often only last a single day. Furthermore, applications that have frequent wireless communication require substantially more battery power than offline applications.

Other Services

In addition to the line-of-business application being developed, are there other services that mobile users will require? This may include access to corporate email, wireless Internet support, or instant messaging.

Application Architectures

In the wireless world, rarely does everything in your architecture perfectly suit your application needs. You will almost always have to make trade-offs between features, convenience for the end user, time to market, range of devices, ease of use, complexity, deployment options, and so on.

Figure 4.1 shows the spectrum of mobile application architectures covered in this chapter. But, note, though the diagram shows five application models, for the purpose of our discussion, the architectures have been divided into three categories: wireless Internet, smart client, and messaging. As we will see shortly, messaging is interesting because it can be used either on its own or in conjunction with the other application architectures. As we move from left to right on the diagram, the complexity of developing and deploying the application increases, as do the capabilities the application provides.

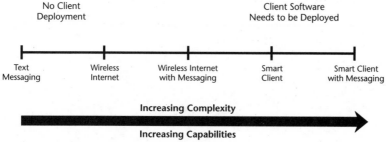

Figure 4.1 Application architecture spectrum.

Finding the right combination of complexity and capabilities for each application being deployed is important. In some cases the benefits of developing a smart client with messaging application will prove to be worth the added development complexity, while in other cases a wireless Internet solution will be more suitable. It is a good idea to enter the design and development phase with an open mind. The purpose of answering the questions in the previous section is to gain an understanding of the user and application requirements, so they can then be matched against the most suitable architecture. Having preconceived opinions about a particular solution may lead to a poor architecture decision.

You may also find that incremental gains in an application can have a large impact on the result. This is often the case when messaging functionality is added to an application. Push messaging can turn a user-driven application into an enterprise-driven application. Rather than requiring the user to request data periodically, important data can be pushed to the user as soon as it is available.

We are now going to take an introductory look at wireless Internet, smart client, and messaging application architectures. The purpose of this overview is to provide enough information for you to understand the high-level benefits of each architecture. Later in this book, detailed information is provided for each application type, as follows: Messaging applications are discussed in Chapter 5, "Mobile and Wireless Messaging"; smart client applications are detailed in Chapters 7 to 10, which make up Part II, "Building Smart Client Applications"; and wireless Internet applications are detailed in Chapters 11 to 15, which make up Part III, "Building Wireless Internet Applications".

Wireless Internet

In most regards, wireless Internet applications have the same architecture as wired Internet applications. The components of the system are essentially the same; the only difference is the way the information is transmitted to the end user. Figure 4.2 depicts the wireless Internet architecture. All of the business logic and enterprise data is stored on servers within the corporate firewall. On the client side is a mobile device that has an Internet browser. These browsers are often called microbrowsers due to their limited size and functionality. No other software is required on the client device. For this reason, wireless Internet applications are also called thin client applications. Each component of the wireless Internet architecture is described in turn in the following subsection.

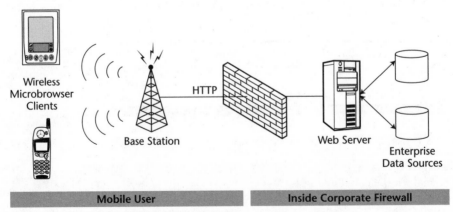

Figure 4.2   Wireless Internet architecture.

Wireless Internet Architecture Components

The following are the key components of the wireless Internet architecture.

Microbrowser client. The browser uses a URL address to contact the specified wireless Web server. The response containing the markup language is then parsed by the browser, and the results are displayed to the application user. In this regard, wireless browsers are very similar to desktop Internet browsers. When the wireless network is not Internet Protocol (IP)-based, the base station contains a gateway that will translate the request and send it as Hypertext Transfer Protocol (HTTP) to the Web server.

Unlike desktop Internet applications, wireless Internet applications may be deployed to devices running a variety of microbrowsers, supporting several markup languages, all having their own unique way of allowing users to navigate through the application. Most wireless devices come with a microbrowser preinstalled, although this is not always the case.

Wireless Web server. The wireless Web server listens to the incoming HTTP request and sends the appropriate response back to the client. This may require some additional steps to gather content from corporate data sources and format the data appropriately for the client in terms of markup language and form factor.

To accommodate the large number of microbrowsers and markup languages used in wireless computing, the wireless Web server often contains wireless-specific frameworks that help with the formatting of the wireless content. In all other regards, this is identical to how an Internet Web server works.

Enterprise data source. The wireless Web server will access the enterprise data source using the preferred access mechanism. Features of your enterprise data source, such as failover, high-availability, and access methods depend on many factors, including how much use you predict your site will see, how critical the service will become, and the existence of alternate methods of accessing your enterprise data source.

Additional Information on Wireless Internet Applications

The entire request/response process happens while the user is connected to a wireless network. If a connection is not available, the user is typically unable to access any part of the application or data. There are exceptions to this with some browsers that provide offline Web content caching. In these cases, cached content may be available, although this content is usually not adequate to perform a complete transaction.

One of the most common protocols for wireless Internet applications is the Wireless Application Protocol (WAP). This protocol was developed and is maintained by the Open Mobile Alliance (formerly the WAP Forum), which is composed of a group of companies to promote wireless Internet development. WAP applications are most common on limited-display devices such as mobile phones.

The majority of consumer application services such as mMode from AT&T (United States) and i-mode from NTT DoCoMo (Japan) are based on the wireless Internet model. Providing access to these services is very straightforward since no software has to be deployed to the device. Essentially, anyone with a wireless device and micro-browser can access the applications provided.

Advantages of Wireless Internet Architectures

The following are the key advantages of wireless Internet (thin client) architecture:

Minimal to zero software deployment. This allows applications to be deployed without any additional client-side configuration. Updates to these applications are also straightforward since only the server has to be updated.

Extends Internet computing model. Many corporate applications are based on the Internet model. Wireless Internet is a natural extension to these applications.

Familiar user interface. Many users are familiar with a browser interface to their applications. Providing a similar interface on mobile devices allows them to be productive immediately; there is no learning curve.

Enterprise integration. If an existing Internet application is being extended, the application logic and enterprise integration layers may already be taken care of. This is a tremendous benefit, as enterprise integration often proves to be the most resource-intensive part of a mobile application.

Broad deployment. If multiple mobile devices are being targeted, wireless Internet application is often the easiest approach since the entire configuration and adaptation of content takes place on the server.

Fresh data. All of the data is assured to be latest version available in the corporate system. There is no worry about stale data.

Security. All of the data is stored on the server behind corporate firewalls. No data is stored on the client.

Disadvantages of Wireless Internet Architectures

Wireless Internet architectures have some disadvantages as well, namely:

Wireless connectivity. To access any data, all of which resides on the server, you need wireless connectivity. This can be problematic when users are moving between multiple locations. The exception is when browsers have content-caching capabilities. That said, even when caching is available, there is still a very limited amount of data and logic available to perform transactions.

Simple user interface. Many microbrowsers have limited capabilities for graphics or other "rich" components. Graphics are also often avoided to minimize the amount of data being downloaded over potentially slow wireless networks.

Application performance. For each request being transferred over a wireless network, performance can be an issue. This is due partially to network throughput and partially to network latency.

Application testing. Controlling, predicting, and testing the behavior of the application is difficult on the full range of microbrowsers. When emulation software is used to simulate devices, it is not always an accurate representation of the end-user experience since it is not executing over a wireless network.

Availability. If a server-side problem occurs, all users will be brought to a halt.

Security. Total control of the environment is not available in most cases, because a wireless gateway exists that may lead to security concerns. This problem is discussed in greater depth in Chapter 6, "Mobile and Wireless Security."

Cost. Wireless airtime fees can become an issue if the mobile user has to constantly be connected to use the application. On circuit-switched networks, where fees are charged based on the time connected, not the data transferred, charges are incurred even when a user is reading Web content or filling in a form.

Application Examples

Nearly any application can be presented as thin client. This section highlights those applications which are almost always best suited as thin client.

Securities Trading

Securities data is extremely volatile and it is imperative that users be assured that the data they are viewing is the most up-to-date and accurate. This data category is an ideal fit for wireless Internet applications. Users viewing stock quotes or portfolio summaries can be confident the information presented reflects the most up-to-date information available on the enterprise server.

Wireless Internet applications are also applicable to financial transactions. When users purchase or sell a security, they need to be updated immediately on the status of the transaction. With thin-client applications, the transaction information is immediately relayed to the appropriate server where the transaction will take place. If the transaction occurs immediately, the result can be transmitted back to the user in real time. If the transaction happens at a later point in time, the user can check the status throughout the day to get real-time updates.

Information Services

Information services comprise one of the leading uses of wireless Internet applications. Any user with a wireless device and a microbrowser can quickly and easily access a variety of informational sites to access many types of content, including news, weather, sports, airline schedules, traffic information, movie schedules, restaurant information.

Entertainment Information

Similar to information services, a variety of entertainment information is available from wireless Internet devices, including interactive games, horoscopes, puzzles, and instant messaging.

m-Commerce

The majority of m-commerce applications are targeted at the general consumer population. These applications are more successful when they are for goods that can be purchased and obtained immediately, such as downloadable games and mobile phone ringtones. Purchases that require pickup or delivery at a later date have not proven to be popular over mobile devices. This is because m-commerce makes the most sense when there is a sense of urgency.

m-Commerce applications, information services, and entertainment information are all targeted at consumer audiences, so supporting the largest possible number of devices, and in turn the largest number of potential customers, is essential.

Smart Client

Smart client applications are a powerful alternative to wireless Internet applications. Instead of using a microbrowser on the client, custom software is developed. This software typically contains a persistent data storage mechanism as well as business logic. This means that smart client applications can be executed at any time, even when a wireless connection is unavailable. Actually, smart client applications are essentially mobile and do not require wireless communication at all. Integration to the enterprise is typically provided via a synchronization process whereby the client application communicates data to back-end data sources through a synchronization server. This communication can occur over wireless or wired connections such as TCP/IP, serial, or USB. The smart client architecture is depicted in Figure 4.3.

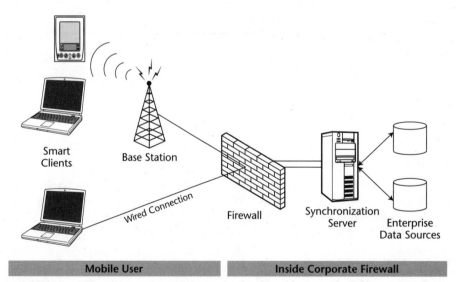

Figure 4.3   Smart client architecture.

In order to deploy smart client applications, the client device has to have a mobile operating system that supports custom applications. This usually eliminates mobile phones from the smart client deployment options, leaving smart phones, PDAs, hand-held PCs, and laptops. The common approach for developing smart client applications is to use libraries native to the mobile operating system, although alternatives such as Java are also available. Let us look at each component of the smart client architecture, in turn in the next subsection.

Smart Client Architecture Components

The following are key components of the smart client architecture:

Smart Client

The smart client application is where client-side business logic is executed. The application itself is either a native executable or Java application that is deployed to the mobile device. To provide offline data access, mobile data store products are incorporated into the application. Many data store options are available, which are discussed in depth in Chapter 9, "Persistent Data on the Client." To communicate to the enterprise systems, a synchronization or messaging layer is also part of the smart client application.

Synchronization Server

Data is sent from the client application to the synchronization server. This can occur over a wireless or wired connection to the server. From there it is then communicated to the enterprise data sources. The synchronization server, with its associated logic, is responsible for ensuring that the minimal amount of data is transferred and that any conflicts are detected and resolved. It also provides the communication layer to enterprise systems. Synchronization is discussed in depth in Chapter 10, "Enterprise Integration through Synchronization."

Enterprise Data Source

The synchronization server will access the enterprise data source using the preferred access mechanism. The access to the enterprise source may occur during the synchronization process if it is imperative for the smart client to receive feedback from the synchronization. This is a simple process and requires that the client connection to the synchronization server remain active until the enterprise is finished processing the data.

It is also possible to synchronize data to a corporate data buffer and return pending status for any operations that need to be completed. The enterprise system can poll the data buffer at its convenience for information to process. On the next smart client synchronization, the status of all pending operations are downloaded. This provides for a much shorter connection time to the server, and a greater number of clients can connect, but at the cost of end-user convenience.

Additional Information on Smart Client Applications

One of the most common concerns about smart client applications is software deployment. Manufacturers of products that specialize in mobile device management have addressed this in the enterprise space. These software packages allow a central server to manage all aspects of application delivery and maintenance. For the consumer market, over-the-air (OTA) provisioning is a key enabler of smart client computing. OTA provisioning enables consumers to purchase and install applications directly over a wireless connection; no companion PC is required. Many wireless operators and device manufacturers are setting up digital malls that users can browse to find a variety of applications. When they find an application they like, they download and install it on the device. A nominal charge [usually between $1 and $5 (U.S.)] will be added to the user's airtime bill.

These malls are very attractive for developers who require a distribution channel for their applications. The most common examples are games, but utilities, music, and other software applications all are benefiting from this type of distribution.

NOTE Smart client applications are sometimes called *terminal-based* since the application runs locally on the device. Other names for smart client applications include rich client and fat client.

Advantages of Smart Client Applications

The following are key advantages of smart client applications:

Always-available data. When no wireless coverage is available, the user is still able to access and modify enterprise data to be synchronized later. This overcomes one of the most significant limitations experienced with wireless Internet applications.

Rich user interface. Sophisticated displays with graphics and charts can be developed. This type of user interface is more similar to what users expect based on desktop experiences. In the case of Pocket PC devices, a subset of the Windows functionality is available, allowing for rich graphical components.

Performance. How the interface is "served up" is determined by processor speed and data access, which are known qualities. No performance is lost in having to continually communicate over wireless networks to access data.

Distributed computing. The execution of business logic and data processing occurs on the client devices, reducing the load on the server machines. This means organizations have to purchase less server-side hardware.

Security. Implementing a solid end-to-end security scheme is easier than with wireless Internet architectures, as you have the ability to control data on both ends of communication. There is no dependence on third-party systems such as WAP gateways.

Cost. Wireless airtime fees can be reduced or even eliminated with smart client applications. Only the data pertaining to the specific mobile user has to be transferred between the smart client and the synchronization server. This results in a minimal amount of data being transferred, thereby lowering wireless airtime charges. The synchronization can also occur over wired networks so that wireless fees can be avoided altogether.

Disadvantages of Smart Client Applications

The disadvantages of smart client architectures are:

Enterprise integration. Because smart client application can provide advanced functionality on the client device, many organizations end up enhancing the client application beyond the original design. While the client-side component may not be complex to develop, the enterprise integration component will often take more time and resources than expected. (In a way, this is both an advantage and a disadvantage since the result is a more capable client-side application.)

Application deployment. Remote deployment and management of the smart client applications may seem a trivial or end-user-driven process, but it can become an IT burden when there are a lot of remote users. Fortunately, software products are available to help with this task; they are outlined in Chapter 16, "Mobile Information Management."

So a simple deployment system is req'd.

Mobile viruses. Introducing a mobile operating system opens the doors to viruses. This can be addressed by software in the application deployment phase; nevertheless, it does add another item of concern.

Development complexity. Programming in the native language of the mobile device requires learning a new client-side API. If you are developing for multiple mobile devices, with different operating systems, this can turn into a significant challenge. Java and other cross-platform tools are on the market to help address this issue. This is discussed in more depth in Chapter 8, "Smart Client Development."

Multiple development cycles. When multiple devices and mobile operating systems are targeted, separate development and testing phases may be required for each device involved. This can be a drain on resources and result in drawn-out development cycles if not planned for ahead of time.

Application Examples

In general, a smart client architecture deserves strong consideration for any application that requires offline data access. Here are a few example applications that are particularly well suited for a smart client architecture:

Sales Force Automation (SFA)

One of the most important assets for a salesforce is current information. The difference between making a sale or not often depends on having immediate access to information while the salesperson is meeting with the customer. The best way to ensure that the sales staff has this information is to deploy a Sales Force Automation (SFA) application with persistent data storage.

SFA applications give sales representative immediate access to important sales information, such as the customer's contact information and order history, product catalogues, price lists, and inventory lists. With this information on hand, the sales rep can put together quotes with delivery dates on the spot. Once the order is placed locally on the device, the rep relays the order to the back office through synchronization.

Field Force Automation (FFA)

Field service workers can benefit from persistent data access in many ways. By empowering service workers with the information they need, they can improve all aspects of their jobs, resulting in higher productivity and increased customer satisfaction.

The benefits start with order entry and dispatching. Orders can be taken at a call center, entered into a central database, and synchronized through to the mobile worker. This eliminates paperwork and streamlines the dispatching process. Once the service worker is at the job site, he or she will have all the related job information on hand. This includes a machine's service history, repair checklists, troubleshooting information, schematic diagrams, and parts inventory levels. Having this information on hand decreases delays that occur when the required information is not readily available.

Once the job is complete, the field service worker can complete the transaction by completing the job report. This involves entering the appropriate work order data, such as placing an order for parts and scheduling a follow-up visit. Once the data is entered into the remote application, it can be synchronized to the enterprise system, reducing administrative costs.

The overall benefit of providing mobile workers with always-available data is improved customer satisfaction, with the ultimate result being increased customer loyalty.

Healthcare Applications

Medical professionals, such as doctors and nurses, can realize many benefits from smart client applications, in particular, the availability of up-to-date, accurate, patient information at their fingertips. With this information, medical practitioners can diagnose conditions more effectively and accurately.

A prime example is prescription writing. In the United States, close to 3 billion prescriptions are written annually, for some 100,000 different prescription drugs, from a variety of pharmaceutical companies, and over a dozen insurance carriers. Managing this data is an arduous task. By automating this process with a smart client application, doctors can deliver accurate, easy-to-read prescriptions that meet the requirements of the patient's insurance plan. This is just one example where persistent data within a mobile application increases productivity and reduces errors. Many other medical processes can be automated in a similar fashion.

Data Collection Applications

An entire breed of applications has been designed for data capture. These applications allow businesses to streamline existing data capture processes. Many of the most recent devices are very well suited for this task. The rich user interfaces and add-on peripherals increase their usability.

For inventory management applications, having a built-in barcode scanner allows for easy capture of merchandise, asset, and document information. This information is recorded without data entry errors and can be synchronized directly into back-end systems. Digital camera add-ons allow the user to take pictures and store them on the device. They can then manipulate the pictures before synchronization. Insurance claims agents can take advantage of this functionality. They can take a picture of an item and make quick edits to the image to illustrate where the damage occurred. This picture can then be attached to the claim as it is synchronized into the central database.

These are just two examples where smart client applications with persistent data storage increase productivity while reducing errors and administrative costs. Other vertical markets that can achieve the same benefits include the financial, retail, transportation, and healthcare verticals.

Mobile Productivity Applications

For professional services staff and other mobile professionals, several applications are available to increase productivity. In the professional services industry, workers' success is often based on billable hours. By providing an application where they can capture the number of hours logged on each task, corporations will have more accurate records of how time was spent. This often results in more billable hours since they can be entered at the job site, reducing the likelihood that they will be forgotten.

Productivity gains can also be accomplished by providing mobile access to other enterprise business applications such as SAP, PeopleSoft, and Siebel. This type of mobile application allows for more accurate data capture and reduced data entry costs.

Messaging

Messaging applications can take many forms, ranging from email to alerts and notifications to application-to-application messaging. In some cases messaging is used as an enhancement to an existing mobile application; in other situations, it is itself an application architecture.

Numerous messaging technologies are available, so it follows that there are many uses for messaging. However, for our purposes here, three categories of mobile messaging are of interest to us. The following is a brief description of each:

User-to-user messaging. Messages can be sent from one user to another using a variety of mechanisms, including email, paging, and wireless text messaging such as the Short Message Service (SMS) or Instant Messaging (IM). Richer messages that include graphics and formatted text can be sent using the Enhanced Message Service (EMS), while multimedia content can be sent using the Multimedia Message Service (MMS). These forms of messaging can also be generated by server-side processes as a means of information dispersal.

Notifications and alerts. Messages that are urgent in nature can be pushed to mobile users on their wireless devices. This allows corporations to ensure that information is received in a timely fashion. These messages can contain a URL link to a wireless Internet site where the user can obtain additional information. These types of messages are called actionable alerts, since the recipient performs an action based on the message content. The two leading technologies for notifications and alerts are Handheld Device Markup Language (HDML) alerts and WAP Push.

Application-to-application messaging. In many cases, user interaction is not required for the message to be successful. Enterprises can communicate data directly from an enterprise server to a client application without user interaction. This can be useful to enhance smart client applications with server-initiated synchronization.

The first two types of messaging in this list do not require additional software to be deployed to the client device. For user-to-user messaging, the client software is typically provided by the device manufacturer; and for notifications and alerts, this capability is part of the microbrowser on the device. From the users' perspective, when they receive these types of messages, they view the information and react accordingly. This type of messaging is relatively straightforward to implement.

Application-to-application messaging is more involved. It requires software to be deployed on a device that understands the communication stream between the client and the server, as well as how to operate over a wireless or wired network. In a way, an application-to-application messaging application falls under the smart client category since a software application is deployed to the device that can be used when disconnected from a wireless network. Additionally, the client application typically contains a message queue that can store information. Figure 4.4 illustrates the application-to-application messaging architecture. Each component of the application-to-application messaging architecture is described in the next subsection.

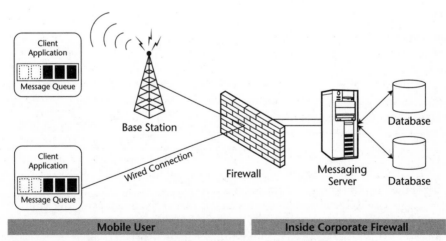

Figure 4.4   Application-to-application messaging architecture.

Application-to-Application Messaging Architecture Components

Here are the main components of an application-to-application messaging system:

Messaging client. The client application contains message queues as well as client-side logic. The message queues can store incoming and outgoing transactions for future access. For example, if an application attempts to send data to the server when a connection is unavailable, it can be stored in an outgoing queue and be sent automatically when the user establishes a connection to the server. This technique is called store-and-forward messaging. The messaging client can be used either on its own or in conjunction with other smart client features such as a persistent data store. The application itself can usually be programmed in languages specific to the operating system or in Java.

Messaging server. The server component is the part of the system that communicates with the messaging client, as well as the enterprise systems. The industry name for messaging servers is Message Oriented Middleware (MOM). Many of these systems are built on the Java Message Service (JMS). JMS provides a reliable and scalable base platform with store-and-forward capabilities. This form of messaging is useful for m-business applications because the sender of the message does not have to wait for the recipient to receive the information, allowing him or her to continue working while the message is routed and acted upon.

Enterprise data source. The messaging server can interact with a variety of back-end systems, including databases, business applications, and other messaging systems. This integration can use the preferred access technique for the enterprise system. The asynchronous nature of messaging systems is well suited for systems that require complex enterprise integration since the user does not have to wait for a response.

Additional Information on Messaging Applications

The store-and-forward capability is not unique to application-to-application messaging. The server components of both user-to-user messaging and notifications and alerts can implement store-and-forward as well. This feature is especially important when communicating over a wireless network as it helps address the issues of intermittent connectivity, dropped connections, and coverage.

Many types of applications take advantage of messaging technology. For corporate applications, messaging is often used in conjunction with other technologies to provide access to enterprise data. Application-to-application messaging excels in this arena, often as an enhancement to smart client applications, whereas, HDML alerts and WAP Push are very valuable for enhancing thin client applications. SMS can be a valuable service for many businesses, but it is not typically integrated into corporate applications. Over 80 percent of all SMS messages are sent from consumer applications. Over time, more business may adopt text messaging, but the initial market focus is clearly on the consumer. After we examine the advantages and disadvantages of messaging, we will take a look at specific consumer and corporate areas where messaging technology is beneficial.

Advantages of Messaging

Here are the advantages of messaging:

Push capabilities. Wireless Internet and smart client applications can be enhanced with push messaging functionality. For wireless Internet applications, the message may contain a URL where the user can go to get more information. For smart client applications the push message could notify the user that he or she should synchronize the client data; or in some cases, the message may actually be able to initiate the synchronization automatically from the server.

Store-and-forward. When working over wireless networks, an important capability is for messages to be stored in a queue and delivered when connectivity is available . On the client, it means users can work as if a network connection is available, since the information will be forwarded to the server at a later time. On the server, it is beneficial because messages will be sent to users when they connect to the network, offering a form of guaranteed message delivery.

Personalized data delivery. The leading messaging products give end users the ability to register their messaging-enabled devices and schedule the delivery of messages to a particular device. For example, the schedule might dictate that between 9:00 A.M. to 5:00 P.M., a message should be delivered to an email account, but between 5:00 P.M. and 11:00 P.M., it should be sent as an SMS text message. In this way, users can receive messages at the location that is most convenient for them.

Wired and wireless communication. Application-to-application messaging can operate over both wired and wireless networks. This gives companies the option of providing wireless access to the application users. It also allows application users to connect through a wired connection when wireless coverage is not available. User-to-user messaging and notifications and alerts typically require wireless networks for communication.

Disadvantages of Messaging

Messaging is an optional component for many applications, so typically it is incorporated into applications only where there is a clear benefit. For this reason, there are not many disadvantages to using messaging, other than when it is simply not required.

However, messaging does add complexity to the development of the mobile application. For application-to-application messaging, development is required on both the client and the server. For text messaging and notifications and alerts, most, if not all, of the development occurs only on the server. In either case, incorporating messaging functionality into an application does require additional development and testing resources.

Sample Consumer Applications

Messaging is extremely popular in consumer applications. Whether using SMS, IM, email, HDML notifications, or WAP Push, consumer applications are usually focused on communication, entertainment, or information dispersal. In all cases, the messages are typically text-based, although this will change as more advanced messaging services grow in popularity.

Communication

Messaging technology is often used for person-to-person communication from something as simple as saying hello to asking a question, arranging a meeting, and to conduct other general interactions. It works well when making a voice call is inconvenient, inappropriate, or too expensive, and when the user wants immediate delivery of messages. SMS is currently the most common mobile messaging technology used for communication, although this may change as IM adoption grows, which will definitely be the case when IM vendors allow for interservice communication between different wireless systems and wired Internet users.

Entertainment

Many entertainment services use messaging technology. Some of the most common include viewing sports scores, reading horoscopes, and exchanging ringtones with other users. SMS can be used for this service, but push technologies such as HDML notifications and WAP Push are better suited. These technologies will include a URL that the user can visit to get more information, usually resulting in higher wireless usage. This benefits both the application provider and the wireless carrier.

Information Dispersal

Perhaps the most useful implementation of messaging in the consumer space is information dispersal. Applications that alert users of, for example, flight information, stock information, and weather forecasts, to name just three, can be very helpful to the mobile user. Essentially, any information that can be incorporated into a message, be it SMS or otherwise, can be dispersed using messaging technology. For urgent messages, push technologies are the best choice, while for general information, a pull-based solution will prove to be less intrusive to the user.

One concern for many people regarding push messaging is the potential of receiving unwanted messages, similar to spam on email. Mobile users are concerned that push messaging will continually interrupt them throughout the day. In addition, push clients have limits on the size and number of messages that can be stored on the device. For example, the Openwave client allows only 1500 bytes to be stored on the device, with each message containing a maximum of 160 bytes. This allows for as few as nine messages before the client reaches capacity.

Already applications are out there that will push information such as advertisements and promotions to mobile users based upon their location. While this is a great service if you are a subscriber, it will quickly become annoying if you are not interested in receiving this kind of message.

Sample Corporate Applications

Corporate applications can reap tremendous benefits by incorporating messaging technology. Whereas consumer applications benefit largely from the communication aspect, corporate applications gain by making enterprise data more readily available. Text-messaging technologies such as SMS and IM do not currently play a large role in corporate applications; application-to-application messaging and some of the push technologies are more conducive to the requirements of enterprise applications.

System Integration

On the server, messaging technology can provide a way to integrate disparate systems. In most organizations, enterprise information is stored in multiple sources, most often with different mechanisms for accessing the data. Messaging systems can provide access to these sources of enterprise data and relay the information to mobile users. The goal is to implement a solution that will require very little, if any, change to the existing systems, while providing a common layer for accessing enterprise data. Systems based on JMS will help meet this goal, at the same time providing integration with other J2EE technologies such as EJBs and JDBC.

Information Exchange

Messaging systems are very effective for exchanging information between mobile and nonmobile applications alike. New mobile applications can be based exclusively on messaging technology. Applications to access corporate email, dispatch work orders, and update inventory levels or pricing information and customer relationship management (CRM) applications are commonplace in the mobile workforce. All of these can be built using messaging as the transport of data between the mobile user and the enterprise system.

In the case of job dispatch, messaging technology is useful both for scheduling new jobs and for retrieving the status of existing ones. The field agent can be notified of an upcoming work order by a text message such as SMS or by a more advanced mechanism using application-to-application messaging, which could add the order directly into a calendar or custom application. Once the service agent has completed the task, he or she can send a message back to the office to keep staff there informed of the current status. This message could also contain notes or other information to keep the customer up to date on the processing of the order.

Application Enhancement

In addition to creating new mobile applications, messaging technology is also useful to enhance existing applications. Many smart client applications can benefit by adding new features such as server-initiated synchronization or store-and-forward messaging, while thin client applications can benefit from HDML or WAP alerts. In most cases, it is possible to add new messaging functionality without having to make major changes to the existing application. In this way, messaging technology can be incorporated into nearly all the smart client and thin client applications that were discussed earlier in this chapter.

Event Propagation

When events occur within corporations that require the attention of some form of service staff (whether it involves repairing a malfunctioning system, fixing a database conflict, or replenishing inventory levels) messaging technology is a great way to notify workers that their attention is required. When a problem arises, the messaging server can take the request, and notify an individual user or a group of users using a variety of technologies. In this scenario, sending out a message to a pager may be the ideal choice for one worker, while an SMS message might be better for another. At the same time, the system administrator may want to receive an email message each time a problem arises to monitor response times. Messaging systems are well suited to handle these types of scenarios.

Other Architecture Selection Considerations

It would be short-sighted to base your device and network selections on a single application. The overall success of mobility in an organization is ultimately based on whether the application users are satisfied with many factors of the solution, even if some of them are totally unrelated to the line-of-business application being deployed.

The best example of this is access to email. Once wireless devices are introduced to an organization, the first request is often whether the selected device will allow for access to corporate email. As the application architect, you may have decided on implementing a smart client solution using a Pocket PC device with wireless access provided by a wireless PCMCIA card. Then you discover that another group has just made the decision that wireless email is a requirement for any solution that is deployed, and in order to get approval for the business application you have designed, a device that is more conducive to email has to be selected. For a variety of reasons, RIM BlackBerry devices are great for accessing email. This is now the device that has to be used, even though there are many good corporate email solutions available for the Pocket PC platform. This decision, made without consideration of the original application has limited the flexibility of the solution you can provide. You might think that this is an isolated case, and that it is not relevant to your situation, but it occurs more often than you might expect.

The next most common request is for Internet access. Even if the solution being deployed uses a smart client architecture, the application users will inevitably want to access Internet content such as stock quotes, weather forecasts, or personal email. So now there is a requirement to select a device that has a microbrowser for wireless Internet access. In most cases this will not be too much of an inconvenience since nearly all wireless devices provide some form of Internet access. This is true of all device classifications (see Chapter 2, "Mobile Devices").

A similar situation is when a company decides that it is going to take advantage of devices already in use. These devices may have been supplied by the company for other applications or they may be devices that employees have purchased on their own and expensed to the company. In either case, you are usually looking at supporting a wide range of devices ranging from WAP phones to high-end PDAs. If you are mandated with developing an application that can take advantage of these devices, then you once again are limited in the architecture that you can choose. In this case, the outcome is usually that a wireless Internet solution is the only option since developing a smart client application for multiple device types running multiple operating systems may go beyond the original scope of the project.

Of course, choosing a device is not the only time when miscommunication between departments causes setbacks or even failure. Talking to the network administrators *before* deciding which method to use to get past the firewall is always good advice.

There are, of course, exceptions to these scenarios. For instance, specialty shops that cater to a certain type of device can usually be guaranteed to provide certain capabilities. Furthermore, if your company has recently mandated a certain device for use in all or certain circumstances, then provided you are quick to market, you can be reasonably assured of your target devices. And, most likely, your company already has in place at least a few methods that enable external devices to access corporate resources, of which you can take advantage.

The overall message to application architects and developers is to expect the unexpected, as there are many unpredictable factors that will impact the application architecture, device, and network selections. It is a good idea to plan ahead so you do not waste resources on decisions that are out of your hands. In general, plan to have email and wireless Internet access as a component of any solution you are deploying, even if they are not part of the specific application you are developing.

Summary

Developing and deploying a solution that provides mobile access to enterprise data sources is challenging. In order to create a successful solution, many factors have to be taken into consideration. These include wireless connectivity, device type, target user base, enterprise integration, power consumption, and security. Three mobile architectures can be used to address these issues: wireless Internet, smart client, and messaging. Each of these systems has advantages and disadvantages that have to be weighed when making any decision. In many cases, the idea solution will include a combination of two, or possibly all three, architectures.

Part II of this book is dedicated to smart client application development, and Part III to thin client development. But before we move on to those sections, in the next chapter we will take a closer look at mobile wireless messaging technologies.

Helpful Links

Web links relevant to the architectures discussed in this chapter are provided in their respective chapters throughout the book. Here you will find links to common mobile development and information Web sites. These sites are a good source for getting up-to-the minute information on the mobile and wireless industry.

Cellular Telecommunications and Internet Association (CTIA)	www.wow-com.com
FierceWireless daily newsletter	www.fiercewireless.com
m-Business Daily	www.mbizcentral.com
MobileInfo Portal	www.mobileinfo.com
ThinkMobile Portal	www.thinkmobile.com
Unstrung	www.unstrung.com
Wireless Business and Technology Magazine	www.sys-con.com/wireless
Wireless Developer Network	www.wirelessdevnet.com
Wireless Internet Daily newsletter	www.wirelessinternetdaily.net
Wireless Week	www.wirelessweek.com

CHAPTER 5

Mobile and Wireless Messaging

The term *messaging* can mean many things. To some people, it may mean email; to others, text messaging; to others still, smart application-to-application messaging. Of course, all of these, as well as others, are valid messaging formats. As discussed in the previous chapter, messaging technology can be used by itself or in conjunction with smart client or thin client applications, but however it is used, messaging technology will be key to implementing a successful mobile solution.

In this chapter, we take a look at the key messaging technologies currently available. We will start with common messaging systems such as email and paging, then move on to SMS, EMS, and MMS, and finish with push and application-to-application messaging. After gaining an understanding of the various messaging systems, we will cover the messaging value chain, ranging from device manufacturers to messaging middleware providers.

NOTE Messaging technology relates equally well to both smart client and thin client applications, therefore it is included in the first part of the book rather than in the parts dedicated to specific application architectures. This is intentional, as messaging technology is a great way to tie together both application types.

In Chapter 4, "Mobile Application Architectures" we answered the question why we would want to use messaging for our applications. Here we will look at how this can be accomplished.

Messaging Basics

We begin by looking at some basics of messaging systems. The two areas of importance at this point are the differences between asynchronous and synchronous messaging, and between push and pull for message retrieval.

Asynchronous versus Synchronous Messaging

One of the fundamental benefits of messaging is its asynchronous nature. Asynchronous operations are nonblocking. This means that the sender of the message only initiates the operation, and does not have to wait for a response before he or she can continue working. If and when a response does appear, the application can then receive that message and respond to it appropriately. Even if the message takes some time to complete, the user does not notice the delay since he or she can continue to perform other tasks while waiting for a response.

This is the opposite of a synchronous operation, which blocks the process until the operation is completed. In these cases, the application is on hold until the operation is finished. Remote Procedure Call (RPC), a term used to describe the distributed computing model implemented by many middleware technologies such as CORBA, Java Remote Method Invocation (RMI), and Microsoft's Distributed Component Object Model (DCOM), is synchronous in nature. These technologies wait for a response before they continue. An example of synchronous communication is wireless Internet applications. Once a Web page is requested, users have to wait until the content comes back before they can continue working. This can lead to a poor user experience if the application is not designed correctly or if the network has high latency or low bandwidth. Unfortunately, both are common in today's wireless networks.

Asynchronous execution of messages is usually superior to synchronous execution, even on unreliable networks; that said, there are times when a synchronous response is required. An example would be an application that requires credit approval before making a purchase. You do not want to continue the purchase until the credit card is confirmed, thereby making synchronous messaging better suited for the task. The point is, when you are looking at implementing messaging within your applications, it may be worthwhile to determine whether you will need either synchronous or asynchronous messaging, and choose a solution that will provide the capabilities you require.

Push versus Pull

One of the common uses of messaging technology is to disperse information, such as notifying a mobile user when a specific event has occurred in the enterprise. The goal is for this type of notification to happen seamlessly, without user interaction. This is often referred to as push messaging. A message is pushed from the enterprise to the mobile user without the client having to request the information manually. In practice, rarely is the information actually pushed to the user without the client doing some form of check to see if new data is present.

Most wireless networks and devices do not support true push technology, but instead use a polling mechanism on the client device. Instead of users requesting new information, as they would in many smart client and thin client applications, the client application periodically polls the server to see if there is a new message for this client. If there is, it will then retrieve the message and react accordingly, often displaying some information to the application user. From the user's perspective, the information was pushed from the server, when in reality a pull mechanism was used.

Using this technique, you can enhance many smart client applications. When a new set of data is available on the server, you can initiate the data synchronization procedure from the server, allowing the data to be updated on the client device without the user having to constantly check for updates. For thin client applications, you can send out HDML notifications or WAP Push messages to notify users of new information, or link them to a wireless Web site that will be of interest to them. (More information on push messaging can be found in the sections "HDML Notifications" and "WAP Push" later in the chapter.)

Types of Messaging

In this section, we are going to take a look at each of the many different forms of messaging technology, giving an overview of how it works and where it fits into the development of the mobile applications. We start with the most common forms of messaging, such as email and paging, and work toward some of the newer and more advanced formats including WAP Push, instant messaging, and application-to-application messaging.

Email

When it comes to messaging, email is the killer application. This is definitely true for PC users and more and more so for mobile and wireless users. Every day, billions of email messages are sent; some users send and receive more than a hundred each day as a matter of course. Email has become the preferred means of communication for many companies, providing a quick and easy way to move information between users, often in the form of standard text or business documents such as Microsoft Word or Excel.

For mobile users, email is also a top priority. Many organizations implement email as their first mobile application, allowing users to get a feel for the mobile device and wireless connectivity before they implement other, more focused business applications. Companies such as Research In Motion (RIM) have built a strong following for their devices due to their capability to retrieve and send email messages. Many other companies have developed technology that enables mobile users to access their enterprise email accounts while away from the office. They provide server software that can integrate with common groupware infrastructure, such as Lotus Domino or Microsoft Exchange, as well as client software that can be used to access the messages. This client software is often available in both smart client and thin client formats.

NOTE Since most readers of this book are familiar with sending and receiving email messages, this section provides only a brief overview of the components that comprise an email system.

Email systems employ store-and-forward messaging. That is, messages are stored in a repository until they are accessed by a client application, at which time they are forwarded to the user. The email client can be either a stand-alone client, such as Microsoft Outlook, Lotus Notes, Eudora, or Pegasus, or it can be browser-based, such as Microsoft Hotmail, Yahoo! email, or any of the other Web email offerings. Each of these client applications has to communicate with an email server to receive messages, and another to send messages. The two most widely used servers are the Simple Mail Transfer Protocol (SMTP) server, used to send messages, and the Post Office Protocol (POP) server, which stores incoming mail. The most recent version of POP is 3, which is commonly referred to as POP3. An alternative to using POP is the Internet Message Access Protocol (IMAP). IMAP is similar to POP in that it stores messages, but it accomplishes this in a more effective manner for users who want to access their email messages from multiple machines.

The process for accessing email from a mobile device is very similar. The only differences are the client application used to interact with the email servers and the type of attachments that are supported. For obvious reasons, some types of email attachments may not be supported from your mobile device, due to either size or format restrictions.

SMTP Server

When you send an email message from your client application, the email client will communicate with the SMTP from your email provider. This SMTP server will then look at the address to which this message is being sent and communicate with the SMTP server at the destination address domain (the part following the @ symbol). When the SMTP server at the destination receives the message, it communicates with the POP3 server at the same domain and puts the message into the recipient's account (the part preceding the @ symbol). From this point on, the POP3 server is used by the recipient to access this message. If, for some reason, the SMTP server at the destination domain cannot be reached, the email message will go into a queue, often called the sendmail queue. This queue will periodically try to resend the message to the destination SMTP server. If the message cannot be sent after a defined period of time, very often set at four hours, the sender may get a return message stating that it was not yet delivered. After a longer period of time, very often five days, the SMTP server will stop trying to send the message and will return it to the user marked as undelivered.

POP3 Server

The POP3 server is used to store received email messages. When users check their email, the client application connects to their POP3 server, providing an account name and a password. Once users are authenticated, the POP3 server allows them to access their stored email messages. The POP3 server essentially acts as an interface between your email clients and the data store containing your email messages. [For more detailed information on the mobile aspects of personal information management (PIM), including email, calendars, and to-do lists, see Chapter 16, "Mobile Information Management."]

Paging

One of the earliest forms of messaging to mobile devices was paging. Paging typically involves a caller dialing a telephone number associated with the intended recipient of the page. Once connected to the paging terminal, the person sending the page can enter a message that will be sent to the pager. The message can be either numeric, alphanumeric, or voice, depending on the system being used. When the message is complete, the paging terminal converts the message into a pager code and sends it to a series of transmitters to which it is connected. These transmitters then send out the message as a radio signal throughout the entire coverage area. Every pager within this area on the particular frequency will receive the message, but only the pager with the proper code (the intended recipient) will be alerted. In essence, the pager works much like an FM receiver.

The majority of pagers are used as one-way communication devices. When the user receives a message, that person responds either by calling the number sent in the page or by some other appropriate means. This is an effective way to alert people that you want to get in touch with them or have them carry out a task. Over the past few years, the capabilities of pagers have increased dramatically. Many of today's pagers provide two-way communication of various forms of data, ranging from short alphanumeric messages to wireless Internet content and email. While the underlying communication mechanism has remained the same, the client uses have become much more advanced. The most common two-way paging devices are the Motorola TimePort and RIM's BlackBerry.

Short Message Service (SMS)

The Short Message Service (SMS) was first introduced in Europe in 1991 as part of the GSM Phase 1 standard. Since that time it has had tremendous success, with more than 1 billion messages sent around the world daily. Though it continues to be more popular in European countries, SMS is catching on in North America as well, as more of the major wireless carriers add support for SMS and as SMS-capable devices start to proliferate in the marketplace. SMS is supported on digital wireless networks such as GSM, CDMA, and TDMA.

SMS makes it possible to send and receive short text messages to and from mobile telephones. The message can contain alphanumeric characters to a maximum length of 160 characters for Latin alphabets, including English, and 70 characters for non-Latin alphabets, such as Chinese. It provides an easy way for individuals to communicate with one another and with external systems. In this way, SMS can be used to turn a cellular phone into a pager. By adding the capability to receive short messages, users can receive the same kind of information that is typically received on pagers. This eliminates the need to carry both a pager and a cell phone, as SMS messages can even be received during a voice call.

NOTE Even though the standard maximum length of an SMS message is 160 characters, many carriers have set their own limitations, which range from 100 to 280 characters.

New technologies such as SMS have gained widespread adoption because they offer compelling benefits to both the user and the network provider, such as:

Guaranteed message delivery using a store-and-forward approach. Even when the recipient is out of the coverage area or does not have his or her device turned on, that person will still receive the message, though at a later time. Note, however, when a user does not retrieve the SMS message for an abnormally long period of time, the message may be removed from the system.

Ease of use, without additional software or hardware. Sending or receiving SMS messages does not require a WAP browser. The ability to communicate using SMS comes pre-installed on the mobile device.

Low-cost method for information delivery. SMS provides an alternative to making voice calls to deliver information.

Revenue source for service providers. Network operators can deliver value-added services using SMS to generate revenue. SMS also helps drive the adoption of mobile phones to younger audiences.

It is estimated that close to 80 percent of SMS messages sent are consumer-oriented. The most common use is peer-to-peer communication, often replacing the need to make voice calls. Other uses include information services, such as stock and weather information, simple email access, and advertising. For corporate applications, SMS can be used for vehicle-positioning applications, job-dispatch services, and remote monitoring.

Unfortunately, when it comes to sending SMS messages from places other than another cellular phone, the process becomes more difficult. This is mainly due to the proprietary interfaces to each operator's short message service centers (SMSCs). In Figure 5.1, you can see the route that a message takes as it moves from its origin to the destination mobile phone. Once the message is constructed, it is sent either over a wired connection or a wireless connection to the SMSC for that particular carrier. If the message originates from another mobile phone, the carrier takes care of the message delivery through the SMSC to the destination device. If however, the message originates from another source, such as an enterprise server, it is the enterprise's responsibility to communicate with the SMSC to have the message delivered. This is not a simple task, as most carriers do not expose their SMSC's APIs. In order to communicate directly with a carrier's SMSC, you either have to establish a relationship with that particular carrier and write to their proprietary interface, or use an aggregator that has established connections to the carrier or carriers you are interested in. The standard protocols used to communicate with SMSCs include Telocator Alphanumeric Protocol (TAP) and Short Message Point to Point (SMPP). (For more information about messaging aggregators, see the section entitled "Messaging Value Chain," later in this chapter.)

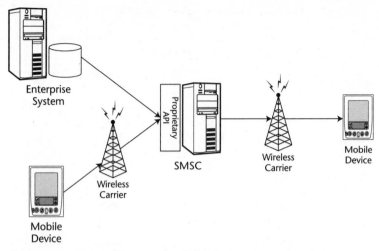

Figure 5.1 SMS architecture for delivering a message.

There is one other possible method to send SMS messages. Many carriers have made SMTP access available to their systems, allowing users to send SMS messages using an email interface. This usually involves sending an email message to an address containing the destination user's telephone number and carrier domain. For example, a message sent to an AT&T phone may have the address 8005551234@mobile.att.net. When this message is sent, it is routed through the SMTP server to the SMSC, where it is then delivered to the mobile recipient over a wireless network.

Communicating directly with the SMSC does make it possible to retrieve the status of a message, which SMTP cannot. This comes at a cost, however, as there is usually a per-message fee for sending messages via the SMSC via a system aggregator.

The many benefits of SMS messages are easy to identify, but there are also some major limitations. The most serious of these is the 160-character maximum message length and the lack of support for more advanced data types such as audio and graphics. Another limitation is the lack of interoperability between network operators. The ability to send a message from one carrier to another has only recently been enabled by T-Mobile (formerly VoiceStream), AT&T Wireless, Verizon Wireless, and Cingular, thereby removing a major obstacle for SMS adoption in North America. Even with the cooperation between North American carriers, there are still limitations that need be overcome, including roaming outside of North America and the ability to encrypt the message content.

Enhanced Message Service (EMS)

The Enhanced Message Service (EMS) adds powerful new functionality to SMS. In addition to being able to send text, EMS allows users to send richer content, including pictures, animations, sounds, and formatted text. EMS can be added to existing SMS infrastructures, saving operators from having to make large investments to add these new features. This should help drive the adoption of EMS until more advanced messaging services, such as Multimedia Message Service (MMS) are rolled out in 2003 and 2004.

The following is an overview of the features introduced in EMS:

Sounds and melodies. Users are able to send and receive either predefined sounds for such things as notifications or melodies to add new ringtones to the phone. It is possible to combine several sounds and melodies into a single message and to combine them with text and pictures.

Pictures and animations. Users can send and receive predefined pictures that are included on the phone, as well as new pictures that they create or download from the Internet. Some phones may also contain the capability to edit the picture directly on the phone using a built-in picture editor.

Formatted text. In addition to plain text, EMS makes it possible to format text. This may include changing font sizes; using text attributes such as boldfacing, italics, or underscoring; and changing text alignment. This feature will help to make news items and information updates more attractive.

Concatenated messages. To help overcome the message size limitations, EMS allows for message concatenation. This can be accomplished directly on the mobile phone for both sending and receiving messages. This is very important for messages containing rich content, since EMS is still limited by message size as defined in SMS.

There is some doubt on whether EMS will truly succeed as a messaging standard. Ericsson is promoting it heavily, but other vendors, including Nokia, are not embracing EMS as warmly—for obvious reasons: Nokia has a competing messaging format called Nokia Smart Messaging, which the company is promoting until carriers and devices support MMS. The bottom line is that it looks as though SMS will continue to dominate for text messaging, and MMS will be the leader for multimedia content, leaving EMS with little hope for widespread adoption.

Multimedia Message Service (MMS)

The Multimedia Message Service (MMS) takes the capabilities of EMS one step further, adding true richness to the message content. In addition to the capability for pictures, formatted text, and sound introduced in EMS, MMS also provides support for voice, audio and video clips, and presentation information. This is accomplished in a manner very similar to SMS: providing automatic immediate delivery for custom content, as well as store-and-forward capabilities when the recipient is unable to receive the message. MMS also adds support for email addressing, so messages can be sent to an email address from the MMS client.

MMS has been standardized by both the Open Mobile Alliance (OMA) and 3rd Generation Partnership Project (3GPP). OMA's MMS specification defines the message encapsulation and application protocols, while the 3GPP specification defines the network architecture and general functions. The transport for MMS is accomplished using WAP transport, making MMS bearer-independent, therefore not limited to GSM or WCDMA. This also makes it possible to use WAP Push features to deliver the message from the server to the recipient. (Note: For MMS to work, WAP 1.2 or above is required, although it will most likely be implemented alongside WAP 2.0 in many cases.)

To enable true multimedia content, the SMS message size limitation and the transport mechanism had to be discarded. And to avoid the problems encountered with SMS, and to enable future interoperability, no maximum size has been specified for MMS messages. This leaves the message size open to the implementation of each operator. That said, the message size will still be defined, but by the bandwidth and mobile device storage capabilities. It is expected that the first generation of MMS messages will typically be between 30 KB and 100 KB in size—a dramatic increase over the levels available in SMS. The obvious drawback to this message size is that many of today's wireless networks do not provide the bandwidth to support it. For this reason, MMS is a technology that requires 2.5G wireless networks, with a minimum bandwidth of 14.4 Kbps.

To help reduce the perceived wait times to download MMS messages, the MMS centers (MMSCs) use a store-and-automatic-forward mechanism to deliver the messages. The MMSC is a similar concept to the SMSC for SMS messages, whereby the MMSC can temporarily store a message for the time required to find the receiving phone. Once the receiving phone has been located, the message is immediately forwarded to the intended recipient and deleted from the MMSC. MMS messages cannot be sent without going through the MMSC. If the MMS message originates at an enterprise server rather than another mobile phone, the application developer will be responsible for integrating with the MMSC API to send the message. As with SMSCs, each vendor's MMSC will have its own API with which to interface, adding additional complexity. Unfortunately, unlike SMSCs that provide an SMTP interface, MMSCs are not expected to deliver this capability.

The first generation of MMS messages are laid out as slide shows. Each slide show will contain at least one slide, divided into two sections, one for text and the other for multimedia. The slides simply define the layout, while the actual content, such as video, audio, and text, are separate pieces sent along with the slides. These files are incorporated to the slide show using the Synchronized Multimedia Integration Language (SMIL). SMIL is an eXtensible Markup Language (XML)-based language specified by the World Wide Web Consortium (W3C); it is used to control the presentation of multimedia elements. Within the SMIL specification is a set of tags that can be used for defining images, text areas, and layouts. It is very similar to HTML, making it an easy transition for Web developers. (If you are interested in learning more about SMIL, visit www.w3.org/AudioVideo for full detailed information on the SMIL 2.0 W3C Recommendation.)

Multimedia messaging on wireless devices is definitely of interest to many consumers and corporate users alike. Unfortunately, as of early 2002, only Japan's NTT DoCoMo was offering commercial MMS support on its 3G network. Trials are underway in other regions in Asia and Europe, but the lack of devices and suitable wireless networks is delaying its availability. In North America, adoption is even further off. There, SMS is just starting to catch on, and few users are willing to pay to download pictures and videos wirelessly when they can download them for free at home. It will still be a few years before MMS is available throughout the United States and Canada.

Instant Messaging

Instant messaging (IM) is well positioned to be the next killer application for the wireless industry. With the monumental growth rate of SMS, and more than 100 million desktop instant messaging users, the potential for wireless instant messaging is incredible. It provides similar capabilities to other two-way messaging technologies, such as paging, SMS, and email, with the addition of one significant feature: *presence*! Presence is so elemental to IM that this form of messaging is often referred to as Instant Messaging and Presence Services (IMPS).

Presence lets users know the current status of the people with whom they are conversing. This introduces a new way of communication. Presence information can include device availability, device capabilities, user status, location information, as well as personal information such as the user's mood. When a user wants to send a message to another party, he or she can first check the status of the intended recipient to make sure that person is available. Based on the presence information, the user may decide to send a message, try another means of communication, or simply wait until later. This is an important concept because instant messaging does not have store-and-forward capabilities. When a message is sent, it goes directly to the intended recipient. If that person is not able to receive the message, it is lost; it is not sent at a later time. This differs from how email, SMS, EMS, and MMS work.

Instant messaging has been available for fixed Internet users for some time and is very popular. Adding the mobile aspect to these services will enable users everywhere to communicate with one another, regardless of their type of connectivity. Unfortunately, before this can happen, and/or mobile instant messaging can meet its potential, the proprietary nature of the leading IM services will have to be resolved. Interoperability between IM services will be a key ingredient to its success. The leading desktop instant messaging services, such as AOL Instant Messenger, MSN Messenger, Yahoo! Messenger, and ICQ, do not allow for cross-service communication. Users can only communicate with others using the same vendor's product, resulting in many users having multiple IM clients on their PCs. In the mobile market, having multiple clients will not be an option, and in some cases, users will be required to use the IM service that comes preinstalled on their device.

Several mobile instant messaging clients are already available, including those from Microsoft, AOL, and Openwave, and the list is sure to grow. To promote interoperability, and in turn drive IM adoption, Nokia, Motorola, and Ericsson are involved in a joint effort called Wireless Village (which is now a component of the Open Mobile Alliance). Their goal is to create a set of standard specifications for handset makers and carriers to follow. This will enable all users to communicate with each other using instant messaging regardless of the device or carrier they are using. If successful, a user at his or her desktop will be able to send a message to a wireless user across the country with little effort, even if the recipient is using a different IM service.

The potential for mobile instant messaging is tremendous. As devices with always-on capabilities penetrate the market, the opportunity for IM will become even stronger. Some believe that IM will be the answer to the slow uptake of SMS services in North America, which would be a welcome relief for wireless carriers who are looking to capitalize on the growing messaging market.

HDML Notifications

HDML notifications were the first form of push messaging available to mobile Internet users. They allow for asynchronous communication, allowing the server to send relevant information to clients in a timely fashion—similar to SMS text message functionality. They are different from SMS in that they interact with the device's microbrowser. This interaction can take on many forms, including:

Alert. A text message sent to the browser that will beep or display a visual signal to notify the user that new information is available. The alert will often contain a URL, which, if selected will load the URL's deck or page and display the content on the device's microbrowser. These notifications are often referred to as actionable alerts.

Cache operation. These notifications can remove certain URLs or all URLS from the microbrowser's cache. This is done to prevent obsolete content from being viewed and enacted upon before the specified time to live (TTL) is reached. This type of operation can occur behind the scenes, without the user's involvement.

HDML decks, images, and digests. These can all be preloaded into the microbrowser's cache to make interaction with the application more efficient.

To send one of these notifications, you are required to know the subscriber ID of the target device. If you do not have this information available, a script is available at http://developer.openwave.com/dev/sdk.40/scripts/whoami.cgi that can help extract the subscriber ID. Once you know the ID of the device to which the message is being sent, you then need to interact with the carrier's HDML gateway. Methods of accomplishing this are covered in the "Mobile Message-Oriented Middleware" section later in this chapter.

There are two delivery channels for HDML notifications: the push channel and the pull channel. On packet-switched networks, all data transmissions are treated the same, allowing for push delivery of information, regardless of the size. Circuit-switched networks, on the other hand, use SMS to deliver asynchronous messages, preventing the HDML gateway from delivering messages that exceed the SMS message size limits. In all cases, the push channel is meant for delivering time sensitive material, using only alert or cache operations. The pull channel is better suited for data that is not critical, and for preloading content into the microbrowser.

Once an alert is sent to the HDML gateway, it is queued for delivery. The length of time it spends in the gateway's queue depends on the following information:

- For all push notifications and for pull notifications on packet-switched networks, the gateway will attempt to deliver the message immediately. If the destination phone is unavailable, the gateway will keep the message in its queue and reattempt to deliver it periodically. If the message TTL is exceeded, it will be removed from the queue.

- For pull notifications on circuit-switched networks, the message will remain in the queue until the destination phone opens up a circuit. At this time, the message will be sent to the user for viewing. If the message TTL is exceeded, it will be removed from the queue.

Any notification that is in the gateway's queue but has not been delivered is referred to as a pending notification. The sender of the notification can request to delete or get the status of any pending notification. The sender can also request the status of completed notifications. This capability is important for applications that require guaranteed message delivery.

HDML notifications provide a powerful way to push content to wireless users. In North America, where HDML is still widely used, these notifications are often the only option available for push services. However, because HDML notifications are a proprietary messaging technology developed by Openwave, they are only supported in Openwave microbrowsers. As HDML gateways are replaced by WAP gateways, and HDML handsets are replaced by WAP handsets, HDML notifications will gradually give way to WAP Push for push messaging capabilities.

FURTHER INFORMATION The best source of information on HDML notifications is the "SDK Developer's Guide," version 3.2, which is available at http://developer.openwave.com/support/techlib.html.

WAP Push

WAP Push, first introduced in the WAP 1.2 specification, is the successor to HDML notifications. You will notice that its push capabilities and delivery channels are very similar to those used for HDML notifications. The main difference is that WAP Push is an industry standard for pushing content to WAP-enabled devices, defined by the WAP Forum. This allows for multiple gateway vendors, microbrowser providers, and wireless carriers to communicate using the same protocol, making it easier for application developers to create wireless Internet applications with push capabilities.

Architecture

Figure 5.2 depicts the WAP Push framework. Three parts of this framework are of interest to us:

Push Initiator (PI). The PI is an application that pushes the content and delivery instructions to the Push Proxy Gateway (PPG). It typically runs on a standard Web server and communicates with the PPG using the Push Access Protocol (PAP).

Push Proxy Gateway (PPG). The PPG does most of the work in the push framework. Its main responsibility is delivering push content to the WAP client. In doing so, it may have to translate the client address into a format understood by the wireless carrier. The PPG is also the location where messages are stored when they cannot be immediately delivered to the client. It also maintains the status of each message, allowing the PI to cancel, replace, and request the current status of a message. The PPG uses the Push Over-the-Air (OTA) Protocol to deliver push content over a wireless network.

To send a WAP Push message, the PI must have two sets of information about the destination: the domain of the PPG and the client address. An arbitrary text string, such as an email address, can be used to identify the client. The PPG is then responsible for translating the string into a format that is understood by the mobile network.

WAP client. The WAP client is typically a wireless device that contains a WAP microbrowser capable of receiving WAP Push content. This is where the user is able to view the content that was pushed from the PPG using the Push OTA Protocol.

The PAP has been designed to be independent of the underlying transport, making it suitable for any protocol that allows for the transport of MIME types over the Internet. Currently, HTTP is the only protocol specified as a transport, although new protocols may be added in the future. The content for a push message is specified using XML with a defined set of tags.

Within the XML document, the PI is able to communicate the following operations to the PPG:

Push submission. Contains the push message itself.

Push cancellation. A request to cancel a previously submitted message.

Push replacement. A request to replace a previously submitted message.

Status query. A request of the status of a previously submitted message.

Client capabilities query. A request for the capabilities of a particular device on the network.

In turn, the PPG is able to respond to the PI with one operation: result notification. This response is an XML document that contains information indicating whether the message delivery was successful or unsuccessful. A successful delivery occurs only after the target application on the WAP device has taken responsibility for the pushed content. If it cannot take that responsibility, it must abort the operation, in which case, the PI will know that the content did not successfully reach its destination.

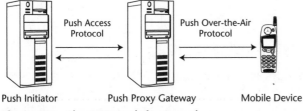

Figure 5.2 The WAP Push framework.

The Push OTA protocol is responsible for transporting the content over the wireless network from the PPG to the WAP client. It can run on top of either HTTP or the Wireless Session Protocol (WSP). The OTA-HTTP protocol is primarily used with IP bearers, while the OTA-WSP is used with WAP bearers.

Operations

The WAP Push specification adds three new MIME types for delivering WAP-specific content from the PI to the WAP client. These are supported in addition to any MIME media type that is already available. The following is a summary of each of these MIME types:

Service indication (SI). Provides the ability to send notifications directly to end users. These notifications can contain information directly in the message or in a Uniform Resource Identifier (URI) that directs the user to a service, often containing WML content. When the user receives an SI message, he or she has the option of visiting the URI immediately or postponing the SI for later handling. Push messages that contain URI references are often called actionable alerts.

Service loading (SL). Contains a URI that points to content loaded by the WAP client, without requiring user interaction. The SL also contains an instruction dictating whether the content should be executed immediately or placed in the cache memory. The PI is able to control the level of user intrusiveness of the SL message.

Cache operation (CO). Removes individual objects or all objects that start with the same URI prefix from the WAP client's cache. This is typically used when the expiry times of the cached content cannot be specified beforehand.

Note, implementing SI is mandatory for clients implementing WAP Push. Both SL and CO are optional. These operations are very similar to those supported by HDML notifications.

FURTHER INFORMATION A good source for more information on WAP Push is the OMA Web site at www.openmobilealliance.org. Here you will be able to find all of the specifications related to WAP and the WAP Push protocol.

Application-to-Application Messaging

Application-to-application messaging technology is different from all of the messaging technologies we have discussed so far. Rather than using standard client software, application-to-application messaging usually is incorporated to custom applications. This provides nearly unlimited flexibility as to the types of messages that can be sent and how messages are handled.

When implementing application-to-application messaging, the application on the mobile device is smart client in nature. This client application communicates directly back to the messaging server, without going through a gateway from the wireless carrier. This eliminates the requirement of having to communicate with SMSC or MMSC servers. Instead, the client application communicates directly to the messaging server—hence the name application-to-application messaging.

The vendor can choose the communication protocols to use, along with the compression techniques and security features. It is recommended that you select a solution that has addressed all of these issues and that has created a solution that is well suited for wireless communication networks. This involves using a suitable protocol, such as the User Datagram Protocol (UDP), that provides built-in message compression and integrated security, including data encryption and user authentication.

If you plan to deploy your messaging-based application to a variety of devices, it also is a good idea to look for a solution that features a messaging client for the mobile OS, device, and programming language that you will be using. This typically involves C/C++ or Java/J2ME support across the leading mobile operating systems, including Palm OS, Windows CE/Pocket PC, and Symbian OS. Java support is available for all of these operating systems, so native messaging support is not necessarily required.

NOTE More information on developing smart client applications, with an overview of the mobile operating systems, is provided in Part II of this book, "Building Smart Client Applications."

Application-to-application messaging technology can be used to create stand-alone messaging applications or applications in conjunction with other smart client technology such as a mobile database and synchronization. When used on its own, application-to-application messaging technology provides a mechanism for the client application to communicate with enterprise data sources. The messaging server can interact with the enterprise data and then send a message to the client with the data relevant to the mobile user. The user of the application can then respond to this message in the appropriate way. For a field service worker, this may involve adding an entry to his or her list of work orders; for a salesperson, it may involve updating inventory levels or pricing information. Whatever the case may be, messaging technology is a fundamental way to disperse information to mobile workers.

In conjunction with mobile database and synchronization technology, application-to-application messaging can add some significant advantages. The foremost is the ability to add push capabilities to applications. In smart client applications, the application user typically initiates the data synchronization process. This occurs either when users want to send information back to the enterprise or when they want to update the information they have in their client application. Users who are concerned that they do not have the most up-to-date information may end up synchronizing with the server multiple times before they get the new data they were waiting for. This can be a time-consuming and costly procedure. In this scenario, application-to-application messaging technology can be used to push the enterprise data to the client device, thereby ensuring the data on the client is always accurate. This process is often referred to as server-initiated synchronization.

Whether you are creating a stand-alone messaging application or enhancing a smart client application, store-and-forward capabilities are important. Often, mobile users cannot send a message immediately, in which case store-and-forward makes it possible for the message to be put into a queue, from where it will be sent at the first opportunity. Users can be confident that the message will be sent; they do not have to be concerned with the wireless connection. This so-called fire-and-forget feature makes mobile applications much more approachable for application users, especially when they are new to wireless communication.

Application-to-application messaging is proprietary in nature. That means there is no industry standard that defines how to create or deliver messages. This can be both good and bad. It is good because it allows vendors to add advanced features into their products, resulting in very versatile solutions. When vendors do not have to adhere to a predefined specification, they can respond quickly to their customers' needs without having to wait for a new specification to be released. The downside is that there is no standard API to communicate with the messaging client or server, therefore vendors create the interface to their systems however they see fit. This can lead to a steep learning curve for developers who are new to a particular vendor's messaging product.

Fortunately, this does not have to be the case. The Java Message Service (JMS) provides an industry standard way that vendors can follow to implement messaging solutions. JMS is a core technology defined in the Java 2 Platform Enterprise Edition (J2EE). JMS is predominantly used to provide an abstract level of access into different message-oriented middleware (MOM) products. It enables messaging vendors to provide a common programming model that is portable across messaging systems. As a developer, you only have to learn the JMS API to be able to interact with all of the leading messaging solutions on the market. JMS can be used on both the server and the client. For mobile applications, we are starting to see the first JMS clients, or JMS-like clients, developed in other programming languages, including C++. This type of product lets developers use their knowledge of JMS to quickly create advanced application-to-application mobile messaging applications.

Messaging Value Chain

Many stakeholders are involved in the implementation of a complete mobile messaging solution; each providing a relevant part of the technology that goes into creating this type of application. The following are the major stakeholders in mobile messaging:

- Device manufacturer
- Wireless carrier
- System aggregator
- Mobile messaging-oriented middleware provider

Figure 5.3 shows how each of these members of the messaging value chain interact with one another. Starting with the device manufacturer, we will work our way through the value chain describing the role of each technology and its importance to an overall messaging solution. Note, however, that not all of these technologies are required for every messaging solution. For example, you will read later in this section that though a system aggregator is not required for a messaging solution, its service is definitely helpful when working with multiple carriers.

Device Manufacturer

From the application developer's standpoint, the role of the device manufacturer is clear: to create wireless devices that incorporate support for messaging technologies. It is the device manufacturer's job to install the appropriate messaging client software for each type of messaging that will be supported. (The one exception is application-to-application messaging, which typically involves a custom application.) From the user's perspective, the device is the most visible part of the solution, so choose its manufacturer wisely!

Recently, the list of messaging-capable devices has grown significantly. In the past you were limited to a small selection of pagers (one- and two-way) and wireless phones. Today you have the following options (and the list is growing):

One- and two-way pagers. These devices are capable of sending and receiving pages, as well as corporate email and often Internet content. Manufacturers include Motorola and RIM.

Wireless phones. These devices are capable of sending and receiving a variety of messages, depending on the unit being used. The most common messaging technology on these devices is SMS, although MMS and IM are becoming more popular. Manufacturers include Nokia, Samsung, Motorola, and Sony Ericsson.

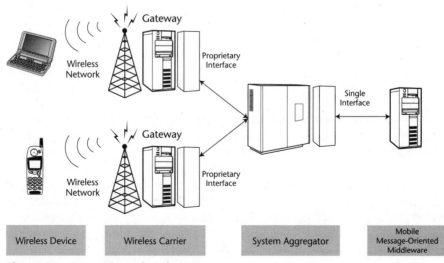

Figure 5.3 Messaging value chain.

Wireless PDAs. These devices allow for custom application deployment, as well as a variety of messaging options. Many PDAs have wireless capabilities using either built-in wireless modems or add-on components. Application-to-application messaging is well suited for the custom applications on these devices. Various forms of text messaging are also supported, depending on the wireless carrier being used. Manufacturers include Palm, Handspring, Casio, Dell, HP, and Sharp.

Hybrid devices. This is a new class of device that incorporates PDA functionality along with voice capabilities. These devices can support the full range of messaging technologies, and are particularly well suited for the more advanced technologies such as MMS and custom messaging applications. This type of device is sometimes referred to as a communicator. Manufacturers include Nokia, Handspring, Samsung, and RIM.

With so many options, it can prove to be a difficult task for end users to determine which device, and over which network, is best for their messaging application needs. In the corporate world, this choice is usually not made by the end user, but rather by the IT development staff, who can choose a device, or multiple devices, that make sense both from a cost and a support and maintenance standpoint. (Refer back to Chapter 2, "Mobile Devices," which gave an overview of device classifications and a list of device manufacturers. This information may be helpful when trying to make a device selection.)

Wireless Carrier

Choosing which carrier(s) to use is a very important decision. It will dictate your wireless coverage, as well as the types of messaging that are supported. In addition, for many forms of messaging, you are required to communicate with a gateway that the carrier provides, such as an SMSC. This often involves setting up a dedicated connection between the carrier and your enterprise, so do your research before selecting the carrier you are going to use.

In many cases, you will make the carrier decision in conjunction with the device selection. This is because certain devices will work only with a particular wireless protocol, which is usually provided only by certain carriers. Each decision you make along the messaging value chain affects other parts of the application, so you need to consider the entire scope of your application before you start. Currently you have the choice of carriers implementing GSM, CDMA, TDMA, CDPD, GPRS, and CDMA2000 1x, along with the 3G protocols. (Refer back to Chapter 3, "Wireless Networks," for insight into wireless networks; it will come in handy when trying to select a carrier.)

In addition to your current messaging needs, you also need to consider the future plans for your messaging applications. Multiple carriers may support SMS messages, but their plans and timeframes for rollout of MMS may vary. This type of knowledge is good to have when making the important decision as to your wireless carrier. It may turn out that you will require more than one carrier to handle your application needs.

This is usually true if your application is being deployed in many countries or across a broad range of geographies. If this is the case, be forewarned that it may add to the complexity of developing your messaging applications, because you may now have to integrate with more than one wireless gateway.

System Aggregator

System aggregators help to alleviate the problem of working with multiple carriers, and in turn, multiple gateways. Rather than having to establish relationships with multiple carriers, you can work with a single system aggregator that manages all of the relationships for you. This aggregator provides a single integration point to multiple carriers, meaning that you have only one point of contact for all of your messaging requirements.

For example, let's say that you are planning on working with three carriers to send out SMS messages. In this case, you will be responsible for, one, establishing a business relationship with each carrier, two, establishing a physical connection to each carrier, and, three, integrating your messaging server with the proprietary interface for each carrier's SMSC. This scenario happens quite frequently. If you are a very large organization, it may not make sense to use a system aggregator, so as to avoid introducing another player to take a piece of the pie. However, if yours is not a large company, or if you want to avoid the hassle of setting all this up, a system aggregator may be the way to go.

In addition to access to messaging systems, many aggregators also provide access to other corporate systems, such as email or Web servers. System aggregators include MobileSys, MobileWay, and RPA Wireless.

Mobile Message-Oriented Middleware (MOM)

The final component of the messaging value chain is the aforementioned mobile message-oriented middleware, or MOM. This technology is used to create messaging systems that allow different software applications to interact with each other. In the mobile world, this can involve a message server interacting with a wireless gateway or directly with the messaging client.

In the enterprise messaging market, several established MOM products are available, including IBM MQSeries, Microsoft Message Queue (MSMQ), and Tibco. These products provide advanced enterprise messaging capabilities such as load balancing, fault tolerance, and transaction support for organizations that need reliable software to deliver a large quantity of messages. When it comes to mobile messaging, the situation is somewhat different. Vendors such as IBM and iAnywhere Solutions provide messaging products geared specifically to meet the challenges of mobile computing. This area of messaging is still relatively new and the market is not completely developed. As more organizations deploy mobile applications, expect to see more of the enterprise messaging vendors expand into the mobile space.

Summary

Mobile and wireless messaging is available in many forms, including email, paging, SMS, EMS, MMS, IM, HDML notifications, WAP Push, and application-to-application messaging. Of all of these messaging options, SMS has so far been the most widely adopted in the mobile space, though in the future it can expect competition from both instant messaging and MMS. For smart client applications, application-to-application messaging is most suitable. Implementing a complete messaging solution will involve many components of the messaging value chain. The only required component of the value chain is the mobile device; a wireless carrier, system aggregator, and mobile messaging middleware provider are not required in every solution.

Now that we have investigated mobile messaging, we are going to learn more about mobile and wireless security before diving into the building of smart client and thin client applications.

Helpful Links

Use the following Web links for more information on the topics covered in this chapter.

JavaSoft JMS	http://java.sun.com/products /jms/index.html
Open Mobile Alliance	www.openmobilealliance.com
Openwave Technical Library	http://developer.openwave.com /support/techlib.html
3rd Generation Partnership Project	www.3gpp.org
OMA WAP Forum Technical Papers	www.wapforum.org/what/technical.htm
Wireless DevNet SMS channel	www.wirelessdevnet.com /channels/sms/
W3C Recommendation for SMIL	www.w3.org/AudioVideo/
Wireless Village (OMA) Mobile IMPS Initiative	www.openmobilealliance.org/ wirelessvillage

Mobile and Wireless Security

One of the major concerns when implementing mobile and wireless solutions is data security. Securing enterprise data in a wired environment is difficult enough; adding wireless data transmission and mobile storage makes the task even more challenging. A number of security technologies are available today that make it possible to create mobile solutions with end-to-end security. These technologies should be incorporated into your application from the initial design through the final implementation.

Giving all aspects of security equal attention is crucial. For example, it's counterproductive to spend hours choosing the right security algorithm only to find out that a user is using his or her surname as the password to the system. Parties with malicious intent will always attack the weakest part of the system, so, clearly, having a single weak link is very dangerous. To implement a truly secure environment, you will require both the right technology and a corporate security policy. This will help ensure that all aspects of your system remain secure.

This chapter provides information on general security concepts as well as the security issues around building WAP and smart client applications. The goal is to provide developers with enough information to make educated decisions when implementing security into their mobile solutions.

Security Primer

Before discussing the security issues surrounding enterprise applications, we are going to take a look at some security concepts and technologies. We will begin by looking at the major components involved in creating a secure environment, followed by the leading security threats that you need to be aware of. The last part of this primer covers the leading security technologies and other security measures that you will want to consider in your solutions.

Creating a Secure Environment

For end-to-end security you have to consider the entire environment, including enterprise access, middle-tier components, and client applications. End-to-end security means that the transmission of data is secure along the entire path from the sender to the receiver—usually the client application to the enterprise server. Contrary to popular belief, this endeavor involves more than just data encryption. In this section we examine five objectives involved in creating a secure mobile environment. Understanding these objectives and the impact they have on mobile application development is crucial for creating secure applications.

Authentication

Authentication is the process of proving that people and organizations are who or what they claim to be. For wireless networks, this is often done at two layers: the network layer and the application layer. The network requires the user to be authenticated before that person is granted access. This can be done implicitly, based on the device or modem being used, or explicitly, using a variety of mechanisms. At the application layer, authentication is important at two levels: the client and the enterprise server. To gain access to enterprise data, the client has to prove to the server that it is what it says it is. At the same time, before a client allows an outside server to connect to it—for example, to push some content—the server has to authenticate itself to the client application. The simplest, and probably least secure, method of authentication is a username/password combination. More advanced methods include digital certificates or digital signatures.

Data Integrity

Data integrity is assurance that the data in question has not been altered or corrupted in any way during the transmission from the sender to the receiver. This can be accomplished by using data encryption in combination with a cryptographic checksum or Message Authentication Code (MAC). This information is encoded into the message itself by applying an algorithm to the message. When recipients receive the message, they compute the MAC and compare it with the MAC encoded in the message to see if the codes are the same. If they are, recipients can be confident that the message has not been tampered with. If the codes are different, recipients can discard the data as inaccurate.

Confidentiality

Confidentiality is one of the most important aspects of security, and certainly the most talked about. Confidentiality is about maintaining data privacy, making sure it cannot be viewed by unwanted parties. Most often, when people are worried about the security of a system, they are concerned that sensitive information, such as a credit card number or health records, can be viewed by parties with malicious intent. The most common way of preventing this intrusion is by encrypting the data. This process involves encoding the content of a message into a form that is unreadable by anyone other than the intended recipient. More information on encryption is provided later in this chapter in the *Security Technologies* section.

Authorization

Authorization is the process of determining the user's level of access—whether a user has the right to perform certain actions. Authorization is often closely tied to authentication. Once a user is authenticated, the system can determine what that party is permitted to do. Access control lists (ACLs) are often used to help determine this. For example, all users may have read-only access to a set of data, while the administrator, or another trusted source, may also have write access to the data.

Nonrepudiation

Nonrepudiation is about making parties accountable for transactions in which they have participated. It involves identifying the parties in such a way that they cannot at a later time deny their involvement in the transaction. <u>In essence, it means that both the sender and the recipient of a message can prove to a third party that the sender did indeed send the message and the recipient received the identical message</u>. To accomplish this, each transaction has to be signed with a digital signature that can be verified and time-stamped by a trusted third party.

Security Threats

Building a secure solution is difficult without awareness of the potential risks, so now that we have looked at the requirements for a secure environment, we will look at four common security threats: spoofing, sniffing, tampering, and theft. Whenever data is being transferred, whether over a wireless or wired network, you need to take precautions against these risks.

> **NOTE** To simplify terminology, any access to data or systems through a security hole will be considered unauthorized access.

Spoofing

Spoofing is the attempt by a party to gain unauthorized access to an application or system by pretending to be someone he or she is not. If the spoofer gains access, he or she can then create fake responses to messages in an attempt to gain further knowledge and access to other parts of the system. Spoofing is a major problem for Internet security, hence, wireless Internet security because a spoofer can make application users believe that they are communicating with a trusted source, such as their bank, when in reality they are communicating with an attacker machine. Unknowingly, users often provide additional information that is useful to the attacker to gain access to other parts and other users of the system.

The process of sniffing, described next, is often used in conjunction with spoofing to get enough information to access the system in the first place. For this reason, implementing both authentication and encryption is required to combat spoofing.

Sniffing

Sniffing is a technique used to monitor data flow on a network. While sniffing can be used for proper purposes, it is more commonly associated with the unauthorized copying of network data. In this sense, sniffing is essentially electronic eavesdropping. By "listening" to network data, unauthorized parties are able to obtain sensitive information that will allow them to do further damage to the application users, the enterprise systems, or both.

Sniffing is dangerous because it is both simple to do and difficult to detect. Moreover, sniffing tools are easy to obtain and configure. In fact, Ethernet sniffing tools come with the Microsoft Windows NT and 2000 installs; fortunately, these tools are simple to detect. To combat the more sophisticated sniffing tools, data encryption is the best defense. If an unauthorized user is able to access encrypted data, he or she will lack a way to decrypt it, essentially making it useless. That said, you must ensure that the encryption protocol you are using is nearly impossible to break. Many wireless LAN users have discovered the hard way that Wired Equivalent Privacy (WEP) encryption is often not enough to protect their data. [For more information on this issue, refer back to the *Wireless Local Area Networks (WLANs)* section in Chapter 3, "Wireless Networks."]

Tampering

Data tampering, also called an integrity threat, involves the malicious modification of data from its original form. Very often this involves intercepting a data transmission, although it also can happen to data stored on a server or client device. The modified data is then passed off as the original. Employing data encryption, authentication, and authorization are ways to combat data tampering.

Theft

Device theft is a problem inherent in mobile computing. Not only do you lose the device itself but also any confidential data that may reside on this device. This can be a serious threat for smart client applications, as they contain persistent data, often confidential in nature. For this reason, you should follow these rules when it comes to securing mobile devices:

1. Lock down devices with a username/password combination to prevent easy access.
2. Require authentication to access any applications residing on the device.
3. Do not store passwords on the device.
4. Encrypt any persistent data storage facilities.
5. Enforce corporate security policies for mobile users.

Authentication and encryption, along with a security policy, are required to help prevent malicious data access from a lost or stolen device. Fortunately, this is not as serious a problem for wireless Internet applications, as they rarely store data outside of the browser's cache.

Security Technologies

Given the security threats just outlined, companies need to understand the technologies that are available to help them minimize security risks. Though the requirements for each company will be different, all companies will benefit by implementing a well-thought-out security plan. This section provides information on the main concepts and technologies that are required to implement end-to-end security for your m-business applications.

Cryptography

The basic objective of cryptography is to allow two parties to communicate over an insecure channel without a third party being able to understand what is being transmitted. This capability is one of the core requirements of a secure environment, as it deals with all aspects of secure data transfer, including authentication, digital signatures, and encryption. On the face of it, cryptography is a simple concept, but it is actually quite complex, especially for large-scale mobile implementations.

Algorithms and Protocols

Cryptography works on many levels. At the lowest level are cryptographic algorithms. These algorithms describe the steps required to perform a particular computation, typically based around the transformation of data from one format to another. Building on these algorithms, is a protocol. The protocol describes the complete process of executing a cryptographic activity, including explicit information on how to handle any contingency that might arise. Making this distinction is important, because an excellent cryptographic algorithm does not necessarily translate into a strong protocol. The protocol is responsible for more than just the encoding of data; data transmission and key exchange are also properties of a protocol.

Finally, on top of the protocol are the applications. Once again, a strong protocol does not guarantee strong security, as the application itself may lead to further problems. Thus, in order to create a secure solution, a strong protocol is required, as well as a good, robust application implementation.

Data Encryption

The core of any cryptographic system is encryption, the process of taking a regular set of data, called plaintext, and converting it into an unreadable form, called ciphertext. Encryption allows you to maintain the privacy of sensitive data, even when accessed by unauthorized users. The only way the data can be read is by transforming it back to its original form using a process called decryption. The method of encryption and decryption is called an algorithm or cipher. Figure 6.1 demonstrates the concept of encryption. As the message is transported over an insecure public channel, it is encrypted, preventing anyone eavesdropping on the line from being able to understand the data being sent.

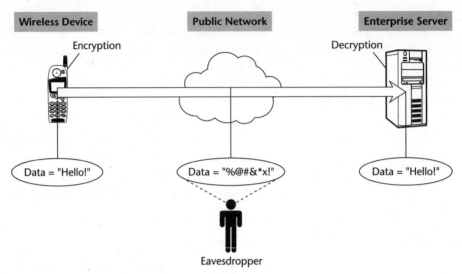

Figure 6.1 Sending a message using encryption.

Modern algorithms use keys to control the encryption and decryption of data. Once a message has been encrypted, it can only be decrypted by users who have the corresponding key. Key-based algorithms come in two classes: symmetric and asymmetric.

Symmetric algorithms are efficient: They use a single key to encrypt and decrypt all messages. The sender uses the key to encrypt the message and then sends the message to the intended recipient. Once the message is received, the recipient uses the same key to decrypt the message. This type of algorithm works well when there is a safe way to transmit the key between users, such as by meeting before the transmission takes place. Unfortunately a substantial problem arises when exchanging data between loosely related parties, such as an e-commerce Web site and a customer. Exchanging the key is a problem that symmetric encryption itself is incapable of solving; and without a secure method of exchange keys, this method is only useful between private parties.

Symmetric encryption is also referred to as secret-key encryption. The most popular form of this method is the Data Encryption Standard (DES), which was developed in the 1970s. Since then more secure forms of symmetric encryption have been developed; leaders among them include the Advanced Encryption Standard (AES), which is based on the Rijndael algorithm; Triple DES; International Data Encryption Algorithm (IDEA); Blowfish; and the Rivest family of algorithms, RC2, RC4, RC5, and RC6.

Asymmetric encryption addresses the main problem that has plagued symmetric key systems: the use of a single key. For years, cryptologists worked to find a solution to the key distribution problem with little success. Just as many mathematicians started to think that a solution was not possible, Whitfield Diffie and Martin Hellman proved them wrong. In 1975, Diffie and Hellman developed a solution using two separate but related keys: one to encrypt the data and another to decrypt it. The key used to encrypt the data is called the public key. This key can be widely distributed over insecure lines, for general public use. The key used to decrypt the corresponding data is called the private *key*. This key is never transmitted, as it is only required by the party that needs to decrypt the data. These keys are related in an obscure way using extremely large prime numbers and one-way functions. This technique makes it computationally infeasible to calculate the private key based on the public key. The larger the key is, the more difficult it becomes to break the system. Sixty-four-bit key systems, such as DES, are capable of being attacked by brute force—that is, trying every single key combination until the attacker finds the correct one. The more common 128-bit systems, such as ECC, so far have proven invulnerable to brute-force attacks.

Here is an example of how asymmetric, or public key, cryptography works. Suppose Alice wants to send a secure message to Bob. Alice can use Bob's public key to encrypt the message, since it is publicly available. She then sends the message to Bob. When he receives the message, he uses his private key, to which only he has access, to decrypt the message. Now Alice is able to send a secure message to Bob without having to do a key exchange. If information is to be exchanged in both directions using asymmetric encryption, each party must have his or her own public key and private key combination.

Encrypting a message with the private key and decrypting it with the public key is possible as well, but this has a different objective. It can be used on nonsensitive data simply to prove that the party who encrypted it actually has access to the private key.

The first, and best-known, asymmetric key algorithm was released in 1977 by Ron Rivest, Adi Shamir, and Leonard Adelman; who are commonly known as RSA. Other popular algorithms include Elliptic Curve Cryptography (ECC) and Diffie-Hellman (DH). RSA is being challenged in the mobile space by ECC, which is much less expensive in terms of processing power and key size, which are essential attributes in mobile computing.

Asymmetric ciphers are not, however, a perfect solution. Choosing a private key is not trivial, as a poor choice can lead to an easily broken scheme. Also, asymmetric ciphers provide a solution to the key distribution problem by using a public key and a private key, but they also are much more complicated, and therefore computationally slower than symmetric ciphers. For large sets of data, this can be problematic. In these cases, a combination of symmetric and asymmetric systems is an ideal solution. This allows you to take advantage of the higher performance of symmetric algorithms, by sending the secret key over insecure channels using public key systems. Once all parties have the secret key, the remainder of the data for that session can be encrypted and decrypted using symmetric algorithms. This is the basis for public key cryptography as used by many of today's leading protocols.

Digital Certificates

Digital certificates provide a way to guarantee that a public key belongs to the party it represents. For this to be successful, the certificate itself also has to be verified to ensure that it represents the claimed entity (a person or organization). This is accomplished using a trusted third party called a certificate authority (CA). The most common certificate authorities include VeriSign, Entrust, and Certicom. Users can purchase a digital certificate from the CA and use this certificate for authentication and to circulate their private key. Once a recipient has their private key and they are confident the recipient is who he or she claims to be (which is largely based on the trust of the CA), they can then send encrypted messages to this person, knowing they can decrypt them using their private key. Digital certificates typically contain the following:

- The name of the holder, as well as other information that uniquely identifies the holder. Additional information can include the URL of a Web server using the certificate or an email address.

- The holder's public key.

- The name of the CA that issued the certificate.

- The lifetime that the certificate if valid for (usually a start and end date).

- A digital signature from the CA to make it easy to detect if the transmission has been tampered with.

Certified users also have the option of self-signing a digital certificate, thereby becoming a CA themselves. This additional party can be considered trustworthy if he or she was signed by another trusted key. In this way, you can continue to navigate to the root CA, thereby determining who provided the initial certificate. If the root CA is not recognized or trusted, then each certificate in the chain is considered invalid.

The leading format for digital certificates is X.509, an industry standard for authentication. These certificates are prominent in Internet applications. In the wireless space, another form of digital certificate, called a WAP Server WTLS Certificate, is available. These certificates, commonly shortened to WTLS certificates, are a simpler version of X.509, created because X.509 certificates were too big for wireless applications. The WTLS certificates are primarily used in WAP applications where the microbrowser wants to authenticate the identity of a WAP server and encrypt information using the Wireless Transport Layer Security (WTLS). (There is more information on WTLS later in this chapter.)

Digital Signatures

Digital signatures are used to verify that a message really came from the claimed sender. It is based upon the notion that only the creator of the signature has the private key and that it can be verified using a corresponding public key. The digital signature is created by computing the message digest of a document, then concatenating it with information about the signer, a timestamp, and any other required information. A message digest is a function that takes arbitrary-sized input data (the message) and generates a fixed-size output, called a digest. This set of information is then encrypted using the private key of the sender using a suitable asymmetric algorithm. The resulting encrypted block of information is the digital signature.

The message digest that was calculated is a bit value that represents the current state of a document. If the document changes in any way, the message digest will also change. By incorporating the message digest into the digital signature, it is possible for the recipient of the document to easily detect if the document has been altered since the digital signature was created.

Public Key Infrastructure

Public key infrastructure (PKI) is the term used to describe a complete organization of systems and rules defining a single security system. The Internet Engineering Task Force (IETF) X.509 Working Group (www.ietf.org/internet-drafts/draft-ietf-pkix-roadmap-09.txt) defines PKI as "the set of hardware, software, people, and procedures needed to create, manage, store, distribute, and revoke certificates based on public key cryptography."

The components of PKI include the following:

- Certificate authorities responsible for issuing and revoking certificates
- Registration authorities responsible for binding between public keys and the identities of their holders
- Certificate holders who have been issued certificates that they can use to sign digital documents
- Repositories that store certificates as well as certificate revocation lists
- Security policy that defines an organization's top-level direction on security

Leading Protocols

The following are some leading protocols that are used for secure data transmission.

Secure Sockets Layer (SSL)

SSL is the dominant security protocol being used on the Internet today. It was developed by Netscape to provide secure and private Internet sessions, typically on top of HTTP, although it can also be used over FTP and other relevant protocols. SSL uses a combination of symmetric and asymmetric algorithms to maximize performance.

There are four phases in an SSL session:

1. *Handshake and cipher negotiation.* Both the client and server agree on the algorithms or ciphers to use.

2. *Authentication.* The server, and optionally the client, is authenticated using digital certificates.

3. *Key exchange.* The client creates a secret key and sends it to the server using its public key for encryption. The server decrypts the message using its private key; and for the remainder of that session, the client and server can communicate using the secret key.

4. *Application data exchange.* Once the secure symmetric session has been established, encrypted data can be communicated between the client and the server.

SSL can be used by many of the more powerful wireless clients, including laptops and Pocket PC devices. You can tell you are using SSL when the URL starts with https://, rather than just http://.

Encrypting the data within the protocol is more efficient than encrypting the data yourself and sending it over HTTP. Within the SSL protocol, the data will be encrypted on a packet level and can be decrypted on a packet level after it has arrived and has passed any sort of integrity test. If you encrypt your data as one large cipher text, then you will not be able to decrypt the data until all the packets have arrived.

Transport Layer Security (TLS)

Transport Layer Security (TLS) is the next generation of SSL. It consists of two layers. The lower layer is the TLS Record protocol, which is layered on top of a reliable transport protocol such as TCP. The two main features of the Record protocol are private and reliable connections. The higher level is the TLS Handshake protocol. This layer provides connection security that authenticates using asymmetric cryptography, negotiates a secret key, and provides a reliable negotiation. Like SSL, TLS is independent and can use a range of ciphers. The overall goals of TLS include cryptographic security, interoperability, and extensibility.

Wireless Transport Layer Security (WTLS)

WTLS is the security layer defined in the WAP specification. It operates above the Transport Protocol Layer, making it suitable for a variety of underlying wireless protocols. It is similar to TLS, but is optimized for low-bandwidth networks with high latency. It also adds new features such as datagram support, optimized handshakes, and key refreshing. It also supports the use of WTLS certificates for server-side authentication, in contrast to SSL/TLS, which typically use X.509 certificates. Overall, WTLS has similar goals to both SSL and TLS, in that it aims to provide privacy, data integrity, and authentication between two communicating parties. (A more in-depth look at WTLS and its authentication models is provided in the *WAP Security* section later in this chapter.)

IP Security (IPSec)

IPSec is different from the other protocols in that it does not operate on the application layer. Whereas SSL, TLS, and WTLS are aimed at providing secure communications over an inherently insecure network, IPSec is aimed at making the Internet itself secure. It does this by providing authentication, integrity, and privacy services at the IP datagram layer. While it is mainly targeted at laptop clients in the mobile space, IPSec-based virtual private network (VPN) products are starting to emerge for PDAs as well (see the upcoming section for more on VPNs). IPSec will become a more prominent solution when mobile devices start to support IP6, which includes IPSec as part of the standard. It is important to know that IPSec supports TCP/IP, not WAP.

Other Security Measures

The following subsections describe other security measures that you should consider when implementing your mobile solutions. You may find that some of these measures are already in place in your organization, as they are commonly used techniques for increasing overall system security.

Firewalls

Firewalls are the most common form of security implemented within organizations. They establish a network perimeter between what is public and what is private. A firewall is a set of software programs, usually located on a separate gateway server that restricts access to the private network resources from users on other networks. As soon as an enterprise installs Internet access to its site, a firewall is required to protect its own resources and, sometimes, to control outside resources to which its own users have access.

At a lower level, a firewall will examine each packet of network data to determine whether it should be forwarded to its destination. Where outside access is permitted, as in the case of a Web server, the firewall will allow outside traffic through the firewall on a specified port, for communication with a specified application. In this case, it would permit access from all outside users. At times, however, access from outside the firewall will be restricted to only known users, usually based on their IP addresses. This is used when a finite number of known users require access to a system.

For mobile devices that have always-on connections, a personal firewall may also be useful. This works in the same way as the firewall just described above, except it prevents unauthorized access to laptops and high-end PDAs. A personal firewall will prevent not only suspicious connections but also inappropriate content. Many devices contain hidden instructions that are used by maintenance professionals to repair a device. Lately, viruses and other inappropriate content have taken advantage of these hidden instructions to cause harm. A personal firewall can help prevent this from happening, although it should not be used a replacement for appropriate antivirus software.

Virtual Private Networks (VPNs)

A VPN allows a company to turn a public network (most commonly the Internet) into a private network. This technology allows remote workers to communicate with the corporate network in a secure fashion. Before VPN technology became available, dedicated leased lines were required to achieve the same result. Actually, VPNs present an additional benefit over leased lines, by providing secure access from many locations— essentially anywhere an Internet connection is available.

VPN technology is currently being used to help overcome the wireless LAN security problems by providing a direct link through a WLAN past the corporate firewall. The drawbacks of this configuration are cost and the inability to roam between WLAN access points. Mobile VPNs for devices on public networks are still in the early phases of adoption. As larger amounts of data are accessed remotely, mobile VPN usage is expected to increase.

Two-Factor Authentication

For some purposes, usually dealing with financial transactions, strong authentication is required. This involves using a two-factor approach, where users have to apply two factors to authenticate themselves. One factor is usually something the user knows, such as a PIN number; the other is something the user has, such a token card to generate a one-time password. This combination makes it much more difficult for unauthorized users to gain to access the system.

Smart client applications inherently provide a form of two-factor authentication: First you must have the device to access the application; second, you must authenticate yourself to gain access to the application, as well as to any back-end system to which it connects. This is not the traditional sense of the term, but it does provide additional security over thin client applications where any device with a microbrowser has the capability to access the content.

Biometrics

Even with the increased security of two-factor authentication, unauthorized users can compromise the system, for example, by obtaining the PIN code and token card, thereby gaining access to the enterprise system.

To avoid this situation, PIN codes can be replaced with a stronger form of authentication: biometric authentication. Biometrics provides a wide range of techniques for authenticating an individual based on his or her unique physical characteristics. Such techniques include fingerprint identification, face recognition, voice recognition, or iris and retina scanning. Using biometric techniques, you can ensure that the identification token is indeed unique. While this use of biometrics does improve security, this type of authentication does have some drawbacks. Many of these systems are somewhat intrusive and therefore not widely accepted by users. Also, the reliability of these technologies varies and so can lead to what's called "false refusal." That said, biometric systems are growing in popularity due to increased security concerns among all users.

Security Policy

The final, and often most important, security measure is the adoption of a corporate security policy. Such a policy will outline all aspects of a corporation's security measures, including both technology and the use and disclosure of confidential information within the enterprise. Even if a corporation has implemented a very strong technical security solution, the overall system will still be insecure if its users do not follow corporate security guidelines. Remember, intruders will always attack the weakest link in a system. Unfortunately, this link is often the users themselves.

Sometimes, very simple measures will dramatically increase overall security. For example, many PDA users do not lock the operating system when it is not in use. If the device is lost, nothing prevents another user from accessing the applications and corresponding data on the device. The same security measures that are in place for desktop users must be extended to remote workers.

WAP Security

The Wireless Application Protocol, or WAP, has been widely criticized by the media and corporations alike for its security shortcomings. What are the security issues with WAP? How can organizations overcome them? This section attempts to answer those questions, by explaining where WAP excels and where it falls short. After we examine the security model of WAP 1.x, we will look at the improvements made in WAP 2.x.

In the WAP 1.x security architecture, two aspects of security need to be addressed:

Transport-level security. This aspect deals with the communication between the client applications and the enterprise servers. This involves two protocols: WTLS is used over the air, while SSL or TLS is used over the wire. This change in protocols is the basis of the major WAP security problem.

Application-level security. This aspect deals with the security of the client application. This involves digital signatures and encryption.

Together, these two security areas will address the security concerns that are typical in any security model, including authentication, data integrity, confidentiality, authorization, and nonrepudiation.

NOTE Though this section focuses on WAP, most of the concepts and issues relate to other forms of thin client applications as well. Depending on the protocols and networks being used, other wireless Internet applications will be similar to either the WAP 1.x or WAP 2.x architecture.

Transport-Level Security

Transport-level security, also known as channel security, deals with the point-to-point communication between a wireless client and the enterprise data source. This involves communication over both wireless and wireline channels. With WAP, data is encrypted during over-the-air transport using Wireless Transport Layer Security (WTLS) protocol, and over-the-wire transport using Internet security protocols such as SSL and TLS. This discrepancy leads to one of the main WAP security issues. But before we discuss that topic, we will examine the features of WTLS.

WTLS

Wireless Transport Layer Security (WTLS) protocol was developed to address the unique characteristics of wireless networks, namely low bandwidth and high latency. It is a variation of the Transport Layer Security (TLS) protocol, which is the IETF standard for security on Internet. Unfortunately, TLS cannot be used directly because it is not efficient enough for a wireless environment. WTLS improved on the efficiency of the protocol while adding new capabilities aimed at wireless users. The following are some of the major features added to WTLS, which are not in TLS:

- *Support for other cryptographic algorithms.* SSL and TLS primarily use RSA encryption. WTLS supports RSA, Diffie-Hellman (DH), and Elliptic Curve Cryptography (ECC).

- *Definition of a new compact public key certificate, WTLS certificates.* These are a more efficient version of X.509 certificates.

- *UDP datagram support.* This impacts many areas of the protocol, from how data is encrypted to extra support for message handling, to ensure messages do not get lost, duplicated, or delivered out of order.

- *A key refresh option.* This is renegotiated periodically, based on the number of messages sent.

- *An expanded set of alerts.* This adds clarity for error handling.

- *Optimized handshakes.* This reduces the number of round-trips required in high-latency networks.

In addition to these changes, WTLS also introduced three levels of authentication between the client and the gateway. They are listed in ascending order:

Class I WTLS. Anonymous interactions between the client and WAP gateway; no authentication takes place.

Class II WTLS. The server authenticates itself to the client using WTLS certificates.

Class III WTLS. Both the client and the WAP gateway authenticate to each other. This is the form of authentication used with smartcards. GSM Subscriber Identity Modules (SIM), for example, can store authentication details on the device for two-way authentication.

The WAP Gap

Unfortunately, at the same time WTLS improved on TLS for wireless communication, it also caused a major problem: Now that both TLS and WTLS are required within the WAP architecture, there is a point at which a translation between the two protocols occurs. It is from this point, not from the WTLS protocol itself, that the security issues arise. The translation occurs on the WAP gateway: From the client device to the WAP gateway, WTLS is used; from the gateway to the enterprise server, TLS is used. At this point, the WTLS content is decrypted and then reencrypted using TLS. The content exists as plaintext while this transfer takes place, creating the so-called WAP gap. Keep in mind that the amount of time that the content is unencrypted is minimal, and that the WAP gateway is not in the public domain, so there is still security in place. However, for many corporations, this risk is still too great, as it presents a vulnerable point in the network, preventing end-to-end security.

There are two options for alleviating the WAP gap:

- Accept that the gateway is a vulnerable point and make every effort to protect it using firewalls, monitoring equipment, and a stringent security policy.

- Move the WAP gateway within your corporate firewall and manage it yourself.

Choosing between these two options is a business decision that will depend on the individual enterprise. It is a trade-off between the extra resources required to maintain a WAP gateway and the potential security threat to corporate data. Fortunately, a solution is available, in the form of WAP 2.x.

WAP 2.x

There are many new features in WAP 2.0, but none is as important as the move to standard Internet protocols. This move to using HTTP, TCP, and IP allows the TLS protocol to be used for data communication, thereby removing the need for WTLS. Once a single protocol can be used from the client device to the enterprise server, WAP can enable true end-to-end security, making the WAP gap a thing of the past. Suffice to say, this is a major change in the WAP, and it will take some time for wireless carriers to move to WAP 2.x gateways. Nevertheless, it provides new life for WAP in the wireless Internet space. (For a complete summary of WAP 1.x and the changes made in WAP 2.x, see Chapter 11, " Thin Client Overview.")

Application-Level Security

With so much attention given to the WAP gap and transport-level security, developers often forget about application-level security altogether. Application-level security is important for two main reasons: (1) when security is required past the endpoints of transport-level security, and (2) when presentation content needs to be accessed but enterprise data does not. This can happen during transcoding, that is, when another markup language (often HTML) is being transformed into WML.

The first scenario can be addressed using the techniques provided in the WML specification. In general, the default settings are set to the highest security, but the following are a few things to keep an eye on:

- Any WML card that requests access to sensitive data should set sendreferer=true in the <go> element.

- The script that handles requests for sensitive information should check the URL specified by the REFERER header HTTP request to make sure that requests being handled are from friendly domains.

- Use HTTPS and require basic authentication. Relying on the phone's identity alone is not sufficient.

The second scenario can be addressed using WMLScript and the Crypto API. Using this signText function in the API, digital signatures can be created, opening the door for wireless PKI to manage and issue public key certificates. This technology allows for end-to-end encryption between the content provider (usually the enterprise) and the client.

Smart Client Security

The smart client architecture does not depend on a gateway for protocol conversion, so it does not suffer from the WAP gap. However, these applications do have security issues that must be addressed. Anytime data is available outside of corporate firewalls, security mechanisms are required to protect sensitive information. With a smart client architecture, it is possible to provide end-to-end security for your enterprise data. This goes beyond what is currently capable with many thin client solutions.

The main areas of security concern for smart client applications include user authentication, encryption of the client data store, and transport-level security. We will take a look at each of these in turn.

User Authentication

Smart client applications store data directly on the device, much like a client server desktop application does on a PC. To restrict access to this data, user authentication is required. A username/password combination is the minimal level of authentication that should be implemented. The password used should not be stored anywhere on the device, as this would jeopardize the system security. Commonly, this form of authentication is implemented as a check between a userid/password stored within the data store against the userid/password typed in by the user. This usually results in a machine-code jump if the comparison is successful, and, unfortunately, this kind of application can be hacked to always jump, thereby giving the hacker access to the data whenever he or she uses a cracked application. For a better implementation, refer to the *Data Store Security* section, which follows.

However you implement user authentication to the device and its data, it should not automatically authenticate the user to the enterprise server's data. At this level, a second form of authentication should be implemented, usually one more sophisticated, such as a VPN or digital certificate. In this way, an enterprise can minimize the amount of data that can be accessed by unauthorized users. They will be able to view the small amount of data stored locally, but be unable to update or retrieve any of the enterprise data.

It is also a good idea to take advantage of the security offering of the mobile operating system. All of the leading operating systems provide a mechanism to lock the machine, requiring users to be authenticated before they can gain any device access at all. This would provide a third layer of authentication and serve as a deterrent for someone who may find a lost device.

Data Store Security

With smart client applications, corporate data is stored locally on mobile devices. This data requires protection from unauthorized access, just as other parts of a mobile solution do. In many cases, requiring users to be authenticated before accessing the data is one step to securing this data. Another step is to encrypt the data store itself, making it impossible to view without providing the proper identification, ideally in the form of a digital certificate.

Implementing both authentication and encryption in a single process is the best way to ensure the data remains confidential. The data store can be encrypted using the password as the seed to a symmetric key algorithm. In this way there is no single check to gain access to the application, as all data being returned from the data store will be decrypted using the password.

The encryption should be strong enough so it cannot be easily broken; 128-bit security is not uncommon. This high level of security comes at a price, however: Every time data is added or retrieved, it has to go through the encryption algorithm. This could potentially have a significant negative impact on performance, especially on devices with limited computing power. Also, if for some reason the encryption key is lost, there is no way to access the data without breaking the key, which is computationally infeasible.

If the performance penalty of strong encryption is deemed to be too high, a weaker form of protection, such as data obfuscation, could be used. Obfuscation can scramble the data so it is not viewable by a text-viewing tool. It does, however, not provide nearly the level of security that strong encryption does, but it may be enough for some types of data. Obfuscation has little or no impact on overall performance.

The leading mobile database vendors, including iAnywhere Solutions, provide built-in encryption technologies for their products. If another form of data storage is being used, such as Palm DB or a custom solution, the developer should ensure that data encryption is implemented in the solution. Many of the mobile operating systems provide cryptography libraries to help with this process.

Transport-Level Security

At the transport level, data encryption is required to secure the enterprise data being synchronized. Whether you are using a packaged synchronization solution or building one in-house, the synchronization may be the most important part of the application to secure. Fortunately, there are many encryption products available to provide this security. Companies including Certicom and RSA provide products that can be used to encrypt nearly any type of data being transferred to and from mobile devices. Many of the smart client application vendors include 128-bit data encryption with their solutions. In this way, you can be ensured that the data being transferred over public networks is private from the time it leaves the device to the time it reaches the enterprise server.

Along with encryption, using a strong form of authentication, such as digital certificates, is recommended. In addition, try to keep the firewall around your corporate data as secure as possible; don't open any ports that are not absolutely required by your synchronization server.

Summary

There are five key components to a secure environment: authentication, data integrity, confidentiality, authorization, and nonrepudiation. When implementing a secure environment, remember that a system is only as secure as its weakest point. Therefore, you have to protect every gap in your solution to ensure that unauthorized users cannot gain access to your system. In order to accomplish this, you may be required to implement a variety of security technologies, including public key cryptography, digital certificates, digital signatures, and PKI. Other measures such as firewalls, VPN, biometrics, and a corporate security policy are also helpful for maintaining a secure environment.

For thin client development, WAP incorporates WTLS for transport-layer security. Just keep in mind that even though it is a strong protocol, WAP suffers from a security problem often referred to as the WAP gap. This gap is on the gateway when WTLS is translated into TLS. WAP 2.x addresses this issue by removing the conversion between protocols.

Smart client applications do not suffer from such limitations. Smart client application developers have full control over the security technology they implement. They can make the enterprise solution as secure (or insecure) as required for the data and applications being developed. When you evaluate smart client technology vendors, make sure security is part of your selection criteria.

The next chapter, the first in Part II, "Building Smart Client Applications," provides an overview of the smart client architecture, highlighting the main components for building a successful smart client solution.

Helpful Links

Helpful links to pursue for more information on any of the topics covered in this chapter are listed here.

Applied Cryptography by Bruce Schneider	www.counterpane.com/applied.html
Certicom Crytography Resources	www.certicom.com/resources/index.html
Cryptography FAQ	www.faqs.org/faqs/cryptography-faq/
Entrust Cryptography Resource Center	www.entrust.com/resources
IETF Public-Key Infrastructure (X.509) Working Group	www.ietf.org/html.charters /pkix-charter.html
WAP 1.x and 2.x Specifications	www.wapforum.org/what/technical.htm; www.openmobilealliance.com

PART

Two

Building Smart Client Applications

One of the first lessons we learn when developing enterprise mobile and wireless applications is that access to data is required at all times. According to the predictions of the wireless carriers and mobile phone vendors, this is supposed to be an easy task. For many years now, these vendors, along with industry analysts, have been forecasting the rollout of high-speed, widespread, wireless data networks. Networks that would provide us with the ability to deliver many forms of data, including multimedia content, to wireless devices around the world. Unfortunately, these predictions have not yet come true, nor does it look like they'll be coming true anytime soon. For some applications, this means that we need to look for alternative ways to implement our mobile and wireless solutions—ways that do not depend on high-speed wireless networks and pervasive wireless coverage.

Fortunately, we do not have to look very far, as the technology to build these applications has been around for many years and has been utilized in thousands of successful enterprise solutions. If we build applications that have persistent data storage, and the capability to synchronize this data with back-end enterprise data sources, we can create mobile applications today that will meet the needs of our users. This type of application is referred to as smart client.

This part of the book covers the technologies that are pertinent to developing smart client applications. It is divided into the following chapters, focusing on the technologies that comprise a complete smart client solution:

- Chapter 7, "Smart Client Overview"
- Chapter 8, "Smart Client Development"
- Chapter 9, "Persistent Data on the Client"
- Chapter 10, "Enterprise Integration through Synchronization"

Smart Client Overview

What do we mean when we say smart client? What is the defining factor of smart client applications versus thin client applications? On which devices can we deploy smart client applications? When developing a mobile solution it is important to learn about the various application models available, and one of the best ways to do that is to ask questions. The goal of this chapter is to provide you with an overview of the smart client application model to answer the important questions. Doing so will help you make intelligent decisions when deciding which application type is appropriate for a given project.

This chapter gives an overview of the smart client application architecture by highlighting the main components that comprise a successful solution. Following the architecture overview, we will take an in-depth look at the major mobile operating systems for smart client solutions. The combination of the mobile operating system and the device hardware often dictate whether it is possible to implement a smart client solution.

As you read, keep in mind that smart client applications are not limited to wireless devices. Many mobile applications run on a laptop environment, in addition to PDAs that communicate via cradles. Actually, until recently, laptop-based applications were considered to be the leaders in the mobile space. As we will see in the remainder of this chapter, smart client applications provide a very flexible way to extend enterprise data to mobile workers, on a variety of devices ranging from laptops to smart phones.

Smart Client Architecture

By now you are probably aware that smart client applications differ significantly from thin client, browser-based wireless solutions. As a general rule, smart client applications enable the user to access data when disconnected from the network. There are many ways that this can be accomplished, but for the purposes of this section, we are going to focus on building applications that incorporate a persistent data store, rather than just using simple caching mechanisms.

The smart client architecture is illustrated in Figure 7.1. On the client, you have the user interface, the business logic, as well as a persistent data store. This application communicates with a back-end data source, often through an intermediate synchronization server. The communication stream itself can run either wirelessly or over a wireline connection. Depending on the technology being used, the connection may require an IP-based network or an additional communication layer for the synchronization process.

When this type of solution is implemented, it quickly becomes clear why it is called a smart client solution. By deploying an application to the device itself, you have the ability to give the client application some "smarts," or logic. This logic dictates many aspects of the application. It determines where the application gets it data from (either locally or from a round-trip to the server), how the data is presented and stored, as well as the set of data that needs to be communicated back to the enterprise systems via a synchronization process. In the wireless space, the impact of having business logic on the device is often overlooked as many vendors and developers focus on low-level technical features of a solution rather than satisfying the requirements of the mobile user.

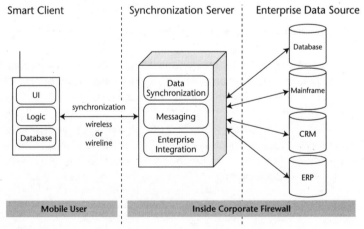

Figure 7.1   Smart client architecture.

The Client

Smart client applications provide many attractive features for end users. Many of these features reside in the client application itself. By providing a rich user interface with persistent data storage, smart client applications are suitable for a large variety of corporate applications. We are going to take a closer look at each of the components that comprise the client of smart client applications.

User Interface

Initially, the user interface may not seem important when developing mobile applications, as most people feel that there is always going to be a trade-off between size and usability. But this trade-off does not always have to take place. Smart client applications can have very sophisticated user interfaces. In most cases, they are programmed using development tools that have extensive component frameworks. Incorporating tables, drop-down lists, radio buttons, and graphs are straightforward operations. Contrast this with the capabilities available for most wireless Internet applications, where the user interface is often text-based with limited support for graphical components.

Whether you are developing for the Palm OS, Windows CE, Symbian OS, or a Java-based system, you will find that user interface development is a very important part of a successful application. When the screen size is limited, as it is for handheld devices, the developer must take full advantage of the space that is available. In addition, putting special focus on the navigation through an application is essential. Mobile application users are often the same people who have been using advanced Windows-based application for years; hence, they have certain usability expectations that need to be met. One of the main reasons given by consumers for their slow acceptance of wireless applications is the complexity of the available solutions. This complexity is usually due to poor application design, rather than technical limitations. Obtaining end-user feedback during the development phase via a prototype can help you address usability concerns at an early stage in the application development process. In the *Mobile Operating Systems* section later in this chapter, we will take a look at the most popular mobile operating systems and review the types of applications that each operating system can support.

Data Storage

When you look at the requirements of typical mobile enterprise customers, you see some trends emerge. First and foremost, data access is a requirement. Users do not really care how the data is retrieved; they care only about its availability. Theoretically, they understand that wireless access is not pervasive; nevertheless, they quickly become frustrated if they can access their data only periodically, depending on the network coverage and/or penetration. By providing data storage capabilities on the device, you can ensure that users will be able to access important data when they need it. (Note: Data storage mechanisms are not covered here, as they are covered in depth in Chapter 9, "Persistent Data on the Client.")

Performance

In addition to providing timely data access, smart client applications also provide performance benefits. Consider how computer technicians work: They travel from location to location repairing a variety of computer systems. As often as possible they carry some inventory with them for common repairs; but just as often, they need to order a special part from the warehouse and schedule a follow-up visit. Before they can schedule the next visit, they have to make sure that the part in question will be available at a time that fits into their schedule. By utilizing a smart client application that contains a persistent data store, technicians can have immediate access to inventory information, as well as their schedule. Imagine if they had to search a database for a specific part number. This task could take a considerable amount of time if a wireless browser-based application were being used. The task could even be impossible if a technician happened to be at a location that was outside of the wireless network's coverage. By implementing a mobile data store, technicians can perform a quick search of the inventory right on the device to provide them with the information that they require.

The one concern with this approach is data freshness. If the data on the device is not updated regularly, then the technician may end up, for example, believing that a part is available, when in fact it is out of stock, leading him or her to make decisions based on inaccurate data. This leads to the data synchronization aspect of smart client applications.

Data Synchronization

In most smart client architectures, the majority of the data synchronization work is executed, not on the client, but on a synchronization server. Nevertheless, the client application is still required to have a certain amount of synchronization knowledge. Minimally, the client has to know the location of the synchronization server, details about the communication stream to the server, what data has to be synchronized, and how to handle incoming data from the server.

If you are developing the synchronization layer yourself, you will have to keep these issues in mind when designing the application. If you are going to use commercially available software, most vendors provide a synchronization client module that incorporates the required functionality with various levels of sophistication. (Chapter 10, "Enterprise Integration through Synchronization" provides a closer look at the options available for data synchronization for both the client and the server.)

Messaging

Smart client applications can also take advantage of intelligent application-to-application messaging systems to communicate data. To do so, a client component has to be able to both send and receive these messages. When receiving a message, it has to be smart enough to relay the information appropriately. This communication may involve notifying the user directly or, possibly, updating a set of the data in the client data store. Application-to-application messaging can be important for applications that require frequent communication with enterprise systems. (For more information on all of the leading forms of mobile and wireless messaging, refer back to Chapter 5, "Mobile and Wireless Messaging.")

The Server

Though the server component of smart client applications is invisible to the end user, it is still very important. The server component is responsible for data synchronization, data storage, and messaging.

Data Synchronization

When we refer to data synchronization, we are talking about how enterprise data is moved from the back-end enterprise system to the mobile device, and vice versa. This data movement can be accomplished in a number of ways, depending on the solution that is chosen. For a general overview, we will look only at the major components that are required in a synchronization server, as shown in Figure 7.2. These components include the interface to the client application, the synchronization logic, and the integration layer to the back-end data source.

Data can be sent between the client application and the synchronization server in a variety of formats. Transferring data over a wired connection is a fairly straightforward process. Transferring it over a wireless network can be a different story. Depending on the requirements of the synchronization software, some wireless networks may work better than others, and some may not work at all. Discussions about wireless data synchronization are saved for Chapter 10; for the purpose of this overview, we will assume a valid connection exists between the client and the server.

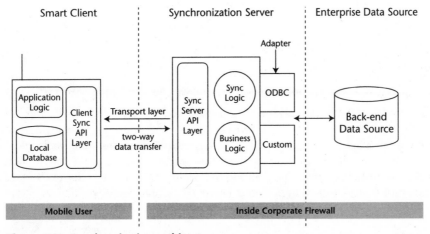

Figure 7.2 Synchronization architecture.

If you are developing an in-house solution, you may choose to create a proprietary data stream format, possibly basing it on XML, or you may decide to use an industry-specified format such as SyncML. The same options are available in commercial synchronization software. In both cases, a stream of data is being transferred that both the client and the server know how to decipher. When the server receives data, it needs to know how to process it. This capability involves understanding the data layout of the enterprise data source. In the most basic form, the synchronization logic will simply take the updates and apply them to the data store, without considering whether they should be applied or not. In more complex cases, the synchronization logic will perform a variety of duties, including conflict detection and resolution.

The integration with the back-end data source is often a simple process, using Open Database Connectivity (ODBC) or Java Database Connectivity (JDBC) drivers for direct database access. At the same time, integration can be difficult, requiring complex logic to interface with more complicated systems such as Enterprise Resource Planning (ERP) systems. In either case, providing the ability to access enterprise data is critical for an enterprise synchronization solution. Rarely does a company use only one enterprise data storage mechanism, so, ideally, the synchronization server will be flexible enough to integrate with a variety of back-end sources.

Enterprise Data Source

The final part of the smart client solution is the enterprise data itself. This data will vary in formats ranging from enterprise databases from vendors, such as Oracle, Sybase, Microsoft, or IBM, to flat-file systems and everything in between. For the more common storage systems, you should be able to find a driver or adapter that will provide an integration layer for your synchronization server. If you are using something that is uncommon or something developed in-house, you will most likely have to develop this integration layer yourself.

Keep in mind that one of the most common reasons for implementing a mobile solution is to provide enterprise data access to the mobile worker, so the enterprise integration is an essential part of the overall solution.

Messaging

The term messaging can be used to describe many different types of systems. Depending on whom you are talking to and the type of system being discussed, messaging can mean email, paging, SMS, voice, text, data, or a variety of other things. For the purpose of this section on smart client applications, messaging will be referred to in the context of application-to-application messaging with store-and-forward capabilities. (If you are interested in text messaging as it refers to thin client applications, refer to Chapter 5, "Mobile and Wireless Messaging," and Chapter 11, "Thin Client Overview.")

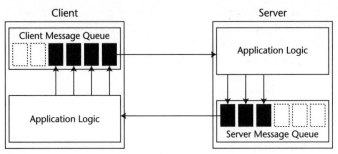

Figure 7.3   Store-and-forward messaging.

As seen in Figure 7.3, the basic form of smart client messaging is composed of a message queue on the client and another queue on the server. These message queues allow you to communicate with messaging system without a wireless network connection. When a message is sent from the server, it is simply placed in the queue that corresponds to a particular client alias. When the client matching that alias connects to the server, the messages in the server-side queue are sent to the client. In the same way, when a client sends a message to the server, it goes into a queue on the client device. When the client connects to the server, the messages from the client queue are sent to the server. This is commonly called store-and-forward messaging, since messages are stored when a connection is not available and forwarded when the connection becomes available.

When building smart client applications you have three options:

- Create a client that uses a persistent data store and synchronization.

- Create a client that uses store-and-forward messaging.

- Create a client that combines both a persistent data store and store-and-forward messaging.

Depending on the type of solution being implemented, any one of these three options may meet your needs.

Mobile Operating Systems

The mobile operating system (OS) market is vastly different from the desktop operating system market. In the mobile world, there is no clear leader in the operating system space. The Palm OS has a large amount of market share in the consumer space; Windows CE has been very successful in the enterprise market; and the Symbian OS is dominant in the European market. In addition, a following is growing for both Java-based operating systems and Linux.

The operating system is an important aspect of mobile application development. It is one of the main factors that will dictate whether you can develop a smart client application. It also plays a role in the development tools, prepackaged software, and device support that will be available to you. We will take a look at five main operating environments: Windows CE, Palm OS, Symbian OS, Linux, and Java.

Windows CE

Microsoft has made great strides in the Windows CE operating system, both in terms of functionality and usability, as well as in market share. Windows CE .NET, Microsoft's current operating system, delivers the most complete mobile operating system available. Windows CE enables multitasking, allowing the user to work on one application while another is executing in the background. For example, you could be working on PocketWord while synchronizing data in the background. Pocket PC 2002 devices use customized versions of Windows CE , and look as though they will be the market leader for the enterprise market, and possibly the consumer market as well. But the situation was not always so positive for Microsoft.

Back in 1995, when Microsoft released the first version of Windows CE, it had very limited success. Microsoft had just released Windows 95 and had a near monopoly in the desktop operating system market. The Windows CE operating system was designed for Handheld PC devices, which, size-wise, are in between PDAs and laptops. Even though many device manufacturers supported Windows CE, it was given a very cool reception by both the consumer and enterprise markets. It did not look as though Microsoft was going to reproduce its desktop OS success with Windows CE.

Unwilling to concede, in 1998, Microsoft released Windows CE 2.x for the PDA market. At this time, Palm dominated the palm-sized device market. Nevertheless, Microsoft was confident that Windows CE was going to be the "PalmPilot killer." This proved not to be the case. Once again, Windows CE was given a cool reception. Many consumers found the user interface too cluttered and complex. Microsoft found itself in an unfamiliar position, trailing both Palm OS and Symbian OS (at the time called EPOC) in popularity. It had to come up with something better.

Finally, in 2000, Microsoft released a version of Windows CE that garnered some attention. Windows CE 3.0 was released when the PDA market was becoming more attractive to enterprises considering building line-of-business applications. For these applications, corporations were looking for an operating system that would allow them to run existing Windows-based software and to create sophisticated data-driven applications. Windows CE fit that description. Even though the popularity of the Palm OS was at an all-time high, Microsoft made some early penetration into the enterprise market during the year. By 2001, Microsoft found itself in a more familiar position, often being the OS of choice for new business application development.

Why did Windows CE 3.0 succeed? For many reasons. It comes with many familiar applications, including PocketWord, PocketExcel, PocketOutlook, and Pocket Internet Explorer. In addition to its office products, Windows CE also features built-in multimedia capabilities for both audio and video. These capabilities make it easy for existing

Windows desktop users to move to the Windows CE operating system. They can maintain the rich level of functionality they are used to, only on a smaller device.

On the development side, Windows CE 3.0 utilizes a scaled-down version of the commonly used Win32 application programming interface (API), allowing developers to quickly refit existing Windows applications for Windows CE 3.0-based devices. This provided the needed link between the Microsoft desktop operating systems and their mobile counterparts. Another major improvement from Windows CE 2.x to 3.0 is enhanced support for real-time functionality made to the OS kernel. This dramatically enhances the performance of Windows CE 3.0 over Windows CE 2.x.

The progression of Windows CE does not end there. At the end of 2001, Microsoft announced another version of the Windows CE family: Windows CE .NET 4.0. This version fits into Microsoft's overall .NET strategy by allowing developers to use the full suite of Microsoft tools, including Visual Basic and Visual C++, for development, as well as updated versions of the Microsoft suite of office products and Internet Explorer. Windows CE .NET also added built-in wireless capabilities, including broad support for WANs, LANs (including 802.11b), and PANs (including Bluetooth).

Then, in the fall of 2002, Windows CE .NET version 4.1 was released. This version added support for IPv6 (the latest version of Internet Protocol), as well as integrated speech recognition. Additionally, devices using Windows CE .NET 4.1 support popular PC file formats including Microsoft Word, Excel, PowerPoint, and Adobe Acrobat. Internet Explorer performance has also been increased by as much as 15 percent from Windows CE .NET 4.0 and 60 percent over Windows CE 3.0. It is anticipated that in 2003, most of the new Pocket PC and Handheld PC devices will take advantage of the Windows CE .NET operating system. Table 7.1 compares the various Windows CE version capabilities.

Table 7.1 Comparison of Windows CE Versions

Version/ Features	Year of Release	Processors Supported	Internet Browser	Wireless Capabilities	Multimedia Support
CE 2.12	1998	ARM, MIPS, PowerPC, SHx, x86	Based on Internet Explorer 4.0	IrDA	No (audio only)
CE 3.0	2000	ARM, MIPS, PowerPC, SHx, x86	Based on Internet Explorer 4.0	IrDA	Yes
CE .NET	2002	X-Scale, ARM,	Based on Internet Explorer 5.5	Bluetooth, 802.11x, Media Sense, OBEX	Yes

Confusion often exists between Windows CE and Pocket PC. To help distinguish between the two, the following extract is reprinted from Microsoft's Web site (www.microsoft.com/windows/ Embedded/ce/evaluation/faq/default.asp):

Pocket PC does not replace Windows CE at all: The former is the hardware device, the latter is the operating system. However, Pocket PC does use Windows CE 3.0 as its underlying operating system. Pocket PC uses a customized version of the Windows CE 3.0 operating system, built by Microsoft and used specifically in Personal Digital Assistants (PDAs) like the Compaq iPaq and the Hewlett-Packard Jornada. While this customized version is only used in PDA-type devices, Windows CE 3.0 can be used in a wide variety of devices including industrial automation, Internet access devices, Web terminals, kiosks, consumer electronics, or retail and point-of-sale devices.

This extract is somewhat dated as the latest versions of PocketPC are now based on Windows CE .NET 4.1, and will soon use Windows CE .NET 4.2.

Even though Windows CE is not the leading mobile operating system, it is definitely a strong platform on which you can build very robust, sophisticated smart client applications. Microsoft has spent the past several years enhancing the early versions of Windows CE to bring consumers the latest Windows CE .NET operating system, which should remain a strong player, if not become the leading mobile operating system.

Palm OS

Palm OS has experienced tremendous success in the PDA market. In the early days of PDAs, Palm established itself as the market leader, capturing nearly 75 percent of the worldwide mobile operating system market by early 2000 (IDC, June 2000). Since then, it has continued to be a dominant player in the market. Lately, however, it has faced some difficult times.

Palm has not had the same level of success in the enterprise market for new application development as it did with consumers. In the early days of PDA application development, many organizations chose Palm OS as a deployment platform because the devices were readily available and many employees were already familiar with them. This situation resulted in a wide range of applications being developed for Palm OS. As the requirements for business applications became more complex, however, Palm OS was found to be unsuitable for many of these tasks. For example, prior to the release of version 5.0, Palm OS had single-tasking capability only; meaning users could perform only one task at a time on the device. Thus, if they wanted to download a file, they had to wait for the download to complete before they could move on to other tasks. Many corporations found this to be a significant limitation. When, with version 5.0, Palm introduced some multi-tasking capabilities to the platform, it improved the usefulness of Palm OS for more complex applications. More enhancements for enterprise applications are also expected in Palm OS 6.0.

It's important to point out that Palm hardware devices and Palm OS used to be considered one and the same. This changed when Palm started licensing Palm OS to third-party device manufacturers. The first licensee was Handspring; many others followed suit, including IBM, Symbol, Kyocera, and Sony. Palm's goal by separating the hardware and software divisions was to maintain the widespread usage of Palm OS while avoiding the same problems Apple confronted as the sole manufacturer of both the hardware and the operating system software for the Macintosh. To date, it looks as though Palm is on the road to success in this regard. Handspring, for example, in 2001, released a device on the market called Treo, which incorporates a phone with wireless modem, Palm OS, email, wireless Internet access, SMS text messaging, and a full thumb keyboard, all in one device. Samsung and Kyocera have similar offerings.

Palm OS 5.0 also includes new features that make it more suitable for enterprise customers. The most significant of these is the move to the ARM series of processors. The result of this move is a significant increase in the overall performance of Palm OS, allowing for the development of more advanced, feature-rich, enterprise-level applications. At the same time, version 5.0 also added the aforementioned multitasking capabilities, additional security for encrypting private data, new expansion slot capabilities, and new wireless capabilities for easier access to the Internet and email systems; it also enhanced the level of color support.

Moreover, the new features of the Palm i705, including wireless support for always-on wireless networks, Bluetooth, and clip-on modems, will make the Palm OS a formidable challenger to Windows CE in the enterprise space. The only concern is that these new capabilities may have come too late; Windows CE had two years to become established before Palm OS delivered these new enterprise-level capabilities. To further distinguish their enterprise offerings, in the fall of 2002, Palm introduced two new product families: Tungsten and Zire. The Tungsten family of products is targeted at the enterprise market, providing powerful solutions for mobile professionals and enterprise work forces; the Zire family is focused on the consumer market, providing affordable options for individuals to organize their schedules and contacts. These new offerings should help Palm regain the momentum it had in 2000/2001.

On the technical side, Palm OS has three main components, listed here (Figure 7.4 shows how these components relate to one another):

- Reference hardware design, consisting of device hardware, third-party hardware, and the hardware abstraction layer.

- Palm OS with embedded system and third-party libraries.

- Application software, featuring the applications included with Palm OS, such as the HotSync conduit and third-party applications.

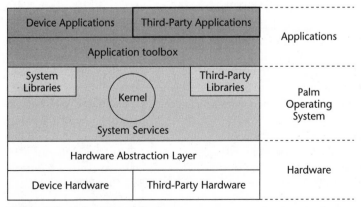

Figure 7.4 Palm OS architecture.

Today, Palm OS has a bright future in the PDA marketplace. Even though other mobile operating systems are closing in on its market lead, it will be some time before another operating system overtakes Palm OS in market share. With the release of version 5.0, as well as innovative new devices based on the Palm OS, the market will remain split between the Palm OS and Windows CE in North America, and among the Palm OS, Windows CE, and the Symbian OS in other parts of the world.

Symbian OS

The Symbian gained its popularity in the mobile phone sector; in fact, Symbian is an operating system developed exclusively for mobile devices. As phone capabilities increased, a more functional operating system was required. Proprietary operating systems from the device manufacturers were no longer practical, so a group of the largest cell phone manufacturers, including Nokia, Ericsson, Motorola, and Matsushita, together with Psion, formed a joint venture known as Symbian. The goal of this partnership was to create a standard operating system for smart phones and PDAs based on Psion's EPOC operating system.

NOTE Symbian OS is the name assigned to the latest versions of the EPOC operating system. Versions prior to 6.0 are called EPOC, while later versions are called Symbian OS.

In June 1999, EPOC version 5 started shipping. It contained support for devices based on a 640×240 screen resolution, with pen and keyboard capabilities. Since then, versions 6.x and 7.x of the Symbian OS have been released by Symbian. These new implementations, described in turn in the subsequent paragraphs, build upon the core EPOC operating system, while adding new capabilities required for next-generation mobile devices.

Symbian OS version 6.0, which made some radical improvements over the previous version, was released in 2000. The design goal was to bring together various forms of communication protocols, including TCP/IP, WAP, GSM, Bluetooth, IrDA, as well as serial connections, while allowing applications to be presented using C++, Java, WAP, and other Web protocols. By linking the communications protocols and applications, Symbian hoped to achieve close integration of contact information, messaging, browsing, and wireless telephony. In 2002, Symbian OS version 7.0 was released. This version was designed for the unique requirements of advanced 2G, 2.5G, and 3G mobile phones. Some of the key features in this release included over-the-air data synchronization using SyncML; full-strength security using HTTPS, WTLS, and SSL; enhanced multimedia capabilities; support for advanced messaging, including MMS, EMS, and SMS; and a complete suite of application engines for contacts, schedules, Web browsing, and system control. These features make Symbian OS a leading choice for smart phone implementations.

In order to target the full range of mobile devices, there are a range of different platform models that can be used on mobile devices. The original two are the reference designs: Quartz and Crystal. Quartz is targeted at PDA-sized devices(320×240 pixels), while Crystal is targeted for communicator type devices (640×200 pixels). More recently with the release of Symbian OS 7.0, two other user interfaces called called UIQ and Series 60 were introduced. UIQ provides a customizable pen-based user interface for media-rich mobile phones. It was designed to take advantage of the large, touch-sensitive color displays found on 2.5G and 3G handsets. UIQ is targeted for screen sizes ranging from 208×320 to 240×320 pixels, the standard size of most PDAs. The first device that takes advantage of UIQ is the Sony Ericsson P800. The Series 60 platform is developed and licensed by Nokia to a variety of phone manufacturers including Siemens, Panasonic, and Samsung. It is designed for mobile phone users with easy-to-use, one-hand operated handsets that have high-quality color displays, rich communications, and enhanced applications.With these innovative new designs, Symbian OS is making a strong move into the smart phone and PDA markets. To make the job easier for device manufacturers, UIQ and Series 60 come with several built-in applications, including email, SMS and MMS messaging, integrated contact lists and calendars, and high-quality Web page delivery. In addition to providing the operating system platform, Symbian also provides a development kit for developers who are building applications for the Symbian OS. This kit offers a wide range of application development choices. For on-device applications, these include EPOC C++, an efficient and purely object-oriented language that provides full access to all of the platform APIs; and Java, which can run on Symbian's Java runtime, which implements the PersonalJava 3.0 JVM and the JavaPhone 1.0 profiles.

For server-based applications delivered to the device, Symbian OS implements WAP 1.2.1 and HTML 4.01 with full frames, HTTPS, and embedded Java applet support. The Symbian OS also supports GSM, GPRS, EDGE, and CDMA networks and full internationalization support with Unicode and international locales.

Historically, the Symbian OS has been dominant in the European markets, with little exposure in North America. This started to change in March 2002, with Nokia's release of the Communicator 9290 device in North America. It is expected that other vendors will also release Symbian OS devices outside of Europe, making Symbian OS a potential challenger to both Palm OS and Windows CE for enterprise application development.

Linux

Linux is a free UNIX operating system that was initially created by a developer named Linus Torvalds. It is an open source operating system developed under the GNU General Public License, making the source code freely available to anyone who is interested. Current versions of Linux are being implemented by developers around the world who submit updates to the original code. The development effort started in 1991 with version 0.02. In 1994, version 1.0 of the kernel was released. The current full-featured release, version 2.4, was released in late 2002, and is still constantly being updated.

NOTE The latest Linux kernels can be downloaded from the Linux Kernel Archives at www.kernel.org.

When it comes to mobile devices, many flavors of Linux are available. Most of the device manufacturers supporting Linux will have their own version of the OS. In addition, many commercially available versions of embedded Linux are available. The major implementations of the Linux OS seem to be based on the handhelds.org kernel. Three companies offer Linux operating systems that are compatible with the Compaq iPaq: TransVirtual PocketLinux, Century Software MicroWindows, and TrollTech. The iPaq is not the only device that has Linux support though; other versions of the Linux OS are available for devices from Casio, Hewlett-Packard, Sony, and Sharp.

Note, however, that many Linux implementations are provided by the developer community and are not officially supported by the device manufacturers. This situation is starting to change. In 2002, Sharp released the Sharp Zaurus device, which is based on the Lineo Embedix Linux platform. Several other smaller manufacturers, including Royal (www.royal.com) and Infomart (www.infomart.info), announced plans for Linux-based PDAs in 2002, although none were shipping commercially at the time this book was written.

NOTE The fact that Linux is freely available does not mean that companies cannot charge for the enhancements they make. Companies such as RedHat have built an entire business around the Linux operating system.

One of the drawbacks of using Linux is the range of applications available. With a small market share and no official support on many devices, it is a challenge to find the applications and support that is required in the enterprise space. This situation will surely change for the better with the release of more commercial devices.

While Linux is growing in popularity on the server market, it has yet to be shown whether it will have a lasting impact on the mobile operating system market. Judging by the number of delayed Linux PDA offerings, it looks as though Linux will fill a niche spot in the mobile operating system market for Linux enthusiasts.

Java and J2ME

When the Java programming language was initially developed by Sun Microsystems in the early 1990s, it was aimed at the embedded device market. The large number of devices with different chipsets made programming difficult with Assembler or C. The idea behind Java was to design a language that could be ported to different architectures and operating systems without reprogramming. It seemed like an ideal fit for the diverse device market, but device manufacturers were not interested. The result was a language that could run on multiple chipsets with multiple operating systems but that did not generate any interest in its target market.

Fortunately, Sun Microsystems did not give up. With the growing popularity of the World Wide Web, the need for such a language resurfaced. At the time, Internet content consisted of static HTML pages that did not generate much viewer excitement. Missing was dynamic content. But how could it be programmed? What was needed was a language that could be run on a variety of operating systems based on different hardware architectures—seemingly, a perfect fit for Java. Using the Java platform, developers could create little Java applications called *applets* that could run inside of a Web browser and provide a rich, dynamic user interface for Web applications. The version of the Java platform that provides this interface is the Java 2 Platform Standard Edition (J2SE). Java applets failed to catch on as many expected. The general feeling was that client-side Java applications were too slow and too large to download.

Around that time, Sun Microsystems was also working on a server-side API for Java, now called the Java 2 Platform Enterprise Edition (J2EE). Rather than developing server-side Web applications using proprietary software, developers can use the J2EE specification to develop applications that can be ported to a variety of servers across a variety of operating systems. The Java platform finally found a large audience of companies and developers ready to take advantage of the capabilities offered by the language. As you are probably aware, most of today's application server vendors support the J2EE specification, and many wireless server-side applications are based on J2EE as well.

With the success of the J2EE specification, in 2000, Sun Microsystems decided to give Java another try on mobile devices. This time the company implemented another version of Java called the Java 2 Platform, Micro Edition (J2ME). J2ME targets the embedded and consumer device space, which ranges from Java Smart Cards to set-top boxes and smart appliances. Java smart phones and PDAs are included in that range. J2ME applications maintain the same core features of J2SE and J2EE: code portability, cross-platform support, along with the Java programming language. In addition, J2ME applications are upwardly scalable, to work with J2SE and J2EE platforms.

That said, it's important to point out that the J2ME architecture is somewhat different from J2SE and J2EE as it incorporates a few new concepts. J2ME is based around configurations and profiles. A configuration consists of a Java Virtual Machine (JVM), core libraries, standard classes, and APIs. Two configurations exist for J2ME today: Connected Limited Device Configuration (CLDC) and Connected Device Configuration (CDC). Each is designed for a set of devices with varying capabilities. The configuration provides the lowest common denominator set of classes and the building blocks on which profiles are created. A profile sits on top of a configuration and provides a complete set of APIs for a narrower set of devices. The combination of a configuration and a profile results in a complete J2ME platform.

The CLDC is based on the K Virtual Machine (KVM) and associated libraries. It is targeted at highly constrained devices with either 16- or 32-bit CPUs and less than 512 KB of total memory available for applications. The current profiles for CLDC include the Mobile Information Device Profile (MIDP) and the PDA Profile. Devices that fit into this category include smart phones, two-way pagers, and low-end PDAs.

For devices with more robust resources, the CDC is the configuration of choice. These devices typically run on 32-bit processors and have over 2 MB of memory available for applications. The base profile for the CDC is the Foundation Profile. On top of the Foundation Profile is the Personal Basis Profile and the Personal Profile. The relationship between these profiles is somewhat unique. The Personal Basis Profile actually includes the Foundation Profile and is a subset of the Personal Profile. The Personal Profile provides a similar level of client support to J2SE, but has a smaller footprint. The relationship between the J2ME configurations and profiles is shown in Figure 7.5.

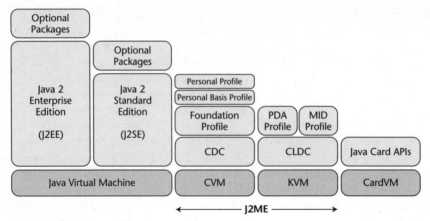

Figure 7.5   J2ME architecture.

Java is listed in the operating system section because you can create smart client applications using Java technology and deploy them to any operating system that has a JVM, which includes all the OSes listed previously. This means that rather than developing a native application that will work only on the targeted operating system, you can develop your application using Java and be able to target a variety of operating systems without having to rewrite the application. It sounds like the ideal solution. However, it is important to keep in mind that J2ME still suffers from some of the same problems experienced over five years ago with the Java on desktop clients; specifically, performance is not always adequate and the portability is not 100 percent. For most applications, extensive testing is required for each platform to which the application will be deployed.

As device manufacturers continue to add more capabilities to handheld devices, thus increasing their performance, it is expected that Java technology will become a viable option for creating sophisticated, data-driven smart client applications.

Proprietary Operating Systems

Before mobile operating systems were readily available, many manufacturers relied on proprietary operating systems for their devices. A proprietary operating system is defined as one that is only used, and available for use, by the device manufacturers themselves. Some examples include the RIM OS, used on the Research In Motion BlackBerry devices, and the Wysdom OS that Motorola used in its TimePort two-way pagers.

The use of proprietary operating systems is becoming quite rare, as it prevents large-scale development. For this reason, both RIM and Motorola have adopted J2ME as the development platform of choice for their most recent devices.

Summary

Smart client applications are an effective way to extend information to mobile devices in an always-available fashion. By incorporating persistent data storage and enterprise synchronization, the application users are ensured of having access to their data whenever and wherever it is required. The deciding factor on whether a smart client application can be developed for a particular device is the mobile operating system. Windows CE, Palm OS, Symbian OS, Linux, and Java/J2ME are the main operating systems that are well suited for smart client applications.

In the next chapter we will continue our exploration of the smart client architecture by investigating the smart client development process.

Helpful Links

The following are some useful links to find out more information on the topics discussed in this chapter.

Windows CE	www.microsoft.com/mobile/; www.microsoft.com/windows/embedded /ce.net/default.asp
Palm OS	www.palmsource.com/
Symbian OS	www.symbian.com/
J2ME information	http://java.sun.com/j2me/index.html; http://wireless.java.sun.com
Linux handheld information	www.handhelds.org

CHAPTER 8

Smart Client Development

Launching a mobile development project can be a challenging task. In addition to the user interface concerns that are always present on mobile devices, you have to worry about enterprise integration, connectivity issues, device constraints, and application deployment and management issues. All of these challenges have to be met using technology that is still in its infancy. It seems like a daunting task, but with proper research and project management these challenges can be overcome, resulting in a very rewarding solution.

The goal of this chapter is to prepare you for some of the technical challenges you will encounter while developing mobile applications, as well as to give you some pointers on how to get started. In this chapter, we will step through each part of the development process, investigating the technology that is available to help you build your mobile solutions. Then we will discuss the pros and cons of developing native versus Java applications.

> **NOTE** It is beyond the scope of this chapter to discuss general software design and development cycles. Many good books are available dedicated to this subject. Two that I recommend are *Code Complete* by Steve C. McConnel and *Design Patterns* by Erich Gamma, et al.

The Development Process

Developing mobile and wireless applications will be a new experience for most business application developers, who usually have not had experience working with limited memory and screen sizes. Moreover, most developers are used to working over reliable networks, commonly TCP/IP or HTTP, and to designing, developing, and testing their applications on the same machine. They have not had to worry about the deployment issues that arise when deploying to several machines, all with different characteristics from the one being used for development. Welcome to the world of mobile application development.

Mobile devices vary in shape and size, thereby adding new complexities for the user interface design and resource management. These devices communicate over a variety of wireless networks, often not even IP-based, let alone TCP/IP or HTTP. Finally, the application development does not take place on the target platform, so emulators are commonly used to help make the development and testing cycle more manageable.

To begin the development process discussion, we are going to take a look at the phases involved in designing a successful mobile solution, as depicted in Figure 8.1.

Needs Analysis Phase

A wide variety of mobile solutions are on the market, including voice-based solutions, wireless Internet applications, smart client applications, and messaging-based applications. Which type is best suited for your current project? The answer to that question must be determined during the needs analysis phase of your mobile application development process.

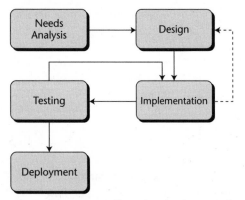

Figure 8.1   Smart client development cycle.

Questions to Ask When Researching User Requirements

While you are researching the user requirements, the following are some questions you may want to ask yourself:

- Who are the end users of this application?

- What is their technical skill level?

- What is the overall goal of this application?

- What data integration is required? Does the user require data access at all times?

- Does this application require wireless connectivity? If so, what type of wireless access does it require, and in which geographies?

- What are the primary usage scenarios for this application?

Try to answer most, if not all, of these questions before beginning to research the device, operating system, and development platform you are going to use. It is very easy to focus on the device and related operating system before determining the actual needs of the end user. If you fall into this trap, be careful not to let the device determine the features of the application. A complete application analysis will lead you to a solution that will be well received by the end user.

Things to Consider

As we consider the issues inherent to gathering the application requirements, we are going to focus on those related to smart client applications. (In Chapter 12, "Thin Client Development," we will look explicitly at the requirements relevant to wireless Internet applications. In Chapter 4, "Mobile Application Architectures," we gave an overview of smart client and thin client applications and the benefits—and drawbacks—related to each architecture type.) Once we gather information about the following items, we can make an educated decision on the solution that is most appropriate, including the operating system, device, wireless network provider, and development platform.

Application Goals

Across the globe, companies are implementing mobile and wireless applications for a variety of reasons. In some cases, the goal is to increase productivity; in others, it is to reduce costs; while in still others, the reason may be simply that it seemed like the next logical thing to do, regardless of the return on the investment. Whatever the reason, it is important that the final solution satisfy the original purpose of development.

During the needs analysis stage, it is important to determine why the project is being undertaken and to produce a result that is in line with expectations. If, say, the project is being implemented to automate order entry, thereby reducing paperwork, the final application should be well suited for rapid data entry, with integration back into the enterprise system. The technical features being implemented should correspond to a business problem that it will solve. By creating a user interface that allows for rapid data entry, we are solving the order entry problem. By having enterprise integration, we are ensuring that the data entered on the mobile device will flow into the back-end system without requiring additional paperwork. These are just a few examples of how technology can help a corporation meet its business goals as it implements a mobile solution.

End User

The end users of the application play a key role in the application design and rollout. Their level of technical ability determines some of the intricacies that may be required. If the user is not comfortable setting up wireless network connection parameters, such details will need to be automated within the application. If the user does not have experience with pen-based input devices, using a device with an alphanumeric keyboard may be more appropriate. Many mobile applications fail to reach their potential because they were not developed with the end user in mind.

Data Access

Extending corporate data to mobile users is essential to many mobile solutions. If this is the goal of the application being developed, then data integration and client storage options will have to be considered. Other factors to consider are how fresh the data must be on the client, how much data should reside on the client, and how often inputted data should be sent back to the server. All of these items will factor into the decision on the data storage and synchronization systems being used.

Wireless Access

Not all mobile applications require wireless access. In fact, many applications are better off without it since real time data access is rarely a requirement. Updates can be made over wireline connections, forgoing the need for wireless access. Other reasons for avoiding wireless access include: complexity of implementation, unreliability of the networks, cost of deployment, or lack of performance.

That said, many compelling reasons do exist for incorporating wireless access into a mobile solution. For field services applications, access to updated work orders, inventory analysis, or routing information can be critical to the success of the solution. Determining the level of wireless connectivity will affect both the implementation criteria and the cost of any given solution.

Usage Scenarios

Predicting how an application will be used is not easy. Nevertheless, it should be attempted, because having some idea of possible use can prove helpful for the application design phase. If, for example, an application is going to be used constantly and is critical to the success of the user's job, then battery life of the chosen device must be considered. It would not make sense to choose a device that has a short battery life that requires constant recharging for such an application. On the other hand, if the application is used infrequently but must perform some complex logic, it might be sufficient to choose a device with better processing power but with perhaps a shorter battery life.

The same type of logic applies to the level of wireless connectivity required. If the application will be used in a fixed location, implementing a wireless LAN solution would be more manageable and cost-effective than relying on a wide area wireless network.

Design Phase

After performing the needs analysis, you will have a good idea of the characteristics the solution requires. You will have limited the number of operating systems, devices, and networks to just a handful that will meet your requirements. From these you will need to make a decision as to which solution best meets your needs and addresses your business constraints. When doing this, review your options with an open mind. Try not to gravitate toward a particular solution immediately, as it is difficult to recognize at this point what will be a successful implementation unless you have an unsuccessful solution for comparison.

One way to accomplish this is to review all the solutions that could possibly meet your requirements, without giving any weight to an individual category. So, for example, if an operating system will meet the requirements, but not perhaps as well as another OS, keep it on the list as a potential candidate. The same goes for devices, wireless networks, and development platforms. After you have gone through this process, ideally you will have two or three possible solutions on which you can concentrate.

Now you can start adding weight to the individual categories to help you choose the option that best meets your criteria. Remember to factor in some "soft" factors as well, such as costs, ease of use, and aesthetics, in addition to the core technical requirements. Once you have chosen what appears to be the ideal solution, keep the other options on file, as you may need to come back to them in the future. (Refer back to Chapter 2, "Mobile Devices," if you require a summary of the device market.)

At this point, you can start to focus on the application design, looking at such issues as what data is required on the device, how you are going to integrate to the enterprise systems, and how the user interface should be designed for maximum productivity.

Client Data Access

Enterprise data sources will contain much more data than you can hope to store on the mobile device. This is true for enterprise databases such as Oracle, Sybase, or IBM; for ERP systems such as SAP or PeopleSoft; and for CRM systems such as Siebel. The amount of data stored on the device will depend on the mobile data store solution being used, device performance characteristics, and physical limitations of the device. You may only be able to store between a few hundred kilobytes to a couple megabytes of data on the client.

When designing a smart client application, start by examining the subset of data that is required for the mobile user. This subset can be determined by looking at many factors, often depending on the type of application being developed. The subset of data that is required can have various levels of granularity. You can partition the data by its structure in the enterprise source—taking only specific tables from an enterprise database. Another approach is to base the partition on specific data, such as a userid, geography, or price range. The more layers of partitioning you add, the more specific the data on the device becomes.

Let's look at an example sales force automation (SFA) application. For such a system, the enterprise database may contain customer contact information, order history, product information, inventory levels, and supplier information. We do not need all of this data on the mobile device, so we can take a subset of it. In this case, only the contact information, order history, and inventory level tables provide a benefit on the device, so we can limit the data to those tables. To go even further, we can say that we only want data for a particular sales representative, perhaps based on a userid, to be sent down to the device. The result is that we will have a subset of the enterprise data that contains everything relevant to the mobile user, and nothing that is not.

The data access design goes beyond just evaluating the subset of data that is required. Before you can go much further, you will need to establish how you want to implement the data store on the client. If you are implementing your own solution, you will want to start laying out your data storage and management logic. If you are going to use a commercial solution, you will want to start examining how storage is implemented in it, and work on the related design issues, such as the database schema. (For more information on persistent data storage on the client device, see Chapter 9, "Persistent Data on the Client.")

Enterprise Integration

Enterprise integration is a term used to describe any communication to systems not on the device. It encapsulates integration with enterprise databases, business applications, XML data, Web content, and legacy data, among other things. For the purpose of designing your mobile solution, you will need to determine to which enterprise systems you require access. This should be based upon the data access requirements that you have already set forth in the client data access stage of the design process. There are two levels of integration that you may require: basic integration and complex integration.

Basic levels of enterprise integration include the ability to access enterprise databases using defined communication protocols. These capabilities may include the following:

- Device communication using the standard synchronization software such as HotSync for Palm devices or ActiveSync for Pocket PC devices
- Communication over IP-based networks
- Direct integration with a relational database or flat-file system
- Limited support for transactions

Usually, you can follow a well-documented API for these types of applications. This integration layer either can be developed in-house, or you can use a commercial synchronization solution. In either case, the enterprise integration is much more straightforward than it is for larger and more complex systems.

When you start incorporating other forms of enterprise data and other communication protocols to the solution, a more sophisticated enterprise integration layer is required. The capabilities for this type of system may include the following:

- Support for a variety of mobile clients including laptops, handheld PCs, and PDAs

- Communication over networks that may not be IP-based

- Support for synchronizing multiple users simultaneously to a central back-end data store

- Communicating with systems that do not have well-defined interfaces, often requiring custom adapters

- Synchronization of complex data models

- Support for very large amounts of data with many transactions

- Conflict detection and resolution

- Administration tools to manage the entire process

If your system requires these advanced features, you should look to a commercial solution so you do not consume too many internal resources on one aspect of the overall solution. Purchasing a solution outright will result in significant cost savings, compared to the costs involved with writing, testing, debugging, and maintaining the conduit that is required to support the features listed. Even with the use of commercial software, the design requirements are much more complex than for a basic system. You will still need to worry about security, data partitioning, enterprise system access, and scalability, in addition to the usual network coverage and bandwidth issues. (Data synchronization is covered in depth in Chapter 10, "Enterprise Integration through Synchronization.")

User Interface

The user interface can account for as much as 80 percent of the total code in a mobile solution. When you have one part of the application accounting for such a large portion of the development effort, it has to be designed correctly to avoid costly changes later in the development cycle.

Be consistent with the user interface (UI). Get a feel for how standard applications that come installed on the device work and stick very closely with their UI styles. Most of these applications provide a very basic entry screen that meets the needs of most users. Typically, these UIs make more advanced functionality accessible via user selection of a button or menu item.

A couple of key items to keep in mind when designing on the user interface include screen size and human interaction. Paying attention to these items will make the application effective for the application users and help you avoid changes later in the development cycle. They are explained in the following paragraphs.

Screen Size

One of the most dramatic differences between desktop applications and those developed for mobile devices is the screen "real estate." When targeting mobile applications, you will have a one-half VGA, one-quarter VGA, or even smaller screen to work with. Using the screen to its maximum benefit is crucial for successful applications. There are many different ways to accomplish this goal, depending on the operating system you are using. Windows CE, for example, supports a tab-based interface, allowing easy navigation to multiple forms with the click of a button. Taking advantage of the menu capabilities in Palm OS applications allows more information to be displayed on a small screen.

Human Interaction

Studying human interaction with your application will prove to be invaluable in determining its overall usability. History has shown that people always interact with the system in unexpected ways. There is no way that these ways can be predicted, so testing is critical.

Before you get to the stage where you can study human interaction, there are application-specific requirements that can lead the direction of the user interface. If the application focuses on data input, then the main input screens should be easy to navigate using the input properties of the device. For example, if the device offers keyboard support, make it possible for the user to quickly tab between entry fields, rather than having to use a scroll-wheel to get there. If the only means of input is a stylus with character recognition, then including radio buttons and drop-down lists are effective ways to improve the efficiency of data input.

By focusing on the screen design and the navigation practices of the general user, you will be able to design effective user interfaces on devices of all sizes. Do not forget that spending some extra time testing the application early in the development cycle can save you much more time later on.

Wireless Connectivity

If you determine that you require wireless data access within your application, keep in mind the state of today's wireless networks. If you are planning on using a public network, keep these key points in mind during the application design:

- *Wireless coverage is not guaranteed.* In fact, many areas do not have adequate coverage for data communication. This situation applies even in countries that have excellent overall wireless coverage.

- *Network penetration issues can arise even in areas that do have coverage.* Network penetration can be problematical not only in obvious places like subways and tunnels but also in many corporate buildings.

- *Wireless networks operate over a variety of protocols.* Some of these are not IP-based, so if your application requires IP for data communication, you may have to add an additional IP layer for connectivity. Many software vendors can provide this layer for you, if required.

- *Limit the frequency of wireless data transfer.* Because there are potential problems with the reliability of wireless networks, this consideration will make the application more effective. Having suitable persistent data storage within the application will make it possible to limit the frequency of network connectivity within the application. Infrequent wireless data transfer also has a positive effect on battery life, as wireless communications require more battery power than local access.

- *Limit the amount of data transferred.* This limitation is urged, once again because of the nature of the wireless networks.

Implementation and Testing Phase

The implementation and testing phase of the development cycle is what most people refer to as software development. It is during this phase that programming takes place and where developers start to see the rewards of their efforts. The preparation phases (needs analysis and design) are often overlooked because concrete results are not apparent. This does not mean that those parts of the development cycle should be minimized; it means the opposite. Having a complete design document is the best way to begin the actual application implementation. It will allow you to stay focused on the requirements; you will be spared having to figure out what is required on the fly.

The implementation of the final solution does not happen all at once. It is an iterative process that requires much testing. Figure 8.2 shows the process in flowchart format. The diagram introduces the concept of device emulators, which are programs that simulate the characteristics of a physical device using software. In order to get to a successful final product, there are some important steps you must keep in mind, the first of which is the development of a prototype.

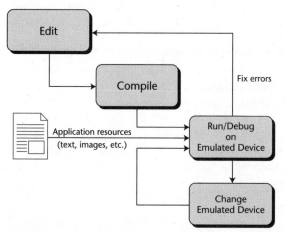

Figure 8.2 Development cycle using device emulators.

Prototypes

Before making any critical decisions about your mobile solution, develop a prototype and do some field-testing. A prototype is a mockup of the final product that simulates the look and feel of the final application, revealing many of the complex interaction problems at an early stage in the development life cycle. It will save you many headaches during the deployment and maintenance stages of the development process.

The following list outlines some of the questions you should ask during your prototype stage:

- Which device and related operating system is most suitable for my application?
- Which wireless network is appropriate for my application, and does it perform as expected?
- Does the user interface provide the most efficient way for the user to operate the application? Does it match the device characteristics?
- Is the appropriate data available in the application? Can the client data store handle the amount of data required on the client device?
- Does the enterprise integration layer work? Is it scalable, to meet the needs for my application?
- Have security concerns been addressed? Are there holes where corporate data is left unprotected?
- Does the application provide an upgrade path for new features? Will it be adaptable for new wireless networks as they arrive?

These are just a few of the questions that need to be answered during the prototype stage of the mobile application development cycle. With the short development times that are now the norm in the software industry, the prototype phase is often omitted from the development cycle. This is a mistake. Each stage of the development process is there for a reason, so avoid skipping phases. More often than not, when you skip part of the development cycle to save time and money, the problems that arise from the omission end up costing you more than if you had done it in the first place.

A prototype does not have to be the complete application. It can represent just a small segment of the overall application, which will allow you to determine some overall application characteristics. For example, to test the network connectivity and bandwidth, only a portion of the application needs wireless connectivity. The information that you gather about this part of the application will be a good representation of what you can expect from the completed product.

Many questions need to be addressed during the implementation of a prototype. In order to answer these questions, you have to actually put the prototype together. To help you accomplish this, we will provide an overview of the emulators and development tools that will aid you in this process. Included in the overview is a discussion of the stages of development as you move from testing on an emulator to a device in a cradle to a device on a wireless network.

Development Tools

Choosing which development tool to use is not a decision to take lightly. Many development tools are available for smart client applications, each with benefits and drawbacks. You need to consider many factors when making your development tool choice. Some of the more important ones include the target mobile operating system, the preferred programming language, the tool's feature set, and the tool's layout. In addition, you should also take into consideration your personal preferences. You will be using this software day in and day out, so if you are not comfortable with the chosen tool, sooner rather than later is a good time to voice your opinion.

Table 8.1 summarizes the development tool options available. Notice that some support multiple operating systems, for example, AppForge Professional, PenRight! MobileBuilder, and Metrowerks CodeWarrior. Cross-platform capability is a nice feature in a tool if you are targeting more than one mobile operating system, but it comes at a cost. Very often, these tools require additional libraries to support the cross-platform capabilities, adding additional overhead. In addition, some areas of development may be limited in functionality, as these tools often work at a lowest-common-denominator level. That said, if you are going to be doing cross-platform development, it's definitely worth your time to take a look at them.

Table 8.1 Leading Development Tools for Mobile Development

MOBILE OPERATING SYSTEM	DEVELOPMENT TOOL	URL
Windows CE	Microsoft Visual Studio. NET with the .NET Compact Framework	www.microsoft.com/mobile /developer
	Microsoft eMbedded Visual Tools (EVT)	www.microsoft.com/mobile /developer
	Microsoft Platform Builder	www.microsoft.com/mobile /developer
	AppForge MobileVB	www.appforge.com
	PenRight! MobileBuilder	www.penright.com
	Sybase PocketBuilder	www.sybase.com/products

(continues)

Table 8.1 Leading Development Tools for Mobile Development *(Continued)*

MOBILE OPERATING SYSTEM	DEVELOPMENT TOOL	URL
Palm OS	Metrowerks CodeWarrior for Palm OS	www.metrowerks.com/products /palm
	AppForge MobileVB	www.appforge.com
	PenRight! MobileBuilder	www.penright.com
	Pumatech Satellite Forms	www.pumatech.com
	PRC-Tools: GCC Development Tools	http://sourceforge.net/projects /prc-tools
Symbian OS	Symbian OS C++ Software Development Kit	http://www.symbian.com /developer
	Symbian OS Java Software Development Kit	http://www.symbian.com /developer
	AppForge MobileVB	www.appforge.com
	Metrowerks CodeWarrior for Symbian OS	www.metrowerks.com /products/symbian
Java	JavaSoft J2ME Wireless Toolkit	http://java.sun.com/j2me /index.html
	Sun ONE Studio	http://wwws.sun.com/software /sundev/jde/index.html
	Metrowerks CodeWarrior for Java	www.metrowerks.com/
	Borland JBuilder MobileSet	www.borland.com/jbuilder /mobileset

In addition to the tools listed in Table 8.1, many other vendor-specific SDKs are available; typically, however, these are optimized for a particular device, using a specific wireless technology, and therefore are not well suited for broad application development. At this time, with support for Palm OS, Windows CE, and Symbian OS, J2ME looks like the best bet for a standard API across the leading platforms. One word of caution: If you decide to go this route, be aware of the current performance limitations that Java may impose on your applications. This subject is discussed later in this chapter in the section entitled *Native versus Java Applications*.

VISUAL STUDIO .NET 2003

With Visual Studio .NET, Microsoft went one step further for making smart client application development easier by adding Smart Device Programmability features to the tool. The overall goal of Smart Device Programmability is to enable developers to use the same programming tool for all forms of development. This means you can use Visual Studio .NET 2003 for building server and desktop applications as well as mobile applications. Smart Device Programmability for Visual Studio. NET is the tool that developers can use to build applications using the .NET Compact Framework. Currently, Smart Device Programmability enables developers to build applications for Pocket PC, Pocket PC 2002, and Smartphone 2002 devices, along with any other device based on Windows CE .NET.

.NET COMPACT FRAMEWORK

A core component of Smart Device Programmability is the .NET Compact Framework (.NET CF). .NET CF is the smart client development component of the Microsoft .NET initiative. It is based on the same programming model as the desktop .NET Framework, allowing developers to easily reuse existing programming skills and code for mobile device, desktop, and server development.

The languages of choice for .NET CF development are C# (pronounced see-sharp) and Visual Basic. C# is an object-oriented language developed by Microsoft. It has many similarities to C, C++, and Java, with some additional features for component development. The source code is compiled into a common language runtime and deployed to the mobile device. *Managed code* is the term Microsoft uses to describe the business logic components in the .NET Framework.

Some of the key features of .NET CF include the following:

XML Web Services. Web Services provide a standard way to provide server-based functionality to a variety of clients. The ability to easily access Web Services is a key component of the .NET CF. Similar to the .NET Framework, accessing a Web Service from .NET CF is as simple as making a function call. (A closer look at how Web Services complement mobile computing is provided in Chapter 18, "Other Useful Technology.")

Corporate data access. To access corporate data, ADO.NET classes are available on the client device. These provide a standard way to access relational databases over a wireless connection.

Performance. Applications running on .NET CF are executed as native code. A just-in-time (JIT) compiler takes the managed classes and compiles them at runtime, resulting in higher performance than if they were interpreted.

Visual Studio .NET integration. Developers benefit from the ability to use Visual Studio .NET to develop smart client Windows CE applications. The same version of Visual Studio .NET is used for mobile clients, desktop applications, and server components. Once Smart Device Programmability for Visual Studio .NET is installed, you can develop your applications using Visual Basic .NET or Visual C# .NET.

Table 8.2 Development Tool Evaluation Questionnaire

FEATURE	QUESTIONS TO ASK
Integration with third-party software	Does the software integrate with other third-party software that you may be using, such as mobile databases, synchronization software, and messaging clients?
Emulation environment	Does the tool provide an environment for emulation so you can quickly test applications without having to deploy them to a physical device?
Debugger	Can you debug the applications? If so, how? Through an emulator, on the device, or both?
Drag-and-drop UI creation	What is the mode of user interface development? Does it provide GUI components or an API?
Multiple-platform support (Palm OS, Windows CE, Symbian OS, etc.)	Which mobile operating systems does it support? Is it limited to a single OS, or can you target multiple OSes with the same tool?
Support programs	Where do you go when you run into a problem? Can you access a developer network or newsgroups? Do the technical support packages meet your requirements?
Training programs	If you require training on the development tool, is it readily available? Is online training or classroom training offered?
Educational books	Are books widely available to help you learn about the tools and how to develop with them?
Online tutorials	Are tutorials available for the development tool? Does the tool vendor or a third-party development site provide them?
Programming language	Which programming language does the tool use? Is it a standard language, such as C/C++ or Java, or is it proprietary to the vendor? Is this language appropriate for the development team's background?

Once you know to which operating system(s) you are going to deploy the application, examine each tool available, to determine if it will meet your needs. To help make the decision easier, you can create a matrix that lists your criteria; then rank each tool as to how well it meets your requirements. You might also want to separate the items into "must-haves" and "nice-to-haves." In this way, you can quickly eliminate any tool that does not meet your must-have checklist. Table 8.2 contains a few features that you may find useful when developing your mobile application.

Once you have decided which development tool is best for your development efforts, you can move on to the programming part of the implementation phase. At this stage you can begin to see the benefits of your needs analysis and design efforts. When you start developing the mobile application, having a complete development environment will make your development and testing efforts go more smoothly.

In the next part of this chapter, we are going to take a look at the development effort as it progresses through various stages of testing. We will start with testing in an emulation environment, then move to a physical device with connectivity via a cradle, on to the final stage of testing on a wireless network.

Device Emulators

An emulator can be your best friend and your worst enemy—at the same time! It will be your best friend when it increases your productivity dramatically. Deploying and testing your application on an emulator is much easier than on a real device. This is especially true if you do not have all the devices for which you are building the application, which is often the case. In such a situation, the emulator allows you to get an idea of how the application will work on the physical device, while you avoid the hassle of actually configuring a real device. On the other hand, an emulator can be your worst enemy if you start to rely on it exclusively. You must never forget that it is just an emulation setup, not the real thing. There is no substitute for testing on the physical device and communicating over the real network on which you will be deploying. In the past, many developers have been embarrassed when their applications did not run properly on the real device, even though these same applications ran flawlessly on the emulator.

Emulators are available free of charge for the leading mobile operating systems. Often, they are included with your development tool; but if not, this section will give you the information you need to obtain them.

The following are some features that are available in most (if not all) of the mobile operating system emulators:

- A window, which can be dragged anywhere onscreen, to display the device's screen, surrounding skin, a keypad, and LED indicators.
- The PC keyboard, to enable additional key input.
- The PC mouse, to enable pointer input.
- A directory tree in the PC's file system, which shows the mobile OS file system.
- The Windows runtime environment, which provides support for communications, Internet, and IrDA through the host PC.

Windows CE

The Pocket PC emulator shown in Figure 8.3 comes with the Pocket PC 2002 SDK. It allows you to test all Pocket PC, Pocket PC 2002, and Windows CE .NET-based applications from your desktop. To create applications for other versions of Windows CE, you will need eMbedded Visual Tools (EVT) along with the SDK for the platform you are developing on. These tools and related SDKs are available as a free download from the Microsoft Web site.

If you are using Microsoft tools, you can configure the IDE to run in emulation mode. When you run the application from within EVT, the IDE opens the emulator on the desktop, downloads the executables files to the emulator, and starts the application. At this point, you can run and test your application on the emulator just as you would on a physical CE device. The emulation environment enables you to test code more quickly than if you had to download files to a device after every compilation. It also gives you the ability to test your applications when a physical Windows CE device is unavailable.

If you are not using Microsoft tools, you can still use the Windows CE emulator. Depending on the Pocket PC SDK you are using, you can either launch the emulator from the Windows Start Menu (Pocket PC 2000 SDK) or from Microsoft Visual Studio .NET or EVT (Pocket PC 2002 SDK). All you have to do is copy your application files to the correct location in the emulation file system. If you want to change the external appearance of the emulator, you can select another "skin" for it. A skin is used purely for aesthetic purposes; it does not change the emulation characteristics in any way.

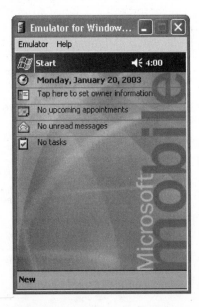

Figure 8.3   Windows CE emulator.

Palm OS

The Palm OS Emulator, shown in Figure 8.4, is software that emulates the hardware of the various models of Palm OS-based devices. It is extremely valuable for writing, testing, and debugging applications, as it creates "virtual" devices by running the emulator on Windows, Mac OS, or UNIX computers.

The Palm emulator is downloadable from the Palm Developer's site. Once you have joined the developer's program, you will need to download the emulator and ROM image (ROM is the software that controls the emulator). The emulator can be loaded with new ROMs, so you can test your application with virtual devices, including those you do not own. These can include foreign-language ROMs and debug-enabled ROMs for extra error checking and debugging features.

The emulator that you download will not include any ROM images; think of it as a computer without an operating system. To emulate a specific device, you will need to obtain a ROM image that is compatible with that device. Typically, the ROM and device must match in processor type and display color depth. In some cases, you will also require some device-specific capabilities. For example, to simulate a Palm VII device with a wireless connection, you would need the Web Clipping components, which are part of the ROM image.

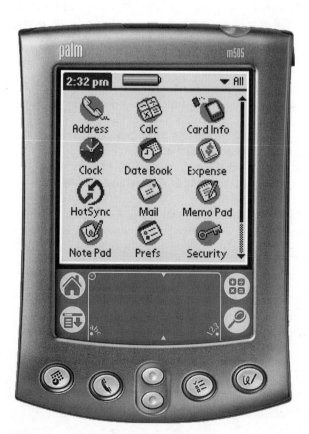

Figure 8.4   Palm OS Emulator with Palm m505 skin.

As new versions of the Palm OS come out, new ROM images are released on the developer's site. When you install a ROM image, the emulator allows you to choose which device you want to emulate. Note that the ROM you use restricts the choices available. Also note that in order to make the emulator appear like the device you are emulating, you will need to download and install the appropriate skins. As mentioned earlier, a skin changes the external look of the emulator; it is not required to get an accurate emulation of the device characteristics.

With the release of Palm OS 5.0, PalmSource has also released the Palm OS Simulator. The Simulator is the Palm OS 5.0 system compiled to run natively on Intel platforms. It does not do hardware emulation like the Palm OS Emulator, but is rather the real Palm OS 5.0 running on top of a device abstraction layer. When developing applications for Palm OS 5.0 it is recommended that you test on both the Palm OS Emulator and the Palm OS Simulator. Testing in both environments is the only way that you can be ensured that your application will run correctly on a physical Palm OS 5.0 device.

Symbian OS

As part of the Symbian OS SDK, you get an emulation environment for the Symbian Quartz platform shown in Figure 8.5. The SDK gives you support for both C/C++ and Java environments. The Quartz emulator provides a full target environment, which runs on desktop Windows platforms. The emulator enables development to be primarily PC-based. You only have to worry about hardware testing in the final development stages. To compile the application code for either C++ or Java development, you are required to use a development environment such as Microsoft Visual C++ for C++ development or a Java IDE for Java development.

The Symbian emulator is an implementation of the EPOC kernel and device drivers, and uses Microsoft Windows and PC hardware to emulate a real EPOC-based device. On the emulator, the application is built as Windows dynamic link libraries (DLLs), containing x86 machine code. Symbian OS programs are rebuilt in native machine code, using the ARM instruction set, to run on a real ARM-based device.

Interpreted languages such as Java use machine-independent data formats, which do not need to be built specially for the emulator or target machine.

There are two Symbian emulator variants: release and debug. The debug emulator is most useful for development as it contains debug information that can be used for stepping though code, and it provides special key combinations for resource checking and redraw testing. This helps developers to trap memory leaks and other errors early in the development cycle.

NOTE Do not worry if the debug emulator does not come up right away, because loading it can be slow. The release emulator is much faster to load as it does not contain debug information.

In addition to the emulation environment in the Symbian SDK, many device manufacturers that support the Symbian OS also have emulators available for their specific devices. Ericsson, for example, has the Ericsson R380 emulator, which is specifically designed to test applications written for that device. Nokia also provides similar emulation environments for its devices.

Figure 8.5 Symbian OS quartz emulator.

Java/J2ME

In the past, many device manufacturers had device-specific emulator environments for the Java platform. More recently, these manufacturers have decided that a better solution is to integrate their devices to a standard environment, such as JavaSoft's J2ME Wireless Toolkit or Metrowerks CodeWarrior for Java, rather than support a proprietary tool. Here, we will focus on using the J2ME Wireless Toolkit offered by JavaSoft, as opposed to examining a variety of implementations. This toolkit offers a complete set of tools that provides developers with the emulation environment, documentation, and examples needed to develop Connected Limited Device Configuration (CLDC) applications using the Mobile Information Device Profile (MIDP).

The J2ME emulator allows you to see how MIDP applications operate on a variety of mobile devices. Figure 8.6 shows the default emulator that is available in the toolkit. (Note: Emulators for other devices are available from the device manufacturers as plug-ins to the J2ME Wireless Toolkit.) Similar to other emulators, the J2ME Wireless Toolkit emulator executes on your local PC so you can test your applications on the same platform you use for development.

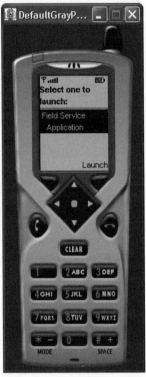

Figure 8.6   J2ME Wireless Toolkit default emulator.

Some of the key emulation features of the J2ME Wireless Toolkit are:

Support for a variety of mobile and wireless devices. You can set up the toolkit to
use emulators for devices with a variety of form factors, including smart
phones, pagers, and PDAs.

Debugging capabilities. You can execute runtime logging of the events pertinent
to your application, including garbage collection, class loading, method calls,
and exceptions. You can also perform source-level debugging with your IDE as
the application runs in the emulator.

The Java emulator comes with the same caution as the other emulation environments; that is, do not use emulator testing as a substitute for testing on the physical
devices. The emulation environment is intended to serve only as an approximation of
the user interface, performance, and other capabilities of the device. You are not running
on the actual target chipset, so the emulator cannot simulate the actual runtime speed.

If you are not using J2ME on the mobile device, then you may want to look at the
specific toolkit offered by the device manufacturer or Java Virtual Machine supplier.
The Symbian SDK for Java offers an emulator that is not J2ME-based, and therefore has
its own development environment for building PersonalJava applications.

Physical Devices

Once you have the application running correctly on an emulator, it is time to test it on a physical device. Ideally, this will be the device that you are planning to use for deployment. This is important because each device has different properties, including the type of processor it uses. For example, if you build a Pocket PC application and test it on an HP iPaq, you are testing an X-Scale processor. If you then want to deploy it to a Casio EG-800 Pocket PC, you have to rebuild it for the MIPS processor. By changing deployment processors, you are changing the behavior of the application, so make sure that you test on the appropriate devices before deployment; do not assume that all Pocket PC devices behave the same way.

Only after you have tested on the real device will you be able to accurately gauge the performance of your application. This will be the first time the application is running on the target processor, with the actual operating system. In most cases, the move from an emulator to the real device will go smoothly, but it is for the times that it does not go smoothly that testing is required.

Cradle

The easiest way to test the communication of the application is through a cradle connection. In this way, you have full control over the network communication and you do not have to worry about establishing the proper wireless connections.

A variety of conduits are available for this cradle communication, including device synchronization software such as Microsoft ActiveSync or Palm HotSync. In addition, you can set up direct serial communication or connect via TCP/IP. Very often, the specific device and its related software dictate the communication method used.

If your application is not going to have wireless connectivity, this could be your final stage of testing.

Wireless Networks

If your application will provide wireless access, the final stage in your testing process will involve the use of a wireless network. Many different types of wireless connectivity can be implemented, depending on your application needs. As discussed in Chapter 3, "Wireless Networks," the two relevant types of networks are wireless wide area networks (WWANs) and wireless local area networks (WLANs).

If the application will be used in multiple areas, possibly in multiple geographies, you will want to utilize a WWAN for wireless access. But note that, currently, for this type of network, wireless communications capabilities are not 100 percent reliable. Therefore, you will want to test your application in a variety of environments and attempt to limit, as much as possible, the frequency and amount of data being transferred in the deployed application. While in the testing phase, verify wireless failure conditions to ensure the application handles wireless connectivity loss in a way that does not hinder continued use of the application.

If you are deploying the application in a controlled area, such as a warehouse or a hospital, then you may be better off using a WLAN for connectivity. A WLAN is not controlled by a wireless carrier, but by the company deploying the solution itself. Because of this, you have full control over the network, hence can often achieve better coverage reliability, while getting higher bandwidth. WLANs are TCP/IP-based, so your testing results should be similar to those over a wired TCP/IP network.

While in the testing stage of the development process, you should set up a staging server for your application. This is a server that has the same environment as your deployment server. Doing this will give you a true idea of what to expect at deployment.

Deployment Phase

Two major drawbacks commonly cited as reasons not to build smart client applications are deployment and management issues. With smart client applications, you have to physically deploy the application to the device, in contrast to wireless Internet applications that run in a browser, and therefore do not require deployment other than downloading the content by accessing a URL. This form of browser-based deployment is one of the reasons that Internet applications have been so widely adopted. Now, with the proliferation of mobile devices, IT departments again have to worry about application deployment and management, and with additional challenges, which include the following:

Wide range of devices that need to be supported. These can range from two-way pagers to PDAs to laptops. Ideally, a way can be found to manage all of these devices in the same manner.

Deployment of applications to these devices. This includes the original application as well as updates as they occur.

Management of mobile assets. This includes keeping an inventory of the devices in the field, as well the software on these devices.

Backup and recovery. Because the primary functions of many of the mobile applications are to retrieve data and to make sure the data is safe in the event of a system crash.

Working with wireless networks. In most cases, the rules that were developed for LAN-based applications do not apply to wireless deployment and management. The bandwidth of the wireless networks makes the efficiency of the solution a top priority.

End users have little access to technical assistance. Very often remote workers do not have direct access to technical support, making troubleshooting and repair difficult.

Participating in business analysis to determine application effectiveness. In today's economic environment, businesses need ways to determine if their mobile applications are meeting their goals of improving employee productivity and cost savings.

For smaller deployments, these concerns may not be as troublesome. In some cases, the applications can be manually deployed to each device before distribution, and support can happen on a one-on-one basis. Also, if the end user has a technical background, he or she may be able to assist with the deployment and management functions.

For larger-sized deployments, manually deploying and managing the software is not practical. This is especially true if the end users are in remote locations, where technical resources are not available. In these cases, companies should coordinate with third-party solution vendors for assistance with device management issues. A number of vendors offer solutions that address these needs.

Deployment and Management Software

Application deployment and management are difficult tasks at the best of times. When it has to be done remotely, on a variety of devices, the challenge becomes daunting. Fortunately, several companies have developed software with the explicit objective of making application deployment and management controllable, even to mobile devices. Most of these products provide a core set of features including software distribution, inventory and asset management, application self-healing, application remote backup, remote device management and enhanced security. All of these features are described in depth in Chapter 16, "Mobile Information Management." Visit this chapter for further information on the capabilities of various mobile device management solutions, and lists of vendors whose products may be of interest to you.

Wireless Deployment Options

In addition to the physical deployment of the software, you may also have to set up wireless connectivity. For wireless networks that are not IP-based, you have two options: establish the connections to the wireless carriers yourself, or rely on a third party for the required connections. If you decide to establish the connection yourself, you will need to establish a business relationship with the wireless carriers that you are using. Typically, you will have to meet some minimum requirements for putting in a direct connection to the carrier; these might include contract length, amount of bandwidth, and number of remote users.

The direct route is often taken for medium to large deployments, where the cost associated with establishing direct relationships with these carriers is less than those of using a third party for connectivity. A diagram depicting this setup is provided in Figure 8.7. As you can see, all of the connections from the wireless carriers come directly into your enterprise, where they access the servers that run your synchronization logic and contain your enterprise data.

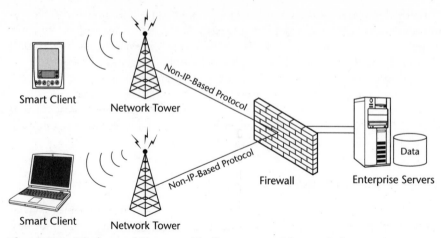

Figure 8.7   Wireless architecture with direct connectivity to wireless carriers.

The other option is to use a wireless Internet service provider (ISP) for your connectivity. The companies that offer this service usually charge a set-up fee for the required connections, in addition to a monthly maintenance fee. The ISP then provides the gateway between the wireless carrier and your enterprise, converting the proprietary wireless protocol to HTTP for communication to your enterprise servers. This infrastructure is all contained within a network operation center (NOC). The benefit of using a wireless ISP is that you can communicate over multiple wireless networks and have all of the technical support, provisioning, and billing provided by a single company. Figure 8.8 illustrates this architecture. As you can see, all of the wireless carriers communicate with the wireless ISP's NOC, which then communicates over HTTP or HTTPS into your enterprise.

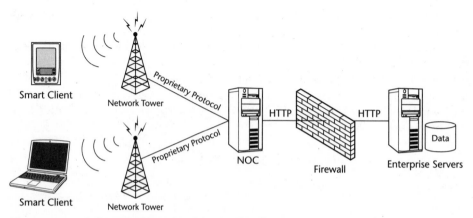

Figure 8.8   Wireless architecture using a wireless ISP.

Keep in mind that connections to the wireless carrier are required only for wireless networks that do not have an IP-based protocol. CDPD, GPRS, and most of the 3G wireless networks are all IP-based, and therefore can communicate to your enterprise using the public Internet, thereby avoiding the requirement for wireless gateways, and hence, direct connections into your enterprise. That said, the benefits that a wireless ISP provides for billing and related issues may still appeal to you.

Native versus Java Applications

Java technology for mobile devices has come a long way. The virtual machines have matured and become more robust and higher performance. J2ME has expanded, now offering two configurations, the Connected Device Configuration (CDC) and the Connected Limited Device Configuration (CLDC), along with several profiles. These developments have made Java a viable alternative to native applications for the leading mobile operating systems.

Deciding whether to develop Java applications over native, compiled applications is not simple. While there are some clear benefits of using Java, there are also some drawbacks.

Benefits of Java

The main attraction to Java is its cross-platform capabilities. When developing native applications, you have to maintain multiple sets of source code, one for each platform targeted. Java alleviates this extra effort. When developing in Java, you can create one set of source code that can be deployed to any of the mobile operating systems that have a Java Virtual Machine (JVM). This allows you to deploy your solutions to additional devices with minimal effort. Of course, if the devices have different form factors, the user interface may require modifications for it to be useful. In addition, the application will need to be tested separately on each device as the JVM implementations differ between platforms.

The programming language itself also attracts developers to use Java; as it is a powerful language for software development. With its pure object model and lack of pointers, developers can reduce the incidence of errors and increase productivity. In addition, Java has built-in communication facilities, with TCP/IP, serialization, and reflectivity all part of the language; it is also the first programming language to seamlessly support communication protocols. This is, of course, an additional benefit for anyone creating wireless applications.

Finally, Java has a single source for the language and API editions that assure developers that Java will stay focused and standardized. The Java Community Process (JCP) controls new features and enhancements, permitting any individual or corporation to contribute to the specification. The downside to this is that it can take longer to release versions of the specification, as they have to go through a community review process.

These benefits, among others, have attracted a large and growing development community to the Java platform. All of the major software companies (with the notable exception of Microsoft) have pledged support for the Java platform in one way or another. In the mobile device space, many device manufacturers are basing their latest offerings on Java. Some of the main vendors in this category include Nokia, Motorola, and Research In Motion, along with other smart phone vendors.

Benefits of Native Applications

Even in the face of Java's growing popularity for mobile application development, building native applications still has some very critical advantages. The primary benefit of developing applications natively for the mobile operating system is performance. Even with the advances made in JVM technology, its performance still is not close to what can be achieved by writing native applications. On some devices, this may not be a concern, but for most devices with their limited resources, performance is the key factor in the decision to write native applications. The Palm platform is a good example. The current Java implementations for the Palm OS limit the complexity of applications that can be deployed due to performance. If you want to build an application that requires anything beyond the basic level of processing power, the application speed, or lack thereof, is quite noticeable to the application user.

In addition to performance, native applications offer increased flexibility. With the ability to directly access the APIs of the operating system, programmers using C/C++ have the ability to take advantage of the low-level capabilities of the operating system without having to go through additional layers. This can be a big advantage for applications that require close operating system interaction. This level of interaction is often required to access low-level telephony or wireless modem APIs.

Finally, development tools are a factor. Most of the leading development tools for any given platform (excluding those that are purely Java-based) are for building native applications. For Windows CE, it is the eMbedded Visual C++ and Visual Studio .NET; and for Palm, it is Metrowerks CodeWarrior. These tools have been around for much longer than their Java counterparts, and have proven to be effective for mobile application development. Few Java development tools aimed at the mobile market are as mature.

Summary

Before getting in too deep with mobile application technology, spend some time focusing on the customer requirements. This will result in an application that solves both business and technical requirements. For developers who have experience with server or PC development, mobile application development will require extra care to address user input, screen size, memory and processor limitations. To aid with development of these applications, there are a variety of development tools and SDKs available for all of the leading mobile platforms.

When it comes to testing applications during the development process, emulators can be extremely effective. They allow you to develop and debug applications on a single machine, without having to obtain physical devices and wireless connectivity. But under no circumstance should emulators replace testing on physical devices over the target wireless networks. Every stage in the development cycle serves a purpose. Not taking the time to execute the complete development process can cost more in the form of changes made late in the development cycle.

In the next chapter, we are going to take an in-depth look at mobile database technology and how it factors into creating smart client applications.

Helpful Links

You will want to go to the Web sites listed here for more information on all the topics discussed in this chapter.

Microsoft Visual Studio	http://msdn.microsoft.com/vstudio/ http://msdn.microsoft.com/vstudio/device/overview.asp
Palm Developer	www.palmsource.com/developers/
Symbian Developer	www.symbian.com/developer/index.html
Sun J2ME Wireless Toolkit	http://java.sun.com/products/j2mewtoolkit/
Wireless Developer Network	www.wirelessdevnet.com/channels/pda/

CHAPTER 9

Persistent Data on the Client

The key component of smart client applications is the persistent data store. This is the technology that allows you to maintain data on the device, removing the requirement for wireless network coverage. Beyond allowing you to work offline, persistent data also makes it possible to build applications with rich user interfaces that have high-performance data access, transactional capabilities and enterprise integration support.

You have a variety of options to choose from for data storage: You can use the device's file system to store data, build your own data storage mechanism, or purchase a commercial solution. In this chapter we are going to take a closer look at some of the reasons why databases are an important component of smart client applications. We will also look at some of the options you have for implementing this type of solution.

Types of Data Storage

Before delving into the requirements for persistent storage within mobile applications, it is worthwhile to step back and take a look at traditional data storage systems. In general, database systems provide a storage mechanism for data. This can be accomplished in a variety of formats, with different storage techniques and relationships between the data. They also provide a means to access that data by using a defined language, like Structured Query Language (SQL), or other proprietary APIs. A complete system that provides the data storage, access mechanism, and administration tools is referred to as a database management system, or DBMS. This is clearly an oversimplified view of what a database is and does, but for our purposes here, it should suffice.

We are going to look at four main database systems: flat-file, relational, object, and XML databases.

Flat-File Databases

The most basic form of database is a flat-file database. This form of database is only able to store data as a single set of records of the same kind. Conceptually, flat-file databases are very similar to a card catalog where each record is kept in a distinct location. For example, a flat-file database might contain a list of contacts, each represented by a string and separated using a common delimiter such as a comma, tab, or other character that is unlikely to be part of the data itself. This type of data storage system is suitable only for the most basic forms of applications that require basic persistent storage, with limited programming capabilities. Since they store data only as a single record, they do not lend themselves to large or complex data sets. If the amount of data becomes too large, performance suffers considerably.

One of the major performance limitations of flat-file systems comes from the lack of tuning and optimizing features that are typically found in relational database models. For example, there is no means of database normalization; that is, there is no way to structure information to reduce redundancy and promote the most efficient use of resources. Normalization is accomplished by using separate tables with foreign and primary keys in the relational database model. In addition, flat-file databases do not allow for joins—retrieving related sets of data based on a common element.

To address some of the performance issues, many flat-file databases provide the ability to designate certain fields as keys, allowing the contents of the field to be indexed. Searches for information in a keyed field are much faster than for those in fields without keys. In a keyed field, the program only has to consult the index to determine which fields meet the search criteria, whereas in nonkeyed fields, every field in the record has to be scanned to achieve the same results.

Flat-file database implementations are rarely found in enterprise applications because of their lack of scalability and poor performance.

Relational Databases

Relational databases have a much more logical structure for storing data than flat-file databases, where you are limited to a single record type. Relational databases allow you to store many records of various data types in a format called a table. Within the table, the data is stored in rows and columns. Each column of data contains elements of the same data type. The table information is related on the basis of a common field (or key), which allows for relationship mapping. In this way it is possible to associate an unlimited number of different record types with one another.

In the late 1970s, Oracle came out with the first relational database; it was soon followed by offerings from IBM and Sybase. Today, relational databases rule the enterprise market. They use SQL for interacting with the database. Using SQL it is possible to perform complex database interactions using simple commands, and it gives users a standards-based way to interact with databases, although it's important to point out that many vendors have extended the SQL standards to create their own proprietary SQL languages.

Over the years, the capabilities offered in relational databases have evolved. The leading vendors such as Oracle, IBM, Sybase, and Microsoft have added the capability to store a wide variety of data types and data processing features, such as support for transactions. These new features are often aimed at making relational databases easier to use for business applications and e-business development. Stored procedures, for example, enable the user to build business logic into the database using either a SQL-based proprietary language or, as of late, using Java. These types of additions make the relational database more attractive than other competing technologies.

In the mobile space, we are seeing many of the enterprise relational database vendors retrofit their enterprise-scale offerings for mobile usage. As well, other vendors have introduced products specifically targeted for mobile use. (More details on this are given later in the chapter in the *Commercial Relational Database* section. The complete relational database system is commonly referred to as a relational database management system, or RDBMS.

Object Databases

Recently, object databases have attempted to challenge relational databases for the enterprise market, but so far have not succeeded. They differ significantly from relational databases in their approach to data storage. This approach is based on the capability to store persistent objects; so, rather than storing tables, as relational databases do, object databases store a variety of persistent objects, which can then be referenced through persistent identifiers (PIDs). PIDs identify the objects and are used to build relationships between them. Instead of using SQL to access and manipulate the objects, common object-oriented (OO) programming languages such as C++ and Java are used.

The first object databases came to market in the early 1990s, more than 10 years after their relational counterparts. Though they have not caught on in the broad enterprise market, they do excel in certain areas. When, for example, you need to be able to store many forms of data, which do not lend themselves to the tabular setup of a relational database, object databases are a good choice. They give you the freedom to create and store any type of information, and provide tight integration with OO programming languages.

Recently, with the proliferation of Web-based applications, object databases are experiencing some surge in popularity because of the many formats of Web content and the requirement for easy access to these data formats. Unfortunately for the object database vendors, the performance overhead and nonstandard design continue to make them unattractive for many business application developers. Whether mobile applications will take advantage of object databases is yet to be determined, but the concept of sending persistent objects, instead of just data, to mobile applications is starting to gain some momentum.

The complete object database system is commonly referred to as an object database management system, or ODBMS. Another term that you may come across for object databases is object-oriented database management system, or OODBMS.

XML Databases

With eXtensible Markup Language (XML) data storage, there are two very different types of databases: one implements XML capabilities on top of another database format; the other uses XML as the storage mechanism itself. It is important to make this distinction, because here, when referring to XML databases, we are referring to the latter. This is because most, if not all, enterprise RDBMS vendors have added some form of XML capabilities to their offerings. These offerings range from the capability to store XML documents in text fields to the capability to interact with the database using XML datagrams. These relational database offerings are not XML databases, as the data storage itself is not XML-based.

The latter type of XML databases, comprise products that have been designed for use with XML from the ground up. They use XML database engines to store large amounts of data, provide concurrent access, and integrated security. The selling feature behind XML databases is that they can interact with XML streams more efficiently than can RDBMS that have XML features, because they do not have the overhead of converting XML to a relational format. In some cases, such as when manipulating an entire document, this can lead to better performance than using an RDBMS. That said, in most cases, storing data in XML formats can be very inefficient, and offers poor performance, due to the lack of indexing capabilities.

Another feature of XML databases is the integration with other XML technologies on mobile devices such as synchronization services, transformation engines, and XML query languages such as XQuery. The premise is that, by having all XML-based technologies on the client device, you should be able to reduce system complexity and cost while increasing the flexibility due to XML's platform independence. Increased flexibility may in fact be possible, but XML databases demand increased processing and storage requirements as well, which are not attractive characteristics for mobile computing.

XML database management systems (XDBMS) are rarely used, in comparison to relational systems, most likely because relational systems are ingrained in most organizations, and not enough compelling reasons exist for moving to a pure XML model at this time.

Reasons for Using a Database

Database technology is the foundation of many enterprise applications. This is true in the world of e-business, and it is becoming more true in the world of m-business. As corporations continue to develop mobile business applications, the requirement for persistent data on the mobile device becomes increasingly prevalent. Just as databases made the client/server era of computing possible, databases are changing the way companies look at mobile computing. As articulated earlier in this book, one of the main goals of mobile computing is to extend enterprise data beyond the confines of the enterprise. In doing so, many lessons have been learned, demonstrating how mobile data access for e-business (or m-business) differs from client/server data access. The following are some of the key differences between the two means of data access:

- *The mobile-device landscape is much more complex.* Compared to the choices in the desktop world, there are many different devices, which use a broad range of hardware, operating systems, and user interfaces.

- *The number of devices and applications that need to be managed in the mobile landscape is increasing.* As mobile computing grows, users may use more than one device, and will definitely use many mobile applications. Desktop computers are a known entity. There are not many new hardware systems or radical new applications coming to market.

- *Mobile users often need to use applications when not connected to a network.* One reason is the unreliable nature of the wireless networks, but it can also be credited to the performance and depth that offline applications can provide. Desktop computers typically have reliable, continuous network connectivity, allowing data access to occur over a network without worrying about prolonged downtime.

All of these points lead the conclusion that mobile data storage is an important aspect of mobile applications.

Key Features

Now that we have looked at the types of databases available, as well as some of the key differences of mobile computing, we are in a better position to evaluate the key features that are important in a mobile environment.

Data Storage Properties

The database engine and the technique used for data storage will play a significant role in the database's storage capacity, performance, and scalability. You want to choose a solution that can store a large amount of data, being limited only by the memory available on the device. At the same time, the solution should have fast data access, to ensure that the database does not cause a performance bottleneck. There is often a trade-off between data capacity and performance; in some products, as the data set grows, the performance decreases.

Another important consideration is the support for enterprise class functionality, such as indexes, referential integrity, and, especially, transactions. Choosing a product that has these features will enable you to produce a robust, high-performance solution. The importance of transaction integrity cannot be overemphasized for mission-critical applications. Transaction integrity means that clients can perform database actions within transactions, so if something goes wrong in any part of the transaction, the entire transaction can be undone. This is an important feature when you are modifying data in more than one table, as you can ensure that internal references remain intact. Since your mobile application will not contain the entire schema of your enterprise data source, the database system should provide an easy way to define the required subset of data. Once defined, the synchronization layer should be able to automatically take care of the data mappings. This leaves you with a minimum amount of data on the device and a means of direct synchronization.

Tool Support

Having tools for the development, testing, administration, and deployment of your mobile application will make the development process more efficient. Support for leading development tools and programming languages enables developers to use products that they know. Ideally, for maximum flexibility, the database system will support development of both native and Java applications. Being able to leverage existing tools and programming language skills will dramatically reduce the amount of training required. When it comes to administration and deployment, the database vendor should provide tools that support all of your target operating systems.

Flexible Synchronization

The database solution should provide an integrated synchronization layer. The synchronization must be bidirectional, so that both the client and server data can be updated. At the same time, the amount of data being transferred has to be kept to a minimum. One way to accomplish this is by synchronizing only the net changes. Imagine the case where a data field changes several times between synchronizations. When synchronizing net changes, the only information sent to the enterprise server would be the final value, rather than all of the intermediate values, which could add considerable overhead to the data being transferred. For most applications, receiving only the final value from the synchronization should be sufficient, but for others a complete transaction history might be required. A solution that offers a choice between the net changes and complete transaction history would be ideal.

Mobile applications run on many devices, using many communication protocols. The synchronization capabilities should be able to run over a variety of both wireline and wireless protocols. The system should also support access to enterprise data being stored in the leading enterprise database products.

Administration Requirements

In most mobile environments, technical staff to troubleshoot and resolve problems is not readily available. At the same time, it is unrealistic to expect the end user to take care of troubleshooting tasks. This means that you need to choose a solution that requires zero client administration. To accommodate this requirement, the database must be easy to install, configure, and manage.

When support is required, having remote device support capabilities can be very helpful. Such support may include the ability to remotely control a mobile device and to remotely deploy application updates to the device. The mobile device management products discussed in Chapter 16, "Mobile Information Management," will be useful to accomplish this task.

Low Resource Requirements

Because mobile devices have constrained resources, the solution has to be able to run efficiently on handheld devices. The database should be capable of being run on slow processors and provide a small footprint reducing memory usage. Choosing a solution that is modular, giving you only the features required for your application will minimize resource usage. Also note that native applications are often more efficient than Java-based solutions, so native support for your target operating systems can be important.

Operating System/Device Support

At a minimum, the database you choose should be capable of running on the leading operating systems and devices, both during development and deployment. Moreover, because the mobile operating system and device market is changing constantly, you need a solution that can adapt quickly to those changes. At a minimum, the operating systems it supports should include Windows CE, Palm OS, and possibly Symbian OS. You are also advised to choose a database that can run in the emulation environments that you have selected. If it does not, the development process will be much more difficult, because physical devices will be required for all stages of the development cycle.

Standards Support

Probably the developers in your organization have done some database development, meaning they have SQL skills as well as knowledge of ODBC or JDBC. Because these standards are used widely and have a broad level of industry support, the database system you choose would ideally support these standards. Even if you do not have previous database experience, choosing a solution that is standards-based will help you to learn the system faster and gain skills that are transferable to other platforms. In addition, technical resources are more readily available for standard systems than for proprietary ones.

Having a SQL relational database on the client can also ease the integration process with enterprise databases, which are commonly SQL relational as well.

Security

The database system should support security throughout all parts of your application. This includes both authentication and data encryption. To prevent unauthorized users from accessing your confidential data, having built-in authentication is important. This authentication should also carry through to your back-end data source during synchronization.

Two levels of data encryption are required: encryption of the communication stream, and of the data stores themselves. The encryption of the data stream will prevent others from listening in when data is transmitted from the data store to the enterprise database. The encryption of the client data store will ensure that confidential data cannot be viewed on a device that is lost, misplaced, or stolen.

Persistent Storage versus Real-Time Access

An argument can be made that direct real-time access to enterprise data is better suited for enterprise applications than persistent client data storage. This solution would involve having a communication layer within the client application that would go over a wireless network to retrieve data from the enterprise server. In the same way, the business logic could reside on the server as well, removing the need for client-side logic. This would leave the presentation layer on the device, along with a communication layer, allowing for enterprise integration, making the result very similar to wireless Internet applications. In this case, the real-time data access solutions would suffer from some of the limitations that real-time browser-based applications do, namely:

Unavailable or unreliable wireless networks. In many cases, mobile tasks such as sales force and field service applications take place in remote locations, where wireless coverage is either unavailable or unreliable. This is also true of workers for whom physical barriers, such as office buildings, cause interference issues.

Insufficient bandwidth. A significant amount of data is being recorded with data capture applications. With the low bandwidth of today's networks, constantly transmitting data is impractical and costly.

Static data. Much of the data retrieved in mobile applications is relatively static, requiring only periodic updates. Examples include customer contact information or product catalogues. In these cases, it does not make sense to be downloading the information over a wireless network when it can be stored persistently on the device.

Battery life. Communicating over a wireless network requires significantly more battery power than accessing data locally. This is a concern for workers who do not have the ability or time to charge their devices throughout the workday.

For all of these cases, developing a smart client application with persistent data storage and enterprise synchronization is a more practical and cost-efficient solution. In many cases, the data can be synchronized via a wireline network, avoiding the wireless network limitations.

And when real-time data is required, wireless Internet applications appear to be a better choice than creating custom client applications with real-time access. Wireless Internet applications do not require a client to be deployed to the device, plus they can communicate over standard protocols such as WAP and HTTP. Another option is hybrid applications that use local data storage for most of the data access, and wireless Internet access for content that is the most critical or time-sensitive.

Database Development Options

Now that we have shown some of the benefits of including persistent data storage in your mobile applications, it is time to take a look at the options available for implementing this functionality.

Several factors will influence which method you decide to use. They include the following:

Required features. Read the features listed earlier. Which are important to your particular application? A few key ones are performance, data storage capacity, transactional capabilities, operating system support, and tools integration.

Application complexity. How complex is the application's data structure? Does it require relationships between data sets? What forms of data manipulations are required?

Available resources. Such resources include both financial and development resources. Often a trade-off occurs between the cost and functionality. In addition, consider the internal development and maintenance costs versus the costs of a commercial solution.

Size of deployment. The number of seats being deployed can have a big impact on your decision. The per-seat cost of developing an in-house solution is higher for smaller deployments. At the same time, the development effort can become much more complex when building an in-house solution that has to support a large number of users. Determining the point at which it becomes more cost-effective to build instead of buy has to be determined on an individual basis

Support for standards. Is a proprietary system suitable for your corporate needs, or do you require support for standards such as SQL, ODBC, and JDBC?

Enterprise integration. What kind of enterprise integration is required? Many of the commercial systems come with built-in database synchronization capabilities that will work over a variety of protocols. Does your application require this, or would it be overkill?

As you read the options available, you will want to weigh the pros and cons of each solution against the requirements of your particular application. This decision will have both near-term and long-term consequences, so be sure to dedicate the appropriate effort to it.

Proprietary Storage

All of the leading mobile operating systems include a form of local data storage. These data storage systems are typically flat-file databases with limited features and storage capacity. The interface to these databases is most commonly at the API level. Let us take a look at how data is stored in each of the leading operating systems.

Palm OS

Any kind of persistent data storage on the Palm OS is contained in a flat-file database. This type of data includes both applications as well as user data. Two kinds of database systems are available: record database and resource database. Each is composed of blocks of memory that can contain any application-defined data. The only difference is the size and contents of the header for each of the objects. In general, record databases are used to store application data, such as contact lists. Resource databases are used to store the application code and user interface objects. For our purposes, we are going to concentrate on the record database, since the resource database is mainly used for managing applications, not user-defined data.

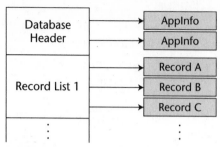

Figure 9.1   Logical layout of the record database.

As the name indicates, a record database stores records of data. These records are essentially segments of memory that are part of the storage heap of the Palm OS. Each database contains a database header that describes the database, AppInfo and SortInfo blocks, and the record list. Both AppInfo and SortInfo are optional blocks that describe the application information and sorting information, respectively. Figure 9.1 shows the logical record database layout.

Within the header, various fields describe the database name (which is a 32-byte string), its attributes, the last modified date, the creatorID, the AppInfo, the SortInfo, and many other items. The header ends with a record list, which has its own header, followed by the record entries. Each record entry references the location of the actual data for the record and contains the attribute and ID information for that record.

From a programmer's perspective, the Palm DB is a quick and easy way to add persistent data to an application. To read or write data, the Palm OS provides functions that can be used to convert data into or out of Palm database formats. (Note: Detailed information on these functions is available in the Palm OS "Programmer's API Reference," which is available at www.palmos.com/dev/support/docs/palmos.)

When using the Palm database for your applications, keep in mind that it does have some limitations. First, the application programmer is responsible for database creation and maintenance. This means that within your application code you will have to use the Palm OS API to open, close, and access the data in any way. In addition, you will have to manage the code to create new records, access or modify them, or remove records. If you require additional functionality, it will have to be created by the programmer.

Second, the Palm database is a flat-file database, with all the limitations discussed earlier in this chapter. The main issues are that the Palm database limits the complexity and size of the data storage capabilities and does not have support for transactions. This may not be a factor for simple applications, but it can be a major limitation when representing a portion of an enterprise system on a handheld device.

The final limitation relates to enterprise integration. Palm databases do not come with very robust synchronization capabilities. You can synchronize blocks of data using HotSync, but this capability is often quite limiting in the enterprise space, where a companion PC is not always available and the data integration is often directly to an enterprise RDBMS. In these cases, you are required to write your own data synchronization conduit. This effort will involve creating your own data synchronization layer, which is not a trivial task. (More on information data synchronization is available in Chapter 10, "Enterprise Integration through Synchronization.")

Windows CE

Windows CE contains a built-in database system in the form of its object store API. It is a flat-file database geared at application programmers, to enable them to include persistent data storage in their Windows CE applications. The database is physically located in what is referred to as the object store, a portion of a device's memory allocated by the operating system for persistent storage of files, the Windows CE Registry, and Windows CE databases.

One advantage of this flat-file database is its ease of use. By using the provided API functions, you can add and remove records, search for a particular record, and sort the records based on four different columns. Another helpful feature is the support for timestamps as a data type. This allows the programmer to see if the data has been altered by another application since its last use. The API also includes functions for finding records. These include the FirstEqual, NextEqual, ValueGreater, and ValueSmaller methods, providing conditional search functionality without much extra programming.

Similar to the Palm database, the object store API has some major limitations for more advanced applications. The first is that it can reside only within the device's RAM, meaning it cannot be placed on an auxiliary storage card. This limitation can pose problems for applications that are shipped on PCMCIA or Compact Flash memory cards. In these cases, before the application can run, the database will have to be written to the device's built-in RAM.

Although flat-file databases work well for simple applications or those with mainly static data, they can introduce performance problems for more complex applications. When searching for a specific set of data or attempting to sort data, extra commands and I/O are often required, thereby inhibiting performance. This will become especially noticeable with large data sets.

As with other flat-file solutions, database synchronization with Windows CE object store is also an issue. You are left to develop your own synchronization routines. As discussed in the Palm OS section, this undertaking can require a significant amount of effort to implement.

Symbian OS

The DBMS in the Symbian OS is significantly more advanced than that in either the Palm OS or Windows CE. Rather than using a flat-file database structure, Symbian OS has implemented a relational database. This means you can have multiple tables in the same database, allowing for more complex schemas. In addition to making the storage capabilities more advanced, Symbian has also added data access through a subset of SQL, as well as using C++ APIs. This too provides a more approachable way to interact with the Symbian DBMS.

For most of the common database commands that you will require, such as opening and closing the database, creating tables, creating indexes, and querying data, you have the option of using either SQL or the C++ API. In most cases, the SQL code is more efficient than using the C++ API, with the exception of inserting rows of data. In this case, using the C++ API provides better performance, especially on large amounts of data. Overall, incorporating the built-in DBMS into your applications will require a considerable amount of custom code.

One of the other nice features of the Symbian OS DBMS is the support for transactions. When database changes occur within a transaction, all changes can be undone if something goes wrong during any part of the data manipulation. This is an important feature when you are modifying data in more than one table, ensuring that internal references remain intact.

As with the other built-in databases, the Symbian OS DBMS also has some limitations. For one, the API is only available in C++, limiting access to the other supported programming languages such as Java. This can be a fairly significant limitation as it is expected that Java and other high-level languages will be frequently used for Symbian OS development. Second, even though it is a relational database, it does not offer functionality that is normally found in commercial products. For example, table joins are not supported. This means that you cannot select, insert, or update data from more than one table in a single statement. Finally, there is the synchronization issue: With this database, too, you are required to implement your own synchronization support, which as explained for the Palm OS and Windows CE databases, is a significant challenge.

J2ME

The Mobile Information Device Profile (MIDP) of J2ME provides a persistent storage mechanism called the Record Management System (RMS). The RMS is a set of Java classes for storing and retrieving simple sets of data. The storage takes place in what is called a RecordStore. Each RecordStore consists of a collection of records that are stored in nonvolatile memory, meaning they are persistent. Internally, the data is essentially stored as a byte array.

The MIDP applications that access the RecordStores are called MIDlets. These MIDlets can be grouped together for deployment into MIDlet suites. Each MIDlet suite has its own space for RecordStores; no MIDlet suite is permitted to access other MIDlet suites' RecordStores. That is, there is no mechanism to share RecordStore data between MIDlets in different MIDlet suites. So, if you want to enable more than one MIDlet to access the same set of data, the MIDlets in question have to be in the same suite.

Each RecordStore can be given a name, which is case-sensitive and has a maximum length of 32 Unicode characters. There are not many naming restrictions, other than that two RecordStores within the same suite cannot use the same name. Within the RecordStore, each record is assigned a unique ID. These IDs are managed by the RMS and are autoincremented; thus, the first record will have an ID of 1, the second an ID of 2, and so on.

The application programmer is responsible for managing the RecordStore. There is a set of APIs to open, close, insert, delete, and move through the records. More advanced classes, such as RecordComparator and RecordFilter, allow you to compare two records and filter records, respectively.

As you can see, the MIDP RMS is a very simple data storage mechanism, for limited amounts of data. As a developer, you are responsible for managing the RMS as well as for implementing any synchronization capabilities that may be required.

For other J2ME profiles, commercial database solutions are available. These are discussed later in this chapter.

Proprietary Storage Summary

The leading mobile operating systems all have a built-in form of persistent data storage. The capabilities of these databases vary, but they all are best suited for simple applications that do not require large amounts of data or have complex data structures. If this is the type of application you are building, then these databases offer you a quick, low-cost way to incorporate data storage into your application. You do not have to worry about licensing fees, because the database is part of the operating environment.

If you are building a more complex application, one that has more demanding data storage requirements, these databases are probably not the best choice. As mentioned before, when the data sets become large, these databases suffer from poor performance. They also do not offer many of the key characteristics that are required for many corporate applications. While they all have low resource requirements, they typically do not provide the security, data integrity, support for transactions, scalability, support for standards, or data synchronization capabilities that many solutions require.

One of the final points to keep in mind about all of these solutions is that they are very specific to the given operating system. If you are planning on deploying your application to multiple operating systems, you will incur additional development and maintenance costs using these solutions, as you are essentially working with a different database for each operating system. Everything about how you store and interact with the data is OS-specific, requiring a significant amount of additional work to port the application to another operating system.

Custom-Coded Databases

If the built-in databases do not meet your needs, you always have the option of creating your own database. In this scenario, you can create as simple or as complex of a database as your application requires. If you are looking for basic storage, but are not interested in the built-in databases due to their proprietary nature, you can create a simple flat-file storage mechanism that can work on many platforms. On the other hand, if you require complex data storage with relational database features, you can hand-code this solution. In most cases, this comes down to a "build versus buy" decision.

Companies face a predicament when they come to the conclusion that the available commercial solutions do not meet their needs for one reason or another. It could be because of cost, features, or because of anticipated performance issues. In any case, they have to make a decision whether building a database in-house will result in long-term benefits. This may not be an easy decision to make.

By building a custom database, you can ensure it has all of the features that you require. If you require a relational format with referential integrity, you can build it into the system. At the same time, to maintain a small footprint, you can omit any extra features that may not be required. When it comes to performance and scalability, the onus is again on you. This is sometimes preferable, as you can ensure that the applications run as expected; that said, development and testing often end up being very time-consuming. Some of the other database features that you may have to custom-code include indexing capabilities to maintain performance, referential integrity to maintain internal data consistency, data-access APIs or SQL support, memory management, and file management, to name just a few. These features alone are often enough to prevent making an in-house solution feasible.

If you require enterprise data integration, you will have to consider building a synchronization layer into your solution as well. This involves building client-side as well as server-side code that can transfer data between your device and back-end server in such a way that you do not lose any data. (More information on what is required in a synchronization solution can be found in Chapter 10.)

Before you start to build a custom solution, make sure that you are aware of the challenges that lie ahead so you are adequately prepared. It is no secret that building a custom database will require significant resources. This is true not only for the development stage but also for the maintenance stage. Remember, mobile operating systems change constantly. Possibly you will have to port your database to new operating systems and new versions of existing operating systems. Each port requires additional development and testing resources to ensure it operates as expected. These cautions are not meant to dissuade you from creating a custom database but rather to make you aware of the effort that will be required.

Of course, there are times when a custom database makes sense. If you are targeting an operating system that is not widely supported by commercial solutions, it may be the only way to get the functionality you require. If you have very specific data storage requirements, which are not addressed by commercial mobile databases, you may find building a custom solution to be to your advantage. These are just two scenarios for which building an in-house solution is beneficial. Just remember that building and maintaining mobile applications is challenge enough; coupled with adding the database, it becomes more so. So if you are contemplating building a solution in-house, research the solutions that already exist to make sure you use your resources to the maximum.

Commercial Relational Databases

Relational databases not only lead the enterprise database market, they are the clear leaders in the mobile market as well. This is largely due to the fact that many of the mobile offerings are from enterprise database vendors. These companies provide a variety of SQL-based solutions with a wide range of capabilities. Some offer key features such as referential integrity, support for indexes, and transaction support, while others do not. Some have open support for operating systems, synchronization integration, and programming languages, while others are more limited.

The point is, when it comes to mobile database implementations, all commercial databases are not created equal. In this section, we examine the leading vendors, summarizing their offerings to give you a place to start for evaluating which technology is best suited for your mobile applications. (Note: The vendors are listed in order of marketshare.)

NOTE This book only describes support for mobile operating systems. Delving into support for desktop environments is outside the scope of this book.

Sybase/iAnywhere Solutions

Product	Sybase SQL Anywhere Studio
URL	www.ianywhere.com/products/sql-anywhere.html
Mobile OS support	Windows CE, Palm OS, Symbian OS, Linux, Java

Key Features:

■ Full-featured Adaptive Server Anywhere (ASA) database for desktop environments and Windows CE

■ UltraLite deployment for mobile operating systems, providing a relational database with a footprint as small as 50K

■ Built-in synchronization to any ODBC back-end data source

■ Strong encryption of both the data stores and communication stream

Summary:
To put more emphasis on its mobile and wireless computing solutions, Sybase launched a new company called iAnywhere Solutions. The flagship product offered by iAnywhere Solutions is SQL Anywhere Studio, which includes a variety of mobile database offerings. This product is the established leader in the mobile database market with 73 percent market share (Gartner Dataquest, December 2002). This lead over the competition is a result of iAnywhere's long-term focused commitment to mobile computing, which has enabled them to develop a set of products that squarely meet the requirements of enterprise customers. In the mobile database arena, three technologies are available: Adaptive Server Anywhere (ASA), UltraLite, and UltraLite Component Suite.

Adaptive Server Anywhere is a high-performance mobile database system that offers complete relational database functionality, including stored procedures, triggers, transaction management, record-level locking, and enterprise data replication. All of this is provided in an easy-to-use package that has a compact footprint (around 3 megabytes on Windows CE environments). This version of iAnywhere's technology is available for a variety of desktop and server platforms, as well as Windows CE.

If you are looking for a more compact solution, iAnywhere Solutions provides the innovative UltraLite deployment option. This technology analyzes the database features that your application requires and automatically generates a customized database containing only those features. You can then link this customized database with your application code to create a single executable for deployment. This reduces the complexity of implementing a mobile database solution. For development, you can use common C/C++ development tools such as eMbedded Visual C++ and Metrowerks CodeWarrior. If you are more interested in Java, iAnywhere also offers a pure Java version of the UltraLite database that allows you to develop with any Java development tool and deploy to any device with a Java VM.

In addition to the application-specific version of UltraLite, iAnywhere Solutions also has a more generic relational database for the same operating systems. This technology is called the UltraLite Component Suite; it provides interfaces for development using AppForge, Visual Basic, ActiveX controls, and Java, in addition to the C++ tools mentioned earlier.

On the synchronization side, iAnywhere Solutions offers its MobiLink synchronization server that provides direct, bidirectional synchronization to any ODBC data source, including Sybase, Oracle, Microsoft, and IBM. The synchronization rules are stored in the back-end database to provide a high level of customization capabilities. Both wireless and wireline protocols are supported for synchronization, including HTTP, TCP/IP, and common conduits such as HotSync and ActiveSync.

For security, iAnywhere provides 128-bit strong encryption of the communication stream as well as the data stores themselves.

With their wide platform support and open synchronization capabilities, the mobile database solutions from iAnywhere Solutions are well suited for companies that require a flexible solution that can integrate with heterogeneous back-end databases.

IBM

Product	DB2 Everyplace
URL	www.ibm.com/software/data/db2/everyplace/
OS support	Windows CE, Palm OS, Symbian OS, Linux, Java

Key Features:

- A relational database management system with a footprint as small as 150 KB
- Support for a wide range of mobile operating systems
- Synchronization to a variety of back-end data sources via IBM DB2

Summary:

IBM incorporates three components in its DB2 Everyplace Enterprise product: a small relational database, a synchronization server, and a tool to develop smart client applications for the Palm Computing Platform and Java. Both native versions and Java versions (through IBM's acquisition of Informix) of the database are available. DB2 Everyplace is a relational database management system with a footprint as small as 150 KB. This RDBMS is composed of the default libraries and the database engine itself. The main goal of DB2 Everyplace it to provide a local data store on the mobile device while attempting to keep many of DB2's enterprise features. Some of the included features are SQL support for adding, deleting, and updating data. Missing is support for transactions, which is notable for many corporate developers. DB2 Everyplace implements a subset of the SQL 99 standard, DB2 Call Level Interface (CLI), Open Database Connectivity (ODBC), and Java Database Connectivity (JDBC).

On the development side, IBM offers its own development tool called Mobile Application Builder for Palm OS and Java applications. Mobile Application Builder offers a graphical rapid application development environment with wizards to aid development. DB2 Everyplace also integrates with Visual Basic, C/C++, and Java development tools.

For synchronization, the DB2 Everyplace Sync Server provides bidirectional synchronization from DB2 Everyplace to DB2 Universal Database. Once the data is in DB2 Universal Database, it can then be transmitted to other ODBC data sources, such as Oracle, Sybase, and Microsoft, by using adapters. The synchronization server is implemented as a Java servlet, so it can be hosted in any servlet engine, although it does work best in IBM's WebSphere application server. Being a Java servlet, the synchronization is HTTP-based, and works for both wireless and wired networks. SyncML is the synchronization protocol used by IBM for its data transfer. (For more information on SyncML, see Chapter 10.)

With its broad platform support and small footprint, DB2 Everyplace is well suited for a variety of mobile applications. However, the enterprise integration is somewhat complex in that it requires DB2 Universal Database as an intermediate step before interacting with other data sources, so it may not be ideal for companies that do not currently use DB2 Universal Database.

Oracle

Product	Oracle9*i* Lite
URL	www.oracle.com
Mobile OS support	Windows CE, Palm OS, Symbian OS

Key Features:

- Lightweight relational database that resides on the mobile device
- Centralized Web management console to manage users, data, applications, and security
- Handles both data and application synchronization
- Synchronization with Oracle enterprise databases

Summary:
The Oracle9*i* Lite database is built upon the Oracle9*i* Database and Oracle9*i* Application server technology. It provides a lightweight relational database with a feature set aimed at being comparable to the Oracle9*i* database server. It incorporates other technologies (such as Web-to-Go and Mobile SQL) into a single product. This streamlines the installation and configuration procedures, breaking them into two groups: Oracle9*i* Lite Mobile Development Kit and Oracle9*i* Lite Mobile server.

The Mobile Development Kit incorporates a set of tools and APIs, along with sample code, to accelerate the development of Oracle9*i* Lite database applications on mobile devices. At the core of the product is the Oracle9*i* Lite database. It is a general-purpose database with support for ODBC and JDBC connectivity on all platforms using common development languages such as C/C++, Visual Basic, Satellite Forms, and Java. For Java development, you can take advantage of Oracle's lightweight JDBC driver from your Java clients, although the Oracle database itself is not written in Java.

As to the Oracle9*i* Lite footprint, it is considered more of a middleweight than a lightweight database. On Windows CE, for example, the combined memory required for the supporting files and engine is close to 5 MB, and this does not include the database itself. On other platforms, the footprint can be as low as 1 megabyte, which is still not small compared to other offerings on the market. In addition, the installation process and feature set vary from platform to platform, so there is no single solution for all mobile devices.

One of the unique features of Oracle9*i* Lite is how it manages mobile applications. The Oracle Mobile Server is an extension of the Oracle9*i* Application Server, making it possible to deploy, manage, and synchronize offline applications from a central console. Rather than having a separate engine for application management, Oracle has incorporated this capability directly into the database. This allows you to synchronize both data and applications from a single console. Unfortunately, the application deployment and management is available only for Win32 clients.

When it comes to synchronization, Oracle does not use standard interfaces such as ODBC or JDBC. This means that you can only synchronize Oracle9*i* Lite clients with Oracle databases. This is fine if your organization uses Oracle technology exclusively; if not, then you will find the synchronization limiting.

Overall, Oracle9*i* Lite is a capable mobile database platform best suited for companies with homogeneous Oracle server installations.

Microsoft

Product	SQL Server CE
URL	www.microsoft.com/sql/CE/default.asp
OS support	Windows CE

Key Features:

- Lightweight relational database that resides on Windows CE-based mobile devices
- Similar interfaces and data management capabilities to Microsoft SQL Server
- Synchronization with Microsoft SQL Server databases

Summary:

The latest offering in the Microsoft SQL Server family is SQL Server CE. This is a scaled-down version of SQL Server aimed at extending enterprise data to offline applications for Windows CE. It has maintained much of the functionality required for mobile solutions, including a wide range of data types and referential integrity. The database footprint varies, but is typically between 1 and 3 MB.

Microsoft SQL Server CE provides a familiar SQL grammar and a consistent development model and API set to the full-blown SQL Server. Development can be accomplished using Visual Studio .NET using the .NET Compact Framework. These tools and technologies allow developers to use their existing skills to build sophisticated Windows CE-based database applications for a wide range of mobile and embedded applications. In addition, Microsoft supports Remote Data Access (RDA) for accessing back-end Microsoft databases in real time without having to configure SQL Server synchronization.

Synchronization is provided over HTTP to Microsoft SQL Server back-end databases via Microsoft Internet Information Server (IIS). Replication can be performed over wired and wireless LANs and WANs. The communication protocol is well suited for wireless transports. Microsoft has also incorporated support for SSL, giving you the ability to encrypt the data stream to protect sensitive data.

Because the Microsoft SQL Server CE solution is targeted directly at Windows CE devices, and synchronizes only to Microsoft SQL Server, it is suitable only for a very specific market. If you are deploying Windows CE devices that only require integration with Microsoft SQL Server, then this solution should meet your requirements, while giving you a familiar development environment.

PointBase

Product	PointBase Embedded Edition and Micro Edition
URL	www.pointbase.com
OS support	Java

Key Features:

- 100 percent pure Java relational database
- Integration with common Java development tools
- Built-in data synchronization to a variety of back-end sources (Embedded Edition)

Summary:

PointBase is the most recent player to enter the mobile relational database market. Its database products are all 100 percent Java, to provide cross-platform development opportunities. There are two versions that will appeal to mobile developers: PointBase Embedded Edition and PointBase Micro Edition. Both are targeted at mobile applications development, but for different versions of the Java platform.

The PointBase Embedded Edition is aimed primarily at J2SE applications that require multiple database connections from within the same JVM. The Micro Edition works in both J2SE and J2ME environments that require only a single database connection. For this edition, the footprint can be as small as 45 KB; however, it is lacking many features that are required for enterprise mobile applications, most notably, synchronization support.

Both versions take advantage of common development tools, such as Metrowerks CodeWarrior for Java, for rapid application development. For deployment, you can use any JVM on the target platform. The Kada Systems' JVM is often used on the Palm OS, while Insignia Jeode is common on Windows CE. For smart phones and Symbian OS devices, you can use the JVM that comes with the device.

The Embedded Edition offers synchronization via PointBase UniSync. UniSync is a middleware server that allows the PointBase clients to synchronize to any JDBC-compliant back-end data source. Even though it is open to many back-end sources, the synchronization is better if you are communicating to the PointBase Server database. In this case, bidirectional, changes-based updates are supported. For other JDBC data sources, only snapshot updates are supported.

Table 9.1 Other Mobile Database Vendors

VENDOR AND PRODUCT	SUPPORTED OPERATING SYSTEMS	WEB SITE
Birdstep Raima Database Manager	Windows CE	www.birdstep.com
DDH Software HanDBase	Palm OS, Windows CE	www.ddhsoftware.com
Land-J Technologies JFile	Palm OS	www.land-j.com
Pervasive Pervasive.SQL	Windows CE	www.pervasive.com
Syware Visual CE	Windows CE	www.syware.com
Thinking Bytes ThinkDB	Palm OS	www.thinkingbytes.com

PointBase's main strength is its focus on Java. The concept of developing an application once and deploying it to any mobile device with a Java VM is very attractive to corporate developers. Unfortunately, Java on mobile devices is still relatively immature compared to server and desktop platforms, and users may suffer through some growing pains. However, on some of the more capable mobile devices, creating Java applications is a promising alternative.

Other Database Vendors

When it comes to mobile database technology, a number of smaller companies have capable product offerings. We are not going to take an in-depth look at them here, because the vendors just described hold over 90 percent of the mobile database market. If you are interested in these alternative technologies, review Table 9.1, which lists the company names along with their Web URLs.

Conclusions

Proprietary databases provide a quick, low-cost way to implement persistent data storage in smart client applications. They have tight operating system integration, but often are not cross-platform. They are best suited for simple applications that store small amounts of data in a straightforward way. If your application is more complex, requiring additional functionality, you will want to consider the alternatives.

If you develop a database solution in house, the feature set and other capabilities will be limited only by your developers' skill and your corporate resources. Given enough time and money, it is possible to create an in-house database that matches the capabilities of a commercial product. However, it is unlikely that you will want to make that effort if a commercial solution that meets your requirements is available. Undertaking an in-house project of this magnitude is recommended only when proprietary solutions or commercial products do not meet your needs; for example, if the other solutions do not support the target operating system, if the performance is unacceptable, or if the initial costs are prohibitive for your organization.

Commercial database systems provide the best features, including built-in enterprise synchronization. Many of these solutions are market proven as well. If cost is taken out of the equation, it is clear that using a commercial database is your best bet in terms of features, performance, and reduced risk, as they support all of the leading operating systems and communication protocols.

When you add cost into the equation, then you must take other factors into account. Keep in mind that total cost includes not just the license fees for the database, but also development, administration, and maintenance costs. When comparing all factors that influence the total cost of ownership, implementing a commercial database system often comes out as the optimum choice.

TIP If you decide to use a commercial database in your applications, it is recommended that you get an evaluation copy of the software to assess some of its capabilities that cannot be learned by reading the product datasheets. Items to focus on are ease of use, stability, performance, and deployment complexity.

Summary

Incorporating persistent data storage into your mobile applications will give users fast, reliable access to their data without having to be concerned with wireless connectivity. There are many forms of databases on the market: flat-file, relational, object, and XML databases. In the mobile database market, most of the freely available databases are flat-file, while the databases available commercially are typically relational. These databases possess key features that make them well suited for mobile application development.

If you decide to incorporate a database into your mobile applications, you have three main ways to accomplish this: using an operating system-specific database, developing your own database in-house, or purchasing a commercial solution. Which option is best for you depends on the application you are looking to deploy.

In Chapter 10 we will complete our look at smart client application development by examining enterprise integration using data synchronization.

Helpful Links

Treat these links as resources for more information on the products and topics described in this chapter.

J2ME Documentation	http://java.sun.com/j2me/docs/
Palm OS documentation	www.palmos.com/dev/support/docs/
Symbian Developer Network Technical Papers	www.symbian.com/technology /whitepapers.html
Window CE technical resources	www.microsoft.com/mobile/developer /techresources.asp

Enterprise Integration through Synchronization

One of the most important questions to answer when implementing mobile enterprise applications is how the remote users are going to access enterprise data. For smart client applications, this integration is commonly implemented using data synchronization. Data synchronization reduces the requirement to constantly transmit data over slow, often unreliable, wireless networks since the data can be stored locally on the mobile device and synchronized periodically at convenient times. It also provides the added benefit of distributing the data access workload down to the mobile device, preventing the server from being overburdened with too many requests.

NOTE In this book, *synchronization* refers to the bidirectional exchange and transformation of data between two separate data stores.

After reading this chapter, you will have a firm understanding of the fundamental concepts of enterprise synchronization, including synchronization architectures and techniques. A comprehensive overview of each technique is provided; their strengths and weaknesses are explained, to guide you as you prepare to implement the solution that is best suited for your applications. In addition, we cover some of the synchronization technologies available commercially, and provide an overview of SyncML with a description of where it fits into enterprise data synchronization.

Synchronization Fundamentals

In the previous chapter, we learned how persistent data storage enables users to access enterprise data without being connected to a network. Without enterprise synchronization, the data on the devices would quickly become stale, hence unsuitable for many corporate applications. The synchronization process allows you to execute bidirectional updates on the required data. Any changes that have been made on the client device can be transmitted to the server database, and any changes on the server can be transmitted to the client device. In this way, you can keep the data on the client and the server synchronized.

> **NOTE** The term *replication* is often used in conjunction with synchronization, leading to some confusion. Replication is the process of making a copy of something (a replica). In database terminology, replication can be used to describe two processes: first, the bidirectional transfer of data between systems, in which use it is a pseudonym for synchronization; second, the one-way copying of data from one system to another, wherein it is quite different from synchronization. To be consistent, we will use *synchronization*, not *replication*, to describe the data transfer process.

Whether you are synchronizing over wireless networks or a wireline connection, synchronization offers many important benefits over an always-connected solution, namely:

- Reduced data transfer over the network, often leading to reduced transmission costs.
- Reduced loads on the enterprise server.
- Faster data access, because the user does not have to constantly wait for data to download.
- Increased control over data availability.

In the following pages, we will take a look at how synchronization works and introduce features you should consider as part of an enterprise synchronization solution.

TYPES OF SYNCHRONIZATION

The term synchronization applies to more than the movement of enterprise data. It is often used to describe the exchange of personal information management (PIM) data and the transfer of application files. This section briefly explains other types of synchronization before we spend the remainder of this chapter on the synchronization of enterprise data.

PIM Synchronization

One of the first applications that many corporations make available to mobile workers is email. This is often enabled at the same time as contact lists, to-do lists, and calendar entries. The combination of this data is commonly referred to as PIM data. Wireless access to email is often considered the killer application in the mobile world. It is a subject to which everyone can relate, and the benefits of which most people by now take for granted.

The leading PIM applications on the desktop are Microsoft Outlook and Lotus Notes. Both vendors provide mobile access to PIM data via laptops, so users are familiar with accessing this data remotely. When it comes to smaller mobile devices such as PDAs, a PIM synchronization solution is required to provide this functionality. PIM synchronization solutions allow users to have offline access to their personal information and to synchronize any changes made back into the enterprise system, often without requiring a companion PC. (For more about PIM synchronization, see Chapter 16, " Mobile Information Management," where the subject is covered in greater detail.)

File/Application Synchronization

Within an organization, a variety of applications and other files need to be distributed to mobile users. In the wired world, this information is often made available through use of portable media such as floppy disks or CD-ROMs or by a corporate server or intranet site. Unfortunately, these techniques do not work nearly as well for mobile workers. Many mobile devices do not have support for traditional portable storage media, and the storage that is available is somewhat expensive, making this an impractical means of distribution. Adding to the challenge is the fact that most users are at remote sites, without access to internal corporate systems.

To overcome these challenges, several file and application synchronization products have come to market. These products allow an administrator to deliver these files to remote workers in an automated way; they require no additional effort by the remote user. This means you can deploy and manage files and applications on remote devices, so you can quickly update remote workers with the applications they need. In most cases, this type of synchronization is one-way: Updates are sent from the server to the remote workers; only an acknowledgment receipt comes back. (More about application deployment solutions can also be found in Chapter 16.)

Data Synchronization

Data synchronization refers to the bidirectional exchange and transformation of data between two separate data stores (and is the form of synchronization that we are going to focus on for the remainder of this chapter). For our purposes, these data stores are most often located on the client device and the enterprise server. In many cases, the synchronized data set is only a subset of the enterprise database. The reason for the limitation is twofold: one, because of the limitations of most mobile devices, and, two, because individual mobile users only require data that is relevant to them.

The data transfer itself occurs between the synchronization layer on the client and the synchronization middleware on the server. The server middleware then communicates with the enterprise data source. This is the most common way that smart client applications are given access to corporate data.

Synchronization Architectures

Enterprise data synchronization is not a straightforward task. Being able to communicate with a variety of enterprise systems over wireless and wireline networks to an assortment of client applications requires some flexibility at the architectural level. Considering that there are half a dozen mobile operating systems, with at least that many networks, and even more back-end systems that have to be accessed, the synchronization layer is often one of the most complex components of a complete smart client application.

After you take a look at the most common synchronization architecture, along with common synchronization topologies and data access methodologies, you will have a good foundation from which you can make informed decisions on implementing data synchronization within your mobile solution.

Architecture Overview

Synchronization is most often implemented in a distributed computing architecture with a client layer, a middle-tier layer, and an enterprise data layer. Each layer can be implemented using varying techniques, all aimed at accomplishing the same goal: providing a way to extend enterprise data to a variety of mobile devices.

We are going to take a closer look at each synchronization component and how it contributes to the overall synchronization process.

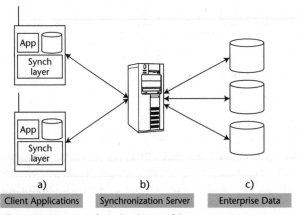

a)	b)	c)
Client Applications	Synchronization Server	Enterprise Data

Figure 10.1 Synchronization architecture.

Client

When synchronizing data between an enterprise server and a persistent data store on the client device, a synchronization layer must be present to manage the two-way data communication (see Figure 10.1a). Ideally, this layer will have a minimal impact on your client application, while still providing a simple, easy-to-use client API for controlling the synchronization process. By implementing a modular, self-contained synchronization layer, you can control the entire synchronization process with little interaction from the client application. In some cases, all that is required from the client is the invocation of the synchronization process; the synchronization layer does everything else from there.

Because so many client devices are on the market, the synchronization client must have support for the leading mobile devices, including laptops, Windows CE devices, Palm OS devices, Symbian OS devices, as well as specialized devices with add-on features such as barcode scanners and other industrial components. Each of these devices can have a different mobile operating system with different network protocol support. The synchronization layer on the client takes care of the network communication from the device back to the synchronization middleware.

Middleware

The synchronization server is where most of the synchronization logic is contained. Figure 10.1b illustrates the role of the synchronization server in relation to the other components of the synchronization architecture. This server is responsible for communicating with the client application to send and receive required data packets. In order to do this, it has to be able to communicate over a protocol that the client application understands. Most of the time, this protocol is IP-based, and often is HTTP. When the synchronization server receives the data from the client, it then has to execute the synchronization logic to determine how this data is transferred into the enterprise data source.

Many of the advanced synchronization features are implemented within the synchronization server. Some of these features include data subsetting, conflict detection and resolution, data transformation, data compression, and security. All of these features have to be implemented while still maintaining server performance and scalability.

Two common synchronization server implementations exist: as a standalone server application, or as Java servlets running in a servlet engine. Both of these methods have benefits and drawbacks. The standalone synchronization server is convenient because it does not require any additional software to execute. These servers are usually programmed using the C programming language, taking advantage of OS-level calls, leading to enhanced performance. This also means that the server has to be available for the operating system to which you are deploying or you are out of luck. In terms of scalability and availability, the server can either have its own built-in load-balancing and failover mechanisms or use third-party load-balancing solutions, such as the hardware-based systems provided by Cisco systems.

For Java servlet-based synchronization servers, you will need a servlet engine for deployment. Since J2EE application servers are now commonplace in most organizations, this requirement does not usually pose a problem. By using an outside servlet engine, the performance, scalability, and availability of the synchronization server now rely on the capabilities of the application server/servlet engine being used. The same goes for the server operating systems that are supported; that is, as long as the servlet engine works on a given platform, the synchronization servlet should work as well. That said, you should give extra consideration to the synchronization vendor's supported platforms and recommended application servers when deciding which application server and operating system to use. (More information on the various features of synchronization middleware is provided later in this chapter in the section entitled *Key Features of Synchronization*.)

Enterprise Integration

The final part of a complete synchronization solution is the enterprise integration layer. While this layer is often part of the synchronization server, we are discussing it separately because it provides different functionality. The enterprise integration layer enables you to communicate with various back-end data sources. If you are using a commercial mobile relational database on the client, you will most likely have integration to enterprise relational databases on the server using ODBC, JDBC, or native drivers.

NOTE Many of the mobile database vendors only have synchronization for their own enterprise database on the server. If you have multiple server databases, this can be a cause for concern. More information on commercial solutions is provided later in this chapter, in the section entitled *Commercial Synchronization Solutions*.

In addition to providing integration to relational databases, you may also require access to other forms of enterprise data, such as ERP systems, CRM systems, or XML data. If this is the case, you will have to look for additional enterprise adapters for the solution you are implementing; or if you have the resources, you can create your own adapter.

Publish/Subscribe Model

One of the most common models of data access being used by synchronization solutions today is the publish/subscribe model, so it is worthwhile to explore how it works.

The publish/subscribe synchronization model is based on the concept of having a master copy of your data, the publisher, and one or more copies of this data, the subscriber(s). The data between the publisher and subscribers is updated periodically to keep the data consistent. The update process, or synchronization, is bidirectional, allowing the publisher to update the subscriber data, and vice versa.

The publisher is responsible for defining which tables (or subsets of those tables) are available for external access. The data sets that are defined as being available are called publications. The application clients that are interested in having access to that information are called subscribers. They define which data they want access to using subscriptions. It is possible to have a number of publications defining the available data. These publications can have parameters, which makes it possible to use different subsets of data for different users. In this way, one publication can be used to send a different subset of data to each subscriber.

Figure 10.2 shows an example of a field service application's data set. In this diagram, you can see that the publisher has the complete set of master data. This data contains all of the information for each field service representative. In this example, the field service representatives are the subscribers. They require only the subset of the publisher's data that pertains to their territory, so rather than keeping a copy of the entire set of data, they have a subset of data that is based on their individual subscriptions. In this way, you can easily define which data is available for mobile users and which parts of the data needs to be synchronized down to the mobile devices.

Common Synchronization Configurations

A variety of network and data management configurations are commonly used with enterprise synchronization. Synchronization configurations (or topologies) are arrangements of publisher and subscriber databases that transmit data to one another. The publish/subscribe model of data access supports both peer-to-peer and hierarchical configurations.

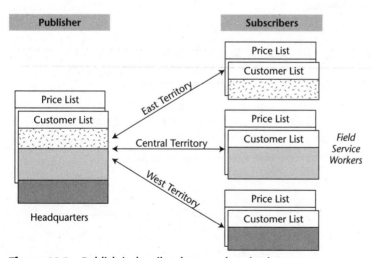

Figure 10.2   Publish/subscribe data synchronization.

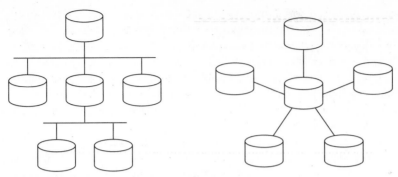

Figure 10.3   Hierarchical database configurations.

The most common of these two is the hierarchical configuration, wherein every database has a single parent database, except for the master enterprise database, which has no parent. A diagram of common hierarchical layouts can be seen in Figure 10.3. In this diagram are two hierarchical configurations: the configuration on the right is commonly referred to as a hub-and-spoke configuration. In both cases, hierarchical configurations are good when a publisher needs to publish data to a large number of subscribers, or, in other words, when there are a large number of remote workers who need to have access to the enterprise data. In this configuration, the master database contains all of the changes made by any of the remote databases. Those updates can then be propagated to the other remote databases during their next synchronization.

Unlike the hierarchical configuration, peer-to-peer configurations do not have a single common enterprise data store. In Figure 10.4 you can see that peer-to-peer configurations enable each database to contain the same information, with no single database acting as a central server. In this configuration, you can still use the publish/subscribe model, but the data updates are not going to be propagated to all databases in the configuration. For this reason, peer-to-peer configurations are best suited for situations in which only two systems need to be synchronized, or when it is not important for each remote user to have the updates from all other users. In general, a peer-to-peer configuration is not recommended, because there is no central server responsible for collecting updates from the remote users and propagating those changes to the other users in the system. Other difficulties of peer-to-peer configurations include maintaining data integrity, implementing conflict detection, and programming synchronization logic.

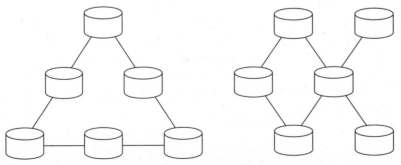

Figure 10.4   Peer-to-peer database configuration.

The Data Synchronization Process

As demonstrated by the synchronization architectures, there are many ways in which a remote database can be synchronized with the enterprise data source. To simplify things, we are going to take a look at a basic synchronization process as the foundation for our discussions on the various synchronization architectures and techniques.

To simplify the synchronization process to a generic level, we have to make a few assumptions about this particular solution, namely:

- We are only synchronizing the changed data, to minimize data transferred.
- We have a synchronization layer on the client.
- We are using synchronization middleware.
- We have a network connection available during the synchronization process.
- The synchronization is bidirectional, and both sets of changes occur during a single synchronization session.

We'll go through each of the steps of the synchronization process shown in Figure 10.5 and explain what happens at each step. In our example, the synchronization process is going to be initiated from the client device.

1. The application user can initiate the synchronization process manually, or it can be programmed into the application. At the point where the synchronization is initiated, the data that has changed since the last synchronization is prepared for sending to the synchronization server. This data preparation often involves compressing the data and, optionally, encrypting the data.

2. Once the data is prepared, a connection is established with the synchronization server. At this point, the user is usually authenticated with the server; the data packets are then sent over the communications network (wireless or wireline) to the synchronization server.

3. The synchronization server receives the data to be synchronized. It then uses the synchronization logic to determine whether the data needs to be transformed before it is sent to the enterprise data source. If it does, transformation can occur at this time.

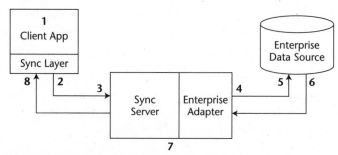

Figure 10.5   Basic synchronization process.

4. The enterprise adapter provides integration to the enterprise data source. This adapter may simply be an ODBC or JDBC driver for enterprise relational databases, or custom code for other more complex data sources.

5. Using the appropriate enterprise adapter, the synchronization server can authenticate the user against the enterprise data source (optional) before it starts the data transfer. Once the user is authenticated, the server can update the enterprise data source with the changes from the client application. At this time, the synchronization server can also detect if there are any conflicts in the data being updated and, if there are, take appropriate action.

6. After the update has been committed, the relevant changes that took place on the server since the last synchronization are prepared for sending back to the client application.

7. The synchronization server takes the enterprise data source's changes and performs any transformation that may be required before sending the updates to the client application. Again, this data is usually compressed and possibly encrypted for additional security.

8. The data is sent to the client application, where it is updated in the mobile data store.

This example uses a straightforward synchronization architecture that is common to many of the synchronization solutions on the market today. It covers only one synchronization methodology, but it can easily be adapted to fit other synchronization architectures and techniques that we are going to cover in the upcoming sections.

Synchronization Techniques

The synchronization solutions available today use many techniques to move data from client applications to enterprise servers. Each of the techniques provides value for certain situations, but not every technique will be right for your particular application. When implementing your synchronization layer, keep these techniques in mind to ensure that your solution is as efficient as possible for your application.

Synchronization Modes

There are two main modes of synchronization: snapshot and net change. To determine which mode is most appropriate, you will need to consider how frequently the data will be synchronized and how many changes to data occur between synchronizations. When using a wireless network for communication, keep the data transferred to a minimum, which will help increase efficiency and reduce costs.

Snapshot

Snapshot synchronization makes it possible to move large amounts of data from one system to another. It involves deleting a table on one system and copying a table from another system in its place. The result is two tables with identical data sets.

This form of synchronization is used when there are moderate to large amounts of data that need to be synchronized or when data is not changed at the remote location. Additionally, due to the large amount of data being transferred, it is best suited for reliable networks with high bandwidth.

Snapshot synchronization is a suitable candidate for instances where a complete, or nearly complete, set of data has to be transmitted. An example might be a salesperson who requires updated product catalogues or price lists. The main office can send these updates via a snapshot so that all the data is updated to the remote worker at once. Since the remote worker will not be making changes to the remote product list, no data will be lost.

Net Changes

For most mobile applications, a net change mode of synchronization is more efficient than using snapshot synchronization. In these cases, only the changed data is sent between the remote and enterprise databases, saving network bandwidth and reducing connection times. In order to accomplish this, the remote database keeps track of the original data and the most recent change that has occurred since the last synchronization. In this way, even if the data were to change several times on the client, only the original data and most recent data have to be transmitted to the server. The changes to the enterprise data source can then happen in a single transaction since there is only one update to be made. The same approach is taken when the server database is updated and synchronized to the remote database.

This form of synchronization is well suited for applications in which the user may synchronize several times in a day, sending only small amounts of data each time. It is ideal for coping with the bandwidth limitations on most of today's wireless networks. It is also a good technique for situations in which larger amounts of data are synchronized less frequently.

The net change mode does have one drawback: If you require knowledge of each individual transaction that occurred to the data, then keeping track of only the most recent change may not be suitable for you. In this case, you might find a transaction log-based synchronization approach to be more appropriate. This approach is discussed in the *Key Features of Synchronization* section later in the chapter.

Data Propagation Methods

The way data is transmitted to and from the remote database can have a major impact on the effectiveness of the application. Therefore, the application designer should take care to synchronize the data in such a way that it is not too intrusive. Ideally, the application will be able to synchronize without user involvement. This capability is possible over wireline networks but often difficult over wireless networks, as most of them require a manual connection to go online.

There are two types of data propagation methods: session-based and message-based. In the following paragraphs, we will take a look at how each of these methods works and the types of synchronization for which each is best suited.

Session-Based Synchronization

Session-based synchronization requires a direct communication stream between the remote and enterprise database systems. The data synchronization happens in real time over a wired or wireless network. During this connection, both the client and server data can be updated, and the updates can be confirmed.

This form of synchronization is similar to a telephone call: Once a connection is made, data is sent in both directions until the conversation is complete; once the data has been communicated, the session is closed. If, for some reason, the communication link goes down during the data transfer, a new connection can be made, and the synchronization can continue from a suitable point. To ensure no data is lost, this point is often at the last committed transaction.

By using a session-based synchronization method, you can make sure that data is synchronized at specified times. For example, if you know that every Friday at 5:00 P.M. you need to have the latest sales information, you can schedule synchronization at this time, and you know that once it is complete, the data on both sides is up to date.

Message-Based Synchronization

Synchronization can also be executed using a message-based system. For this method of synchronization, the data updates are communicated asynchronously using messages that contain a destination address and other control information, eliminating the need for a direct connection to the enterprise system. Instead of requiring a connection over protocols such as TCP/IP, HTTP, or other wireless connections, the data can be transferred over common messaging protocols such as Simple Mail Transfer Protocol (SMTP), File Transfer Protocol (FTP), or any other proprietary messaging system.

This allows the data synchronization to use a store-and-forward mechanism to get the data to and from the database systems. This capability is often provided by the underlying messaging protocol and does not have to be implemented directly into the synchronization layers.

Message-based synchronization works in the same way as an email system. When you want to send a message to someone, you specify the destination address and write a message to send. Once it is sent, you no longer have any control over what happens to it. At some point in the future, maybe a minute, maybe an hour, maybe several days, the email message is read and acted upon. For synchronization, the action is to take the message and perform the required updates to the target database system. Once this is complete, a response may be sent containing a synchronization update in the other direction.

Message-based synchronization systems are well suited for architectures that do not allow for a direct connection between the client and the server databases, that support an extremely large number of remote users, or where a network connection is not available or is unreliable. Direct connections are not possible in many security configurations due to firewall implications. In these situations, you can implement message-based synchronization such that it offers guaranteed delivery. In this way you can be sure the data will reach its destination. The obvious drawback to message-based synchronization is that you do not have control over when the synchronization takes place. If you require a solution that enables data to be synchronized at specific times, session-based synchronization may be more appropriate. (For more information on messaging within mobile and wireless applications, see Chapter 5, "Mobile and Wireless Messaging.")

Key Features of Synchronization

Now is a good time to take a closer look at some of the key features that will prove useful in a synchronization solution. Not all synchronization solutions will contain all of these features, but the solution you choose should, minimally, contain those that relate to the application you are building. Alternatively, if you decide to build your own synchronization system, this section will give you an idea of some of the challenges that lie ahead.

Data Subsetting and Partitioning

The database on the server will contain much more data than is required for the mobile application. For this reason, only a subset of the database should be able to be synchronized to the mobile device. This capability can successfully be accomplished in two ways. First is by limiting the mobile application to include only the portion of the database schema that is relevant to the mobile user. This process will remove any tables that are not applicable, thereby reducing overhead. Once this reduction is accomplished, the second way is to minimize the amount of data being synchronized to those tables.

Data partitioning is the technique used to minimize the synchronized data. There are two forms of data partitioning: column partitioning and row partitioning. Column partitioning means that you will only take the required columns from the table being synchronized. You determine which columns are required by looking at the data displayed to the user and the data the application uses internally. Once you know which columns are required, you may be able to further reduce the data by implementing row partitioning. This means that you will only want to synchronize the rows of data that are required by the specific user doing the synchronization. This can be accomplished by setting up a filter to user subscriptions. For example, if I am a field service representative, I only want to synchronize the data that pertains to the work orders that I will be handling myself. I do not require the rows of data for all of the other technicians in my company. By doing this, I will dramatically reduce the amount of data that has to be stored on my device and that I will have to synchronize to the server database. Side benefits of this technique are that there is a smaller chance of running into conflicts when remote users have distinct subsets of data, plus improved synchronization performance.

Data Compression

Another way to reduce the number of bytes being transferred is by implementing a compression routine. Data compression optimizes the data stream being sent over the network. In a wireless environment, this will reduce synchronization times and costs. The one caveat is that compressing and decompressing data adds extra overhead to the synchronization process on both the client device and synchronization server. This is only a concern, however, on handheld devices that have minimal resources. It is important to note that some data types can be compressed much more than others, so the benefits of compression are dependent on the data being synchronized.

Data Transformation

At times, the data in the enterprise database may not be in the ideal format for your mobile application. In these cases, the data may need to be transformed from one data type to another during synchronization. You may want to transform the data to save space on the client device, by, for example, changing text fields from fixed-length to variable-length. In other cases, you may need to transform the data because the mobile database does not support a given data format. For example, if the client database does not have support for floating-point numbers, these numbers may have to be changed to character representations so they can be displayed on the client.

Transactional Integrity

Databases attempt to maintain what are called atomic, consistent, isolated, durable (ACID) transactions. Maintaining transactional integrity means that the execution of a transaction will ensure that either all of the changes or none of the changes are committed at the same time to maintain database consistency. In addition, the transaction should occur in isolation, that is, not be affected by other transactions that may be occurring. Finally, once the transaction is committed, its effects are persistent, even in the event of future system failures.

Maintaining ACID transactions in mobile and wireless computing is even more difficult. With unreliable wireless networks, and a potentially large number of remote users, it is important for the synchronization layer to maintain transactional integrity at all times. If any part of a remote transaction fails, the entire transaction must be rolled back, and the data has to be kept in a consistent state. The client has to then reapply the transaction during the next synchronization.

With many of the systems implementing a net change approach to database updates, the net changes have to be synchronized in the context of a database transaction. If an organization requires each individual transaction to be committed in the order it occurred, then a log-based system may be required. In this case, all of the transactions performed on the remote database are stored in a transaction log, and during synchronization, the transaction log is read in chronological order, executing each database change individually. While this may appeal to some organizations, it does add additional overhead to maintain and transmit what could potentially be a large file.

Conflict Detection

A conflict occurs when two users attempt to update the same data field with different values. For example, let's say that two sales representatives attempt to update a contact's email address to a new value. Both users have access to the contact information remotely, in offline mode. The first sales representative can update the data to the enterprise server without any issue. When the second sales representative executes his or her update, a conflict will arise because the data was altered since the last time the data was synchronized. When you have multiple databases updating the same fields of data from different locations, a conflict is sure to happen when the data is synchronized to a central location.

The best way to handle conflicts is to design the database schema and data subsetting and partitioning configurations in such a way that a conflict will rarely occur. If this is not an option, the synchronization middleware has to handle them in an acceptable manner. The first step in solving a conflict is detection. For this to happen, the application or, more likely, the synchronization layer, must store both the original and new values of data for each row being updated. In this case, when data is being updated, the synchronization logic can check to see if the original value from the client matches the current value in the enterprise database. If they do, there is no conflict, since the data being updated has not changed since the last synchronization. If, on the other hand, the values do not match, it means that another user has updated the data since the last synchronization, and a conflict has occurred.

Most synchronization solutions do offer conflict-detection capabilities; but it is what happens next that really matters: conflict resolution.

Conflict Resolution

Once a conflict has been detected, it has to be resolved. This can be accomplished automatically, by implementing some predefined rules, or programmatically, by allowing the developer to apply custom business rules to resolve the conflict.

The most common resolution methods include *last-in wins* and *first-in wins*. Both of these can easily be implemented based on the user performing the synchronization. When a transaction is rolled back because of a conflict, a mechanism should be in place to tell the user why the transaction failed, so he or she can correct the problem or, possibly reapply the change.

In the case of custom conflict resolution, the system should allow developers to program the logic in a language familiar to them. When the enterprise data source is a relational database, the most obvious solution is for the synchronization logic to use SQL, the same language that is used for other database interactions. Custom resolution logic may involve notifying an administrator of the conflict, or executing some predefined logic in a stored procedure.

A conflict can also be resolved by evaluating and working with the current data. Let's use the example of the sales representatives again. In this scenario, the representative (let's call him Rep 1) has sold 100 units of product A, and another representative (Rep 2) has sold 50 units of the same product. When Rep 1 synchronizes his data, the remaining inventory value in the enterprise database will be reduced by 100 units. When Rep 2 synchronizes her data, a conflict will occur because her remaining inventory value will not be the same as that in the enterprise database. This conflict can be resolved by reducing the remaining inventory value in the enterprise database by 50. In this scenario, last-in and first-in wins do not make sense, so custom logic is required. The only situation where a problem with this system might arise is when the inventory level goes below zero, in which case either more products have to be ordered or the sales representatives have to be notified.

Network Protocol Support

Data synchronization can occur over one of many different networks, so having support for multiple network protocols is important. The client application may synchronize over a wireless connection, using CDPD, GPRS, CDMA or TDMA, or it may be over a wireline network using TCP/IP. In addition to being compatible with the given protocol, a synchronization solution should also be able to manage the connections. For example, if a wireless connection is dropped, the synchronization layer has to be able to either reconnect and continue the synchronization or notify the user to let him or her know that the transaction will have to be restarted. In the case of a dropped connection, the less data that must be re-sent, the better.

For any synchronization system, at a minimum, you will want to have support for the most common protocols such as TCP/IP and HTTP, or, in the case of message-based synchronization, FTP, email, or even physical media such as floppy disks or CD-ROMs. If your particular application will be deployed over networks using other protocols, then you will want to ensure that they are supported as well. Bluetooth is a good example of a protocol for which you may require support in the future.

Multiple Transport Mechanisms

Smart client applications do not require a wireless connection for synchronization. They can use other data transport mechanisms that may be available. In Figure 10.6 we can see some of the possible synchronization scenarios that may be required. They include the following:

- Direct synchronization through an integrated wireless modem.
- Synchronization through a cellular phone. The device may communicate to the cell phone using a cable infrared or Bluetooth.
- Through an integrated wireless LAN card, possibly using 802.11b or Bluetooth.
- Through a mobile device with a modem that is connected to a regular telephone line that uses a regular ISP to connect to the Internet.
- Through a device in a cradle that is using a companion PC for synchronization. The device may communicate with the PC using IR, a serial cable, or a USB cable. The PC can then communicate to the synchronization server using TCP/IP or HTTP.

Enterprise Integration

To access the enterprise data in your organization, the synchronization server has to be able to communicate directly with your enterprise data sources. In some cases, this may only involve accessing a single enterprise database, but very often, there is a need for access to multiple databases. Even if you do not currently require this, having the capability to synchronize with other databases gives you some protection that your investment will still be useful if other data has to be accessed. For relational databases, the synchronization server should have either JDBC or ODBC access to the following leading databases:

- Oracle
- IBM DB2
- Microsoft SQL Server
- Sybase Adaptive Server Enterprise (ASE)
- Sybase Adaptive Server Anywhere (ASA)

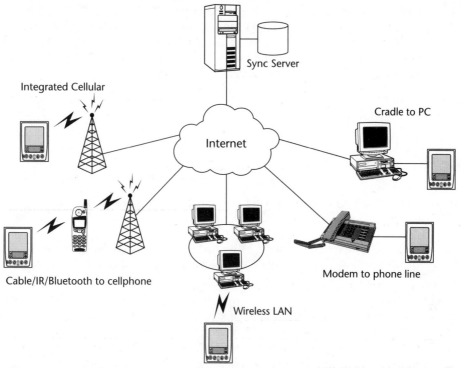

Figure 10.6 Synchronization over a variety of transport mechanisms.

In addition to relational databases, you sometimes need access to other enterprise data sources, such as ERP or CRM systems. To accommodate these enterprise applications, the ability to implement custom adapters is necessary.

Security

As in any enterprise solution, security is always a concern. Within the synchronization component of smart client applications, there are several security layers that need to be addressed, namely:

Authentication. A user can be authenticated at several locations during the synchronization process. These include the client application, the synchronization server, the database server, and the operating system. You may not require all of these authentication layers, but you should at least consider one that is managed at the server, such as using the database or the server operating system.

Authorization. Once a user is authenticated, you can set up access control lists based on individual users. In this way, you prevent users from accessing data that they are prohibited from accessing. This can be accomplished at a user level or by implementing authorization groups such as user, manager, and administrator. Very often authorization is implemented via synchronization logic as well.

Encryption. The data being transported should support some form of encryption to prevent unwanted parties from "listening" to your data stream. For many smart client applications, this can be accomplished using dedicated security libraries such as Certicom's 128-bit Elliptic Curve Cryptography (ECC). If this is not available, then SSL or other encryption routines are available.

For more information about security in mobile and wireless computing, refer back to Chapter 6, "Mobile and Wireless Security."

Synchronization Development Options

Your synchronization options often will be dictated by the choice you make for your persistent data store. Many of the relational database vendors have synchronization solutions as part of their offerings. These solutions are usually integrated extremely well with the mobile database, making them an obvious choice. On the other hand, many of these synchronization solutions integrate only with the vendor's enterprise database, possibly imposing a limitation that is not suitable for your environment. If this is the case, you can either build your own synchronization solution or find a vendor that offers generic synchronization for many client and server databases.

If you use the database built into the operating system, you may also be able to use the OS's synchronization conduit; but more often than not, you will have to create your own synchronization layer. The built-in synchronization software is usually not targeted at accessing enterprise data, and is more suitable for deploying applications and related files to the device.

Mobile OS Synchronization Conduits

Most of the mobile operating systems on the market come with proprietary conduit software for synchronizing data and applications between the mobile device and a companion PC. We will take a look at the leading operating system synchronization conduit software here, focusing on the capabilities for enterprise data synchronization.

Windows CE ActiveSync

Microsoft ActiveSync gives you an easy way to synchronize files and applications between your desktop and Windows CE device. You can connect to your PC through the device's cradle or wirelessly over a wireless LAN or infrared. ActiveSync provides built-in support for synchronizing email, contacts, calendar, tasks, and notes with Microsoft Outlook. In addition, you can browse the Internet directly on your device using the Internet connection on your desktop computer.

ActiveSync is based on a client/server architecture. It consists of a service manager and a service provider. The service manager coordinates the synchronization process between the desktop and the device. It provides connectivity, detects changes, resolves conflicts, and transfers data.

If you are looking for more advanced features, such as synchronizing enterprise data to a client application, you will have to either build or purchase a custom ActiveSync service provider. A service provider is a custom conduit used in conjunction with the ActiveSync application to execute data synchronization. By creating or purchasing an ActiveSync service provider, the users of your application will be able to synchronize their enterprise data using the same method they use to synchronize applications and PIM data.

Building a custom service provider requires the creation of two modules, one for the client device and another for the desktop PC. Both modules are implemented as in-process COM objects, which have exposed interfaces allowing the service manager and other service provider modules to request the tasks for which the service provider has been designed. (Note: Describing the development of these components is beyond the scope of this book. Consult the Microsoft Web site, www.microsoft.com, or a Windows CE programming book for more detailed information.)

Palm OS HotSync

Similar to ActiveSync, Palm OS HotSync provides an out-of-the-box solution for synchronizing files and applications between your desktop and a Palm OS device. You can establish the connection using a cradle or wirelessly using infrared. Additionally, the Palm OS desktop software allows you to synchronize various forms of PIM data, including email, calendar entries, contact lists, and to-do lists. It's important to point out here that the Palm OS HotSync Manager is *not* a synchronization conduit; rather, conduits are plug-ins that can be managed by the HotSync software, and are executed when you press the HotSync button on a cradle or modem. In general, you are required to develop or purchase custom conduits that you require for your applications. The exception is the basic backup and installation conduit that comes with the Palm OS HotSync software. Anything beyond this basic level of functionality has to be developed as a custom conduit. For example, if you want to transfer data from the handheld to your company database, a conduit must be developed to perform the translation, access the appropriate database record, write to it, and save to the database.

Conduits for the Palm OS are usually written using Visual C++, Visual Basic, or Java, using the Palm Conduit Development Kit (CDK). The CDK provides access to the HotSync Manager APIs, and comes with complete samples and documentation. The most useful sample in the CDK is a generic conduit that includes a portion of the required synchronization logic that can be customized for your use.

Symbian OS Connect

Among the standard features included in Symbian OS Connect are file management, backup and restore, remote printing, document conversion, contacts and agenda synchronization, email synchronization, clipboard transfer, machine management, and application installation. This is indeed an impressive list of features, but notice that it does not cover bidirectional enterprise data synchronization. If you want to use Symbian OS Connect for this type of synchronization, you will have to develop it yourself using the converter APIs provided in the Symbian OS SDK. (More information on using Symbian OS Connect can be found on the Symbian Developer Network at www.symbian.com/developer/index.html.)

Synchronization Conduit Considerations

The synchronization conduits included with mobile operating systems have advanced application and PIM synchronization capabilities, but they do not include any functionality for enterprise data synchronization. Hence, in all cases, you will be required to develop a custom conduit or service provider to have these capabilities.

Since most device users take advantage of these conduits for application and PIM synchronization, it is often requested that data be synchronized using these conduits as well. Doing this gives users a seamless way to synchronize their applications and data using the same process. If you are developing your own data synchronization conduit, integration with the operating system synchronization software may be a good feature to include. Many of the commercial data synchronization solutions have support for ActiveSync and HotSync synchronization.

One final point to keep in mind about OS conduit software: It is very specific to the operating system. A conduit written for ActiveSync will have to be rewritten to work with HotSync, and vice versa. If you are going to build a custom synchronization layer, you may want to take a more generic approach.

Custom-Coded Synchronization Solution

Now that you know that the synchronization software that comes with an operating system will not provide any level of database synchronization, you are left with two choices for your synchronization solution: build it in-house or purchase a commercial solution.

Let's take a look at some of the issues that you will have to resolve in a custom-coded synchronization solution. Many of these topics/features have been discussed earlier in the chapter, but are readdressed here to refresh your memory.

In addition to having the knowledge to write your own C/C++ or, possibly, Java-based server application, you will also have to know some specifics about data synchronization requirements:

Database issues. You will have to understand the structure of your database and know how to extract the required information so it can be synchronized to the mobile device. You will have to be concerned with transactional and referential integrity during the data transfer process. Once you understand the enterprise data structure, you have to define a layer to map this data to your client side data source, possibly requiring a transformation layer. And, of course, you will require a data access layer. This may be ODBC, JDBC, or a custom adapter for nonrelational data sources.

Mobile operating systems. Which mobile operating system do you have to support? Each operating system has its own I/O layer with which you will need to integrate. In addition, you may end up having to develop separate applications for each client since each has its own SDK.

Network protocols. Which protocols are required for the data transfer? Is supporting TCP/IP enough, or do you need support for other non-IP wireless protocols? This may require knowledge of how to communicate using the wireless modems that are available for the devices to which you are deploying.

Handling conflicts. Once you are at the point where data can be synchronized, what do you do when a conflict arises? You will have to develop a methodology for resolving conflicts in an acceptable way. This most likely will require a mechanism for custom resolution logic.

Security. What level of security is required for your application? You may need to implement encryption of the communication stream, as well as authentication support for remote users.

Support for SyncML. Do you want this solution to adhere to the SyncML specification? If you are reselling the solution, this may provide a good selling point. Even if you are not reselling this solution, SyncML can offer you a standardized way to handle the most common synchronization tasks that you will encounter. Information on SyncML is provided later in this chapter in the section entitled *SyncML Overview*.

In addition, if you want your synchronization layer to work in conjunction with the synchronization software provided with the operating system, you will need to build specific conduits for each operating system to which you are deploying. To aid with the conduit development, each operating system has a development kit that provides integration APIs.

On top of the development efforts, you also have to address ongoing maintenance of the solution. Each time an operating system is updated or a new network protocol is required, you will have to perform additional maintenance work. In our current mobile environment, where new devices and networks are constantly emerging, maintenance alone could require significant resources.

If you think this sounds like a lot of issues to consider, you are right. Developing a robust, enterprise class synchronization solution in-house is not an easy task. It requires in-depth knowledge about databases, mobile operating systems, network communications, and security. In general, if there is a commercial solution available that meets your requirements, most likely you will end up ahead in the long run if you use it. On the other hand, if you have decided to implement the mobile database in-house, then building the synchronization layer may make sense. It may also be your only option, as many of the commercial synchronization solutions are tied to a specific mobile database offering.

Commercial Synchronization Solutions

The commercial synchronization solutions can be broken into two categories: those offered by mobile database vendors, and general-purpose synchronization solutions. In both cases, the commercial synchronization solution should address many of the key synchronization requirements listed earlier in the chapter. As you learn more about the available products, you will quickly be able to determine whether a solution meets your needs.

Database Synchronization Vendors

The synchronization products in this section are from vendors who also sell mobile databases. In all cases, the synchronization software is only available for each vendor's client database offering, although often it can integrate with other databases on the server side.

Sybase/iAnywhere Solutions

Product/Technology	MobiLink and SQL Remote
Mobile OS support	Windows CE, Palm OS, Symbian OS, Linux, Java
Server OS support	32-bit Windows, Solaris, Linux, AIX, HP-UX, Novell Netware, Mac OS X

Summary:

Sybase/iAnywhere Solutions has two offerings in the synchronization area: MobiLink and SQL Remote. Both are implemented as proprietary servers that sit between the client and sever databases, providing flexible synchronization solutions. Let us take a look at where each technology is best suited.

MobiLink synchronization works with both Sybase Adaptive Server Anywhere and UltraLite as the client databases. It provides session-based bidirectional synchronization to all of the leading enterprise databases, including Sybase Adaptive Server Enterprise (ASE), Oracle, IBM DB2, Microsoft SQL Server, and Sybase Adaptive Server Anywhere (ASA). The integration with these databases is provided in a direct manner using ODBC, without requiring any intermediate data storage. MobiLink uses the publish/subscribe synchronization model, storing all of its synchronization logic in the enterprise database. The conflict resolution logic is stored in the same manner. This gives the developers a flexible way to modify the synchronization logic using the SQL variant implemented by the enterprise database vendor. Additionally, the synchronization logic can be programmed using Java or any .NET language.

MobiLink synchronization is designed to support a large number of remote databases; it can support thousands of remote databases with a single installation. Both wireless and wireline protocols are supported, including HTTP and TCP/IP, as well as common operating system conduits such as HotSync and ActiveSync. For security, MobiLink offers the option of encrypting the data stream to prevent unwanted parties from viewing the data.

The SQL Remote synchronization technology is somewhat different. Though it, too, has support for thousands of remote users in a single installation, it is suitable only for Sybase databases, supporting Sybase ASE and Sybase ASA. SQL Remote is a message-based synchronization layer, permitting asynchronous communication between the client and the server. For each update made on the database, an entry is made into a transaction log. When synchronizing, the changes in the transaction log are sent to the server, using one of many available protocols, including email, FTP, or any other file transfer mechanism. The messages can be processed whenever they are received, allowing replication to occur incrementally.

Overall, the two iAnywhere Solutions offerings are very capable for data synchronization, providing flexible options for the server database. They do, however, require a Sybase client database (either Adaptive Server Anywhere or UltraLite). By providing both session-based and message-based synchronization, iAnywhere Solutions should have a solution that is suitable for your environment.

IBM

Product	DB2 Everyplace Sync Server
OS support	Windows CE, Palm OS, Symbian OS
Server OS support	Windows NT, AIX, Linux, and Solaris

Summary:

The IBM DB2 Everyplace Sync Server is a relatively new technology. At the end of 2001, it supported only the Palm OS client and Windows NT server. In 2002, IBM released the Sync client for Windows CE and Symbian OS, and the Sync server for AIX, Linux, and Solaris. Compared to the other commercial solutions, this product has not been on the market for very long.

The synchronization server is implemented as a Java servlet. This allows it to run within any servlet engine, although it works best in IBM WebSphere. The client components are native for each client operating system. The architecture is based on the SyncML protocol. In accordance with this protocol, the synchronization messages are WAP binary-encoded XML (WBXML). This provides a compact way to send and retrieve data over wireless networks.

User data is defined using the publish/subscribe model. IBM provides a tool to define the server publications and the subscriptions for the mobile users.

Data synchronization can either be one-way or two-way, using HTTP as the transport protocol. When synchronization occurs, the data is sent from the client to the Sync server using SyncML. Once it reaches the Sync servlet, it is temporarily placed in an IBM DB2 Universal database. From there, custom JDBC adapters can transfer the data to the enterprise database of choice. There is currently support for Oracle, Sybase, and Microsoft databases. Once the server has received the client data, the server sends its changes back to the client using the same process.

The IBM synchronization solution is obviously a great choice for those companies using the DB2 Everyplace database. It provides tightly integrated synchronization options for the DB2 Universal database and adapters for other relational databases. The fact that the DB2 UDB is required for temporary storage may be seen as a limitation for some organizations, as it may require additional hardware and maintenance resources.

Oracle

Product	Oracle9i Lite
Mobile OS support	Windows CE, Palm OS, Symbian OS
Server OS support	32-bit Windows

Summary:

Oracle provides a bidirectional synchronization solution between the Oracle9i Lite client database and the Oracle9i enterprise database. There are two components to Oracle Lite's synchronization solution: the Mobile Server and the Mobile Sync client. The Mobile Server sits on top of the Oracle9i AS application server and provides integration to the Oracle enterprise database. The Mobile Sync client works with Oracle Lite on all of its supported platforms.

The Sync Server automatically generates the synchronization logic. This provides a quick way to get an application going. However, if you require custom logic (which you often do), you will have to control the synchronization programmatically, meaning that the database administrator also has to be a developer. There is a similar limitation for conflict resolution logic. The Mobile Server will automatically detect a conflict; but from there, only basic conflict resolution is offered. If these are not sufficient, the conflict data is put on a queue where an administrator can perform manual reconciliation of the data.

For security, Oracle offers data encryption directly from the synchronization client to the synchronization server, thereby prohibiting any access by malicious parties.

The Oracle synchronization solution is targeted only at corporations using both the Oracle9*i* Lite and Oracle9*i* server databases. If this does not describe the technology being used in your organization, then you will want to look at other solutions with more extensive enterprise support.

Microsoft

Product	SQL Server CE
OS support	Windows CE
Server OS support	32-bit Windows

Summary:

Microsoft SQL Server CE provides bidirectional synchronization between the SQL Server CE client database and the SQL Server enterprise database. There are two means of data synchronization: Remote Data Access (RDA) and merge replication. Both provide two-way data synchronization. Merge replication, though it is more complex, has some benefits over RDA for mobile scenarios. For example, merge replication allows data to be updated autonomously on mobile devices and the server. The data is then merged when the device is reconnected to SQL Server. Microsoft SQL Server CE uses the publish/subscribe synchronization model.

Synchronization is provided over HTTP to Microsoft SQL Server back-end databases via Microsoft Internet Information Server (IIS). Replication can be performed over wired and wireless local area networks (LANs) and wide area networks (WANs). Using IIS as the synchronization server provides several benefits, one of them integrated security. IIS provides authentication, authorization, and encryption to the synchronization process. These are tied to the IIS security layer and do not have to be visible to the mobile user.

Microsoft's synchronization layer does have some limitations. Besides working only on Microsoft platforms with Microsoft products, another weakness is the product's support for conflict detection and resolution: Most conflict resolution work has to be done programmatically by the application developer. In addition, the overall synchronization architecture is rather complex, and will take significant effort to set up properly, especially if you do not have experience with SQL Server or IIS.

PointBase

Product	PointBase UniSync
OS support	Java clients
Server OS support	Platforms with a Java Virtual Machine

Summary:

Synchronization between the PointBase Embedded database and other enterprise databases is accomplished using PointBase UniSync. UniSync is a middleware server that provides bidirectional data synchronization. If you are using the PointBase Server or Oracle databases, you can synchronize the net changes. If you are using other enterprise relational databases from Sybase, Microsoft, or IBM, only snapshot synchronization is supported.

PointBase UniSync uses the publish/subscribe synchronization model. It allows data subsetting at the user level. It also has built-in conflict detection with standard resolution schemes, and enables users to implement custom schemes in Java.

PointBase UniSync offers a capable synchronization solution for companies using the PointBase embedded database on the client, and the PointBase Server or Oracle databases on the server. If you are using other enterprise databases, being limited to only snapshot synchronization may prove to be problematic. And note that PointBase Micro version, discussed in Chapter 9, "Persistent Data on the Client," is not currently supported by PointBase UniSync.

Other Synchronization Vendors

The enterprise data synchronization market currently is dominated by companies that offer a corresponding mobile database product. Most of the other synchronization offerings do not focus on the enterprise database market, with one notable exception: Synchrologic.

Unlike the other vendors listed in Table 10.1, Synchrologic also has adapters that allow bidirectional data synchronization between a variety of client and server databases. For mobile client databases, Synchrologic does not currently support IBM DB2 Everyplace or Sybase UltraLite. It does, however, support Sybase Adaptive Server Anywhere, Microsoft SQL Server CE, and Oracle Lite. On the server is support for all the leading enterprise data sources. In terms of features, the Synchrologic synchronization server supports data subsetting, conflict detection and resolution, multiple network support, and security.

Table 10.1 lists other leading synchronization technology vendors. Notice that many of their solutions focus on PIM synchronization, which is discussed in more depth in Chapter 16, "Mobile Information Management."

Table 10.1   Additional Synchronization Vendors

VENDOR AND PRODUCT	TYPE OF SYNCHRONIZATION	WEB SITE
Extended Systems XTNDConnect	PIM, file	www.extended systems.com
FusionOne	PIM	www.fusionone.com
PumaTech Intellisync	PIM, file	www.pumatech.com
Starfish TrueSync	PIM	www.starfish.com
Synchrologic iMobile Suite	PIM, file, data	www.synchrologic.com

NOTE All of the synchronization vendors covered in Table 10.1 support the SyncML initiative and have SyncML-compliant products.

Commercial Solution Conclusions

You have two options for enterprise data synchronization: building it in-house or purchasing a commercial solution. In most cases, building a custom solution will require significant resources for development and maintenance, ultimately making it more costly than licensing a commercial product. If, however, you are using a database that comes with the mobile operating system or a custom-built database, building your own synchronization layer may be the only available option.

All of the leading mobile database vendors offer synchronization to their corresponding enterprise databases. In the case of Sybase and IBM, you also can synchronize to other enterprise databases from their mobile clients. These commercial solutions have rich, flexible synchronization features, making them well suited for large-scale enterprise-wide deployments.

TIP Rather than choosing a synchronization solution based on the mobile database, you might find it easier to choose the mobile database based on the synchronization solution. Since many of the commercial synchronization solutions are more limited than the commercial mobile databases, making the decisions in this order will result in a solution that will directly integrate with your enterprise databases, while still maintaining persistent data storage on the client device.

SyncML Overview

The way data is synchronized varies dramatically from one vendor to the next. Each implementation has its own proprietary way of exchanging data, sometimes making it difficult for device manufacturers, service providers, and application developers to integrate these solutions. Furthermore, each synchronization solution has support for different protocols, devices, and data types, making the task of data integration an even greater challenge. This is particularly the case when you need to synchronize applications from multiple vendors.

In February 2000, a group of companies, including Ericsson, IBM, Lotus, Motorola, Nokia, Palm, Psion, and Starfish Software, joined forces to create a new industry initiative to develop and promote a single, common data synchronization protocol that could be used throughout the industry. This protocol is called SyncML.

To date, the objectives of SyncML have not yet been realized for enterprise data synchronization. The specification is still relatively new and not many vendors have implemented SyncML-compliant products. This is largely because SyncML is not yet proven for corporate applications. Many of the existing synchronization solutions have proven market success, so there is no need for these vendors to rush to implement new SyncML solutions. The one exception is in the PIM space, the first market targeted by SyncML. As you'll learn in Chapter 16, many of the PIM solutions are now SyncML-compliant.

NOTE The information in this section is largely based on the SyncML specifications and related documents from the SyncML Initiative (now a component of the Open Mobile Alliance). Consequently, there is some level of bias in the favor of SyncML. As more vendors implement SyncML solutions, we will be able to take a more objective look at SyncML. For now, we take a preliminary look at target audiences for SyncML and explain how SyncML works.

What Is SyncML?

What exactly is SyncML? The best answer to that question can be found at the SyncML Web site (www.syncml.org) where SyncML is defined as follows:

SyncML is a specification for a common data synchronization framework and XML-based format, or representation protocol, for synchronizing data on networked devices. SyncML is designed for use between mobile devices that are intermittently connected to the network and network services that are continuously available on the network. SyncML can also be used for peer-to-peer data synchronization. SyncML is specifically designed to handle the case where the network services and the mobile device store the data they are synchronizing in different formats or use different software systems.

As the definition reveals, SyncML is an XML-based specification for synchronizing all forms of data between remote devices and enterprise systems. Its goal is to provide an industry-standard protocol for synchronizing data between all types of devices, such as PDAs, handheld PCs, smart phones, pagers, and cell phones, over any network protocols, to any data store. To achieve this lofty goal, the SyncML initiative decided to start with personal information management (PIM) synchronization, and then move to other forms of enterprise data.

Why Use SyncML?

The overall benefit of SyncML is that it is an industry initiative aimed at bringing together synchronization technology for many devices, networks, and data sources. Many of the current synchronization solutions use proprietary conduits that are often complex and limited in the data sources they can communicate with. As the number of mobile computing devices increases, and the expectations for access to corporate data grows, the SyncML Initiative hopes to provide enough benefits to make it the leading protocol for data synchronization.

SyncML Target Audiences

Four main groups are expected to benefit from the adoption of SyncML as an industry standard, each in its own specific way:

End users. SyncML promises to provide a single way to synchronize email, calendar, to-do lists, contact lists, as well as corporate application data. This would be a welcome change to the current landscape where different synchronization software is required for each task.

Device manufacturers. For devices that have built-in synchronization capabilities, SyncML offers a protocol that will allow the device to access a broader range of data sources, over any network and transmission technology. This will enable device manufacturers to provide a wider range of features with reduced overhead.

Service providers. Companies in the application-hosting field currently have to limit their offerings due to the numerous forms of synchronization technologies on the market. As it stands now, service providers already have a difficult time managing the few technologies they do support because they all have different requirements for devices, networks, and back-end data sources. SyncML should alleviate these issues by providing a single protocol that will be able to work with a broad range of applications, devices, and networks.

Application developers. Developers want to focus their efforts on technology that is in demand and will result in successful applications. SyncML should provide a single technology that will allow them to build a variety of data-driven applications without having to learn a number of synchronization technologies. This should result in developers being able to build applications that will work on a larger number of devices, over many network protocols.

SyncML Advantages

In addition to the advantages for the specific groups just listed, SyncML provides seven other advantages that should prove to be beneficial in a synchronization solution. Many of the existing synchronization solutions already include many of these advantages, but again, SyncML is attempting to provide them in an industry standardized way:

Designed for wireless and wireline networks. SyncML works over all networks used by mobile devices, including both wireline and wireless networks. Some of the challenges in this effort come from the characteristics of today's wireless networks: high latency, low bandwidth, low reliability, and high costs. To address these issues, SyncML has implemented a robust protocol that uses WBXML, to reduce the data being transferred; uses a single request-response model, to minimize network traffic; and has support for dropped connections during synchronization.

Support for multiple transport protocols. SyncML works over the following leading network protocols: TCP/IP, HTTP, Wireless Session Protocol (WSP), and OBEX (Bluetooth, IrDA). It can also be deployed over SMTP, POP3, IMAP, and proprietary wireless communication protocols. SyncML does not require any additional capabilities that are not addressed by these protocols.

Support for arbitrary data. The capability to support various forms of data is a requirement for any general-purpose synchronization solution. Since SyncML is targeted at providing synchronization for all forms of enterprise data, it has built-in support for the following data formats, and can add new formats if the need arises:

- Personal data formats, such as calendars, to-do lists, and contact lists
- Collaborative formats, such as email and network news
- XML and HTML documents
- Relational data
- Binary data formats, binary large objects (BLOBS)

Data access for multiple applications. To work effectively with the largest number of possible applications, SyncML is not tied to any specific programming language. It does not make any assumptions about the programming language being used, and will work effectively when a different language is used on the client and the server parts of the application.

Optimized for constrained devices. Since many of today's mobile devices have limited memory and processor capacities, SyncML is designed to work in very constrained environments. By keeping the footprint and processor requirements to a minimum, SyncML-based application should be able to operate on a broad range of devices with varying processors, memory capacities, and operating systems. In addition, the data exchanged using SyncML is in a binary format, reducing memory requirements for storage.

Takes advantage of existing Internet technologies. Since many of today's corporate applications are Web-based, it is important for a synchronization protocol to be able to work with Web technologies. XML has emerged as the preferred way to represent structured data on the Web, so SyncML is XML-based, ensuring easy implementation and quick interoperability testing.

High level of interoperability. SyncML has been designed to work over any network, with any device and any data source. Even though this goal has yet to be realized, SyncML does work very well with highly constrained devices, allowing them to communicate with synchronization servers that are capable of executing complex synchronization logic.

NOTE Again, keep in mind that these are the proposed advantages of the SyncML protocol, which are just starting to be realized with commercial applications.

How SyncML Works

As we take a look at how SyncML works, we are going to focus on the guidelines that have been laid out in the specifications, not individual vendor products. Once you have an understanding of the underlying SyncML protocols, you will be able to quickly evaluate vendor offerings to determine if they meet your specific requirements.

Specifically, we will examine the following components of the SyncML specification:

- SyncML Representation Protocol
- SyncML Synchronization Protocol
- SyncML Transport Bindings

The SyncML Representation Protocol specifies a common synchronization framework and format that can be used to create different data synchronization models. This protocol is defined by a set of well-defined messages (either XML documents or MIME content type) that are transported between the devices participating in a synchronization procedure. It supports both the request/response and the blind push command structures as data synchronization models.

SyncML synchronization is an exchange of packages; each package contains one or more messages. The SyncML representation protocol specifies the expected result of various synchronization operations, based upon the SyncML synchronization framework and associated data synchronization models. In this way, the SyncML representation protocol defines the structure of a SyncML message. Each message within a package contains a single data synchronization operation.

A SyncML message is a well-formed XML document consisting of a header and a body. The header contains metadata, which conveys information about the SyncML message. This metadata may include versioning, session, routing, and security information. The body of the SyncML message contains a list of commands, which allow for inserts, updates, deletes, and other actions to be executed to make the two datasets the same.

Figure 10.7 shows an example of the SyncML framework; notice the dotted line that defines its boundaries. It is in this framework that the SyncML Representation Protocol is applied. When you look at the overall interactions between the SyncML client and the SyncML server, the SyncML Sync Protocol is used. This protocol is essential for achieving interoperable data synchronization. It defines the protocol for various synchronization procedures, which occur between the SyncML client and the SyncML server in the form of message sequence charts. The SyncML Sync Protocol allows for both one-way and two-way data transfer in a variety of modes. One-way synchronization is an important feature because it allows clients to save time by sending only their changes to the server; clients do not have to receive the server changes at the same time. It also is appropriate where only one set of data has changed, in which case two-way synchronization is not required; by default, only the changed records are synchronized between the client and the server. If the full set of data has to be synchronized, there is a slow-sync mode that will allow this.

Within the SyncML Sync Protocol, two roles are defined: a client and a server. Most of the time, synchronization is initiated by the client application, although some of the always-on networks allow the server to initiate the transfer. Do not be fooled into thinking that the server role within the protocol has to be performed by a powerful server-class machine: Both the client and the server may be mobile devices.

When it comes to the transport layer, SyncML defines a set of transport bindings. These bindings are required to ensure that there will not be any problems using the underlying protocols. The three transport bindings currently supported are HTTP, WSP, and OBEX. Each of these is supported for a reason. HTTP is widely used, and usually is able to pass through corporate firewalls; WSP is a part of WAP, allowing WAP phones to support SyncML; finally, OBEX provides support for short-range connectivity, possibly using Bluetooth, IrDA, or USB.

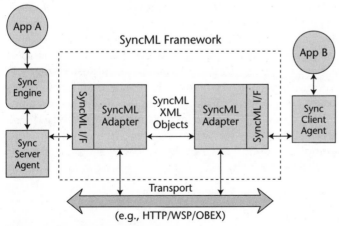

Figure 10.7 SyncML framework.

Now let's see how this works. Figure 10.7 shows how two applications communicate with each other using SyncML. Application A represents a server application that can synchronize with multiple client applications, such as Application B. The communication between these two applications can occur over any network transport, such as HTTP, WSP, or OBEX. The client application (Application B) uses the Sync Client Agent to access the network and send messages to the server. For this to happen, the client agent communicates with the SyncML Interface (SyncML I/F), which in turn uses the SyncML adapter to send the XML message object. The server (Application A), then receives the message through the Sync Server Agent and manages the entire process using the Sync Engine. All of the Sync Engine's network access and data synchronization operations to and from the client application go through the Sync Client Agent.

This example is at a conceptual level. As vendors bring commercial products to market, these products may not contain all of these concrete components. However, the functionality that is contained in these components will have to be incorporated into their solution in some fashion.

> **NOTE** If you are interested in learning more about SyncML, you can download the SyncML specification from www.syncml.org or www.openmobilealliance.org. It will give you in-depth information about the SyncML architecture and protocols.

Future of SyncML

The first products to support the SyncML initiative have focused on the synchronization of PIM data such as email, to-do lists, contact lists, and calendar information. Companies like Starfish Software, Pumatech, Synchrologic, and FusionOne have added support for SyncML at this level. There has been some adoption by other SyncML sponsors, including Palm, Symbian, and Nokia; all of which have added SyncML support to their latest device offerings.

In contrast, SyncML has not been widely adopted for enterprise data synchronization. To date, only IBM (a founder of the SyncML initiative) has released SyncML-compliant products. Oracle, Sybase, and Microsoft have all decided to stay with their own synchronization solutions, which are not based on SyncML. These products have been on the market for some time, and it is unlikely that they will support SyncML anytime soon.

It is still too early to tell whether SyncML will become an industry standard. SyncML is currently in a prestandards phase, where it will remain until it gains sufficient momentum and market acceptance to justify its adoption by an international standards body. At this time, more than 700 companies are supporting the SyncML initiative, although many of them are taking a wait-and-see approach. If these companies start to release SyncML-compliant products, we could very well see plug-and-play integration between any device, on any network, and be able to synchronize with any data source.

Summary

Enterprise data synchronization is one of the key components of smart client applications: It is the technology that integrates mobile databases with enterprise data sources. A variety of synchronization and data propagation techniques can be used to accomplish this integration. Whenever possible, synchronizing only the changed data will reduce the amount of data transferred, resulting in faster response times and reduced costs.

Synchronization solutions either can be developed in-house or purchased from software vendors. In either case, many key features are required for a complete synchronization solution. These include data subsetting and partitioning, transactional integrity, conflict detection and resolution, support for multiple networks, and data security.

The next chapter marks the beginning of Part III, which is focused on building wireless Internet applications. It provides information on the thin client application architecture, and highlights the main components required for a complete solution.

Helpful Links

The following are some useful links for finding out more information on the topics discussed in this chapter:

Palm OS Conduit Development	www.palmos.com/dev/tech/conduits
Microsoft ActiveSync service providers	http://msdn.microsoft.com/library /wcedoc/wceintro/cestart.htm
Symbian OS connectivity whitepapers	www.symbian.com/developer/techlib /papers/pci.html
SyncML	www.syncml.org; www.openmobilealliance.org; http://xml.coverpages.org/syncML.html

Building Wireless Internet Applications

Over the past five years, many corporations have spent countless resources moving their client/server applications to a three-tier Internet architecture. Doing this has made it possible to provide enterprise data access to anyone with a desktop Web browser. You no longer have to deploy client applications on CDs or other media, because the presentation logic is all designed for Internet access, most commonly using HTML as the foundation. In this way, users simply have to enter a URL to access the applications and content that they require.

We have now reached a point where many enterprise applications have been Web-enabled. The result is a vast number of applications making all forms of data available over the Internet. These applications range from business applications on secure corporate intranets to consumer applications available to anyone with a Web browser and an Internet connection. Due to the broad range of applications and content available, the Internet is frequently used for many everyday tasks such as making appointments, checking email, and making purchases; and the corporate world relies heavily on it for information dissemination and business application access. So it follows that, as more people require mobile access to their data, it be distributed wirelessly using the Internet.

In this Part we will take a look at the technologies relevant to building wireless Internet applications with a focus on the enterprise space. The part has been broken into five chapters, providing insight into the requirements of building enterprise wireless Internet applications, as follows:

- Chapter 11, "Thin Client Overview"
- Chapter 12, "Thin Client Development"
- Chapter 13, "Wireless Languages and Content-Generation Technologies"
- Chapter 14, "Wireless Internet Technology and Vendors"
- Chapter 15, "Voice Applications with VoiceXML"

Thin Client Overview

By now most likely you have realized that thin client refers to wireless Internet applications. We call this type of application "thin client" because no software is required on the wireless device except for an Internet browser. The browser accepts a markup language, parses it, and displays it to the application user. Any response from the user is sent back to the server, where it can be handled appropriately. For true thin client applications, all of the application logic resides and is executed on the server platform. Hence, the client does not require much processing power or memory to be able to run these types of applications, making them suitable for very small, resource-constrained devices.

Keep in mind that thin client applications are not limited to cell phones. Many devices on the market support thin client applications. Actually, any device that has a microbrowser and a wireless connection is suitable. This includes laptops, PDAs, smart phones, and cell phones, among other more specialized devices.

This chapter gives an overview of the thin client application architecture by highlighting the main components that comprise a successful solution. Following the architecture overview, we take an in-depth look at one of the leading wireless Internet protocols, WAP.

Architecture Overview

As mentioned previously, the thin client architecture is very similar to the architecture of Internet applications. It is based on a three-tier architecture, as depicted in Figure 11.1.

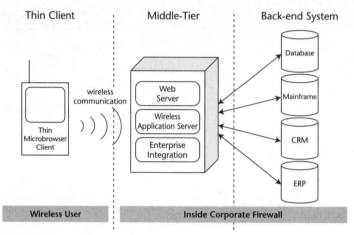

Figure 11.1   Wireless Internet architecture.

The following are the key components of the thin architecture:

Thin client. The client is a microbrowser focused on the presentation of the markup language. It is the part of the application that the end user sees, although it does not execute any of the application logic. The design of the client user interface is crucial to the overall success of the application.

Middle tier. The middle tier is where most of the work is performed. A couple of main components run in this tier: the Web server and the wireless application server. The Web server receives the inbound HTTP requests and routes them to the wireless application server. The application server then takes these requests and executes the appropriate logic. This execution may include session management, content management, as well as integration to the back-end system.

Back-end system. The back-end system is where the data and related services reside. This system may be a relational database, an email server, business applications, an XML content source, or any number of other enterprise systems. In any case, it is important that this data can be reached, so that it is available to the mobile worker.

NOTE ON TERMINOLOGY The term *wireless application server* refers to the middle-tier server that contains both an application server and a framework designed for wireless content delivery. These two components may be supplied either together or independently.

Figure 11.1 shows a very simple three-tier architecture. In reality, several other layers need to be addressed: the security layers, including firewalls, as well as the actual location of the wireless networks and related wireless gateways. In the rest of this section, we will cover each part of the architecture in more depth.

Client

The client interface for thin client applications is an Internet browser. In the desktop world, browsers have many advanced capabilities, with sophisticated user interfaces, often supporting client-side scripting languages such as JavaScript or VBScript. Most desktop browsers also have advanced graphics and multimedia support. Many of these features, however, require high bandwidth and powerful client-side processors, a luxury not available on most wireless devices. The client for most wireless applications is not nearly as sophisticated. The browsers have limited screen size, restricting the richness of the user interface. In addition, there is no de facto standard for wireless browsers, meaning that, to reach a broad audience, application providers have to support a variety of browsers and many markup languages.

With these challenges in mind, we will take a look at the various aspects related to the client layer of wireless Internet applications.

User Interface

The user interface may be the most important part of a wireless Internet application. People may disagree, but if users cannot quickly find what they are looking for, they will stop using the application, in which case, everything else becomes irrelevant. The problem is, even the best-designed user interfaces can cause user frustration. Trying to cram useful data and navigation controls onto a small screen such as a cell phone is a daunting task. Even when the screen size is larger, such as on a PDA, the task is not much easier. Much care has to be taken to ensure that the required information is easy to find, without having to navigate through a dozen screens to get it.

One of the difficulties in this endeavor is that the way we navigate desktop Web applications is fundamentally different from how we navigate wireless devices. A wireless device is not commonly used for general information gathering and Web surfing; rather, the wireless device is used for a specific purpose, and the user interface should make accomplishing that purpose as efficient as possible. Therefore, applications have to be targeted for specific tasks, such as placing an order or looking up a defined set of information. Examples in the corporate space include a salesperson looking up price lists or inventory levels, or a field service employee finding his or her next work order. In the consumer space, such a task may involve looking up a stock quote, purchasing a movie ticket, or viewing the latest sport scores. In all these examples, the application is being used for a specific reason; the faster users can accomplish their goal, the more satisfied they will be. The most effective user interfaces do one job and do it well. (In Chapter 12, "Thin Client Development," we will take a closer look at some design patterns that are useful for creating effective wireless Internet user interfaces.)

Browsers and Content Types

Another challenge in creating wireless thin client applications is the variety of markup languages and microbrowsers currently being used. Whereas in the wired world you can address an overwhelming majority of your target audience by creating HTML applications that run in Microsoft Internet Explorer and Netscape Navigator, the same is not so with wireless Internet applications.

Each device typically has its own browser or, in many cases, set of browsers. Each browser understands a particular markup language. The common markup languages include Wireless Markup Language (WML) as defined in WAP 1.x; XHTML, as defined in WAP 2.x; Handheld Device Markup Language (HDML), which is a predecessor to WML; and the subsets of HTML, such as Compact HTML (cHTML). Each of these markup languages has nuances that have to be addressed. Figure 11.2 shows the common microbrowsers on a variety of device classes.

Moreover, each browser handles markup languages differently, meaning you have to test each browser/markup language combination to be sure that the application is displayed correctly. For example, WML content may display correctly on a cell phone running the Openwave browser, but not function properly on a RIM device with a GoAmerica Go.Web WML browser. For such a situation, you have to create content with slight variations to accommodate different client browsers. The good news is that many wireless application servers incorporate this functionality, so it does not have to be handled manually. (In Chapter 13, "Wireless Languages and Content Generation Technologies," we take an in-depth look at the leading wireless markup languages and content-generation techniques for creating device-specific applications.)

Finally, there are voice browsers that use VoiceXML as the markup language. These applications do not display content visually, but rather speak the response over any telephone; the interaction with the system is through voice as well. (Note that because voice-based applications do not use the typical wireless gateways, they are covered separately, in Chapter 15, "Voice Applications with VoiceXML.")

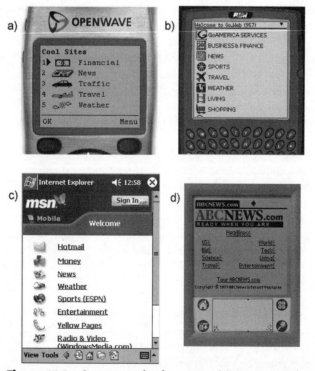

Figure 11.2 Common microbrowsers: (a) Openwave browser, (b) Go.Web browser on RIM 957, (c) Pocket Internet Explorer, (d) Palm Web Clipping.

Wireless Networks

In order for thin client applications to receive data from the server, a wireless network connection is required. This is different from smart client applications, which store data locally, allowing for offline access. With thin client applications, each request goes over a wireless network to a wireless gateway, at which point it is transferred to an HTTP request and forwarded to the Web server/wireless application server.

In most cases, the browser takes care of the communication with the wireless network. This saves developers from having to know the underlying network protocol, allowing them to focus on the application logic. This is a significant advantage, as a large number of wireless network protocols are being used. A complete step-by-step analysis of a wireless request procedure is included in the section *Processing a Wireless Request* later in this chapter.

Middleware

In wireless thin client applications, most of the work is performed using middleware specialized for wireless applications. In this section we take a look at the two main types of servers: wireless application servers and wireless gateways.

Depending on the corporate requirements and resources, these servers can either be hosted by a third-party vendor or be kept on premise. In the most common configuration, the wireless service provider hosts the wireless gateway, and the wireless application server is located within the enterprise.

Wireless Application Servers

Wireless application servers provide the core infrastructure upon which wireless Internet applications are built. They are the engine driving the wireless Internet architecture, providing many of the core feature requirements for this type of application. To meet this rather tall order, wireless applications servers must have a strong server foundation on which the wireless components can be built. For this reason, many vendors have chosen to use J2EE application servers as the underlying technology for their wireless offerings. Using J2EE enables wireless vendors to focus on the core requirements for wireless applications, while leaving the generic server capabilities to the application server vendors.

Initially, many vendors created their own proprietary servers, which were often targeted at a limited number of platforms, thus requiring the developer to use proprietary markup languages and APIs. Realizing that developers preferred to use standards-based technology, most vendors since have ported their frameworks and tools to standard J2EE servers produced by BEA Systems, IBM, Sybase, and Sun Microsystems. In doing so, wireless application server vendors have increased the number of platforms they can support and have reduced the complexity of their offerings. Later in this section, we will investigate the core features of applications servers and how they relate to the J2EE standard and the Microsoft .NET platform. But first we will examine the features provided by commercial wireless server products that are designed for the unique requirements of wireless applications.

The first implementations of many wireless server products were targeted at specific content/device/network combinations. An example would be WML content for the Openwave browser on Nokia devices over a GSM network. While this generated a positive result for the specific combination being targeted, it did not scale well for the multitude of browsers, devices, and networks that were being used as wireless technology became more mainstream. In order to succeed in this rapidly changing marketplace, wireless server technologies had to adapt. This involved creating frameworks that allowed for the development of a variety of wireless applications that could be deployed to a wide range of devices over an assortment of wireless networks.

These frameworks became the basis for the wireless server products on the market today. Most commonly built using J2EE technology, including Java servlets, JSPs, and EJBs, these frameworks provide the core features that are required for wireless application development. These features include device and browser identification, content transformation, voice technology integration, session management, and enterprise data integration. In addition, framework-specific development tools are often available to lower the learning curve associated with using this new technology. Each feature serves a very specific purpose that contributes to the overall success of the wireless application being developed.

A few features on this list generate the most interest. The device and browser identification capabilities are key to providing the appropriate content to the user. By identifying the client device and browser, we can determine the required markup language, the size of the user interface, the types of graphics supported, and the preferred language, among other things. Once we know these factors about the client, the content can be created accordingly, using a variety of techniques and technologies. (In Chapter 13, we take a closer look at each of the leading markup languages, as well as the most common content-generation techniques. In Chapter 14, "Wireless Internet Technology and Vendors," we take a closer look at the features of wireless applications servers and provide summaries of the commercial solutions available.)

Application Server Basics

Here we will take a quick look at the basic services that application servers provide as they apply to wireless Internet applications. (An in-depth look at the capabilities of an application server is beyond the scope of this book.)

The application server is the cornerstone of the distributed Internet architecture. It connects different components of complex applications, while allowing the application to remain modular. It provides the glue to pull together enterprise data, business logic, and the client presentation layer. All of these capabilities are provided in such a way that frees developers from being concerned with the implementation details, allowing them to focus on building the application logic and user interfaces. In general, application servers provide the following three distinct layers:

- Web integration
- Component transaction server
- Enterprise connectivity

Because many commercial wireless servers utilize other application servers as their base platform, we also have to examine the core technology on which they are built. That endeavor involves investigating the core features of the two leading specifications implemented by application server vendors: J2EE and .NET. They both provide the same core functionality, but in significantly different ways; hence, we will review each technology individually.

J2EE

The Java 2 Platform, Enterprise Edition (J2EE) is a specification delivered by Sun Microsystems using the Java Community Process (JCP), which is used to develop and revise Java technology specifications with input from the international community of Java technology experts. In other words, it allows the greater Java community, rather than a single entity, to govern the J2EE specification (and other Java specifications).

The J2EE specification defines a standard platform for building multitier enterprise applications. It builds upon capabilities provided in the Java 2 Platform, Standard Edition (J2SE), such as the core Java libraries, the promise of code portability, and Java Database Connectivity (JDBC). In addition, it defines an enterprise component model in Enterprise JavaBeans (EJBs), Internet capabilities through Java Servlets and JavaServer Pages (JSPs), and XML integration support. Vendors cooperate on developing the specification, then compete on the implementation. This allows a broad range of enterprise systems to use a single standard, allowing for portability between technology vendors and preventing corporations from being tied to a single vendor. Many corporations have desired this type of specification for some time.

NOTE Keep in mind is that J2EE is not a product; it is a standard to which products can be written.

Wireless Internet applications usually take advantage of four main J2EE technologies: EJB, Java servlets, JSPs, and JDBC. They may also use others such as Java Naming and Directory Interface (JNDI), JavaMail, and Java Transaction API (JTA), but usually to a lesser extent, so we will not discuss them here.

NOTE The Java Message Service (JMS) is also an important technology for mobile and wireless applications. It was discussed in more depth in Chapter 5, "Mobile and Wireless Messaging," and is mentioned again later in this chapter in the *Messaging Servers* section.

Let's take a look at each of these four technologies in relation to the functionality they provide to Internet applications. Figure 11.3 provides a logical diagram of how these core layers may be implemented. As you can see, servlets and JSPs act as the main interface between the client applications and the enterprise business logic contained in Enterprise JavaBeans. The EJBs can then communicate with the enterprise data source using JDBC or other adapters.

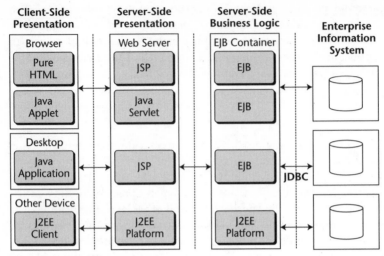

Figure 11.3 J2EE architecture.

Java servlets and JavaServer Pages provide the capability to create dynamic, data-driven content. Java servlets enable developers to easily implement Java server-side code that takes advantage of the other J2EE APIs. They act as an interface between the client application and the enterprise business logic. JSPs provide the same interface, but using a scripting language such as HTML or WML. The JSP specification supports static templates, dynamic markup language generation, and custom tags. Both Java servlets and JavaServer Pages play a prominent role in most wireless application server products.

Business logic is encapsulated in Enterprise JavaBeans components. EJBs make it possible for application developers to create reusable segments of logic, without having to be concerned with the underlying application complexity. The EJB container within the application server handles the various details such as threading, component pooling, and garbage collection. Developers can implement two types of EJB components: Session Beans and Entity Beans. Session Beans typically contain the business logic that maintains workflow and transactions. Entity Beans represent collections of data. They encapsulate data sets and enable operations that can be used to interact with that data. Both component types allow for rapid development and deployment of enterprise logic. Most wireless application servers support the use of EJBs for business logic, but they often do not utilize them for the core wireless capabilities.

The final layer is the enterprise integration layer. If the data resides in a relational database, then JDBC will most likely be the technology used to interact with the data. Most enterprise data vendors have JDBC drivers for their offerings, providing a standard way to manipulate their databases. JDBC offers support for user-defined types, rowset manipulation, connection pooling, and distributed transactions.

If the data does not reside in an enterprise database, then a custom adapter most likely will be required. Depending on the data source, a commercial adapter using EJB technology may already be available. In most cases, these adapters are built to the EJB specification, so they should be able to work in the application server you are using (assuming it is J2EE-compliant). If a commercial adapter is not available, then you may be required to implement an adapter in-house. This undertaking may not be as difficult as it seems. Many enterprise applications have defined APIs that you can use to access the systems, often available for the Java programming language. Additionally, the Java Connector Architecture (JCA) provides an industry standard way to integrate with legacy or nonstandard data sources from J2EE applications.

NOTE All of the leading J2EE application server vendors are compliant with the J2EE specification. This means that J2EE components such as Java servlets, JSPs, and EJBs written to the appropriate specification should be portable across application server offerings. We say "should," because in reality, testing is usually required on each platform to adjust for vendor-specific nuances.

Microsoft .NET

Microsoft .NET is the overarching technology around which Microsoft is building its entire platform. This includes its Visual Studio development tools, programming languages, mobile operating system, messaging architecture, and database servers. In the future, most Microsoft products will be part of the .NET initiative. A major component of .NET is XML Web services. Web services are geared to allow applications to communicate and share data over the Internet, independent of the operating systems or programming language being used.

To help highlight the .NET platform, we will compare the J2EE standard and the .NET products. Notice that J2EE is a standard to which products can be developed, whereas .NET is a set of products offered primarily by Microsoft. We will take a look at these products in relation to how they compare with the J2EE standards.

Let's start with the programming language of choice, C#, the base language of the .NET platform (although other languages can be used as well). In fact, Microsoft is promoting the Microsoft Intermediate Language (MSIL), a set of byte codes into which any supported language (such as VB.NET and C#) can be compiled. These byte codes are then interpreted and translated into native executables using the common language runtime. A strong correlation exists between MSIL and the common language runtime and Java byte codes and the Java Runtime Environment (JRE).

Beyond the programming language, the three main aspects of .NET that we will investigate are the Web integration layer, the component transaction layer, and the enterprise integration layer. Figure 11.4 illustrates the main technologies that are of interest to us in the .NET platform. These include (Active Server Pages) ASP.NET, .NET-managed components, and ADO.NET.

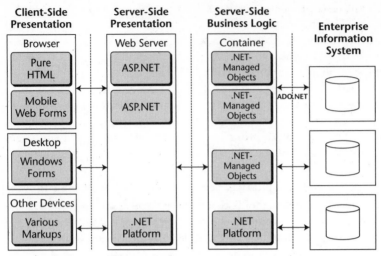

Figure 11.4   .NET platform architecture.

As the name suggests, Active Server Pages .NET are a scripting technology similar to JavaServer Pages. They provide dynamic content to a variety of clients, including desktop clients using Windows forms and browser clients using a variety of markup languages including HTML and WML. As depicted in Figure 11.4, ASPs provide the integration layer between HTTP-based clients and the server-side business logic.

Business logic is encapsulated in .NET-managed components. These are the successors to COM and COM+ components and provide interoperability with them. Essentially, .NET-managed components provide the business logic and data processing layer of the .NET platform, similar to what EJBs provide to J2EE platforms. The business logic for .NET-managed components can be written in a variety of programming languages, since the language will be compiled into the MSIL and executed using the Microsoft common language runtime.

The final layer that we will look at in .NET is the enterprise integration layer. ADO .NET is the technology that provides connectivity to enterprise data sources. It is the latest in a long line of database access technologies that started with Open Database Connectivity (ODBC) many years ago. Since then we have seen OLE DB and ActiveX Data Objects (ADO). One of the main differences of ADO.NET over its predecessors is that it is not as database-centric. ADO.NET is required to support other data sources, as it is the main integration layer between the .NET-managed components and enterprise data. For instances where ADO.NET does not provide adequate enterprise integration, custom adapters can be created using the .NET-managed component architecture. In addition, because .NET is focused on Web services, a variety of XML-based protocols are available for data integration.

J2EE versus .NET: Summary

Debate is ongoing as to which platform is best: J2EE or .NET. Resolving this debate is not easy, as arriving at an answer involves detailed technical and business analysis. The capabilities of both platforms are comparable, but the ways they are implemented are distinctly different. J2EE is a standard that is widely supported throughout the software industry. There are well over 50 vendors that offer J2EE support on a variety of software and hardware platforms. This results in a wide range of companies providing tools, servers, and other products related to J2EE technology, giving the application developer many options for developing and deploying J2EE applications.

.NET, on the other hand, is a product suite offered by Microsoft on Windows platforms. It provides strong tool and server offerings, but not much flexibility outside of Microsoft's product realm. Companies will either see this as a limitation, since they are locked into Microsoft technology, or as a benefit, since it does provide a tightly integrated solution. Having everything provided by a single vendor does have positive aspects: You will have some assurance that the various technologies will work well with each other and with the operating system used for deployment.

Choosing which platform to use comes down to your specific application and business requirements. Both J2EE platforms and Microsoft .NET are enterprise-class offerings. If you are looking for an open solution, with wide industry and vendor support, J2EE is most likely a better choice for you. If you are primarily using Microsoft technology, and are looking for a well-integrated solution from a single vendor, the .NET product suite should suit your needs well.

Wireless Gateways

Wireless gateways are the link between wireless and wired networks, providing integration between wireless communication protocols and the enterprise. This allows all wireless applications to access enterprise applications and data through the same entry point, independent of the wireless technology being used. Wireless gateways allow users with varying needs (devices, applications, coverage) to select the best network for their situation.

As discussed in Chapter 3, "Wireless Networks," wireless networks use many different protocols for communication, often not based on IP. This means that wireless gateways have to obtain a direct connection to the particular wireless carrier for communication, since it cannot travel over the Internet, which is IP based. With this connection in place, the gateway then transfers the information from the proprietary wireless protocol to an IP-based protocol, such as TCP/IP or HTTP(S) allowing for direct communication to the enterprise system. In this way, the wireless gateway is able to provide connectivity from a wide range of wireless networks to enterprise systems.

Due to the complexity and time required to implement and maintain the connections, wireless gateways are usually hosted by third-party vendors outside corporate firewalls. Here's a concrete example. Let's assume the client is a Go.America browser running on a RIM device in North America. The wireless protocol being used is Mobitex, which is not IP-based. The wireless application server is listening using HTTP for any incoming requests. The browser on the RIM device is not using HTTP, so it cannot send a request directly to the enterprise wireless server. At some point between the client and the wireless server, the request has to be adapted from the Mobitex protocol to HTTP. This is one of the roles of the wireless gateway. Next assume that a request is made by the user to retrieve content from www.ianywhere.com. Once made, the request is transferred wirelessly to a base station in the user's vicinity. From there, it will be routed using a direct wired connection to the network operation center (NOC), where the wireless gateway resides. The wireless gateway then takes the request, coverts it to HTTP, and forwards it to the appropriate enterprise Web server. The Web server can then output the appropriate WML content and send it back to the wireless gateway, where it is changed back to the Mobitex protocol and sent back to the user. In this scenario, the wireless server and the gateway are located on separate servers in different locations. (Note: Content transformation capabilities are not always handled this way. In many cases, they are executed by the wireless gateway as well.)

Other features offered by the wireless gateway include optimized communication streams, support for IP over non-IP networks, WAP push messaging, and enhanced security. For some of these features, an application must be deployed to the client device, but not always. It depends on the specific feature, as well as the vendor supplying the technology.

One common wireless gateway is a WAP gateway. WAP gateways provide the link between clients using WAP and enterprise systems. Later in this chapter, in the section called *Wireless Application Protocol (WAP) Overview* we will take a closer look at the role of the WAP gateway within the WAP architecture.

Messaging Servers

Recall that in Chapter 7, "Smart Client Overview," we looked at messaging as it pertained to smart client applications. Here we focus on the aspects that are relevant to thin client applications. Included in this category are the Short Message Service (SMS), HDML notifications and alerts, and WAP Push. (Note: Wireless email will not be covered here, as it is discussed within the context of mobile information management in Chapter 16, "Mobile Information Management.")

Let's start by examining the types of messages that thin client applications can receive. As you may recall, the only application on the mobile device is the microbrowser. All of the input and output for wireless Internet applications goes through the browser interface. This is often the case for thin client messaging. There is no persistent data on the device, meaning the messages sent are intended to be viewed by the user, not to be consumed by an application, as in smart client messaging. The most common form of text messaging is SMS, particularly in Europe and Asia, and more often now in North America.

SMS is based on a packet-based architecture that sends data over the control channel. The messages are typically limited to 160 characters. Typical uses include weather reports, stock quotes, calendar reminders, and communication between two users. SMS messaging is controlled using SMS centers, or SMSCs. Each carrier typically has its own SMSC, with its own proprietary interface. In order for server applications to send messages using SMS, they must have interfaces to each of the SMSCs being used.

Implementing this in-house can be a huge burden. For this reason, companies have created a common interface to the various SMSCs to make server integration easier. Using this type of service prevents you from having to get your own connection to each carrier, and from having to interface directly with each carrier's APIs.

The size limitation of SMS messages restricts the content being sent. Text is the only practical messaging format that can be sent in 160 bytes. Because of this limitation, sending more feature-rich messages using this protocol is not possible. The Enhanced Message Service (EMS) changes this. It adds powerful new functionality on top of the SMS standard. With EMS, mobile phone users can add some "life" to SMS text messages, in the form of pictures, animation, and sound. This gives users new ways to exchange pictures and ringtones with other users. One of the great things about EMS is that it is based on SMS standards, meaning that wireless carriers can support EMS with little effort.

In addition to SMS and EMS, WAP 2.x provides a new messaging service called Multimedia Message Service (MMS) that allows for even richer messaging. MMS is aimed at providing a feature-rich messaging solution, in both instant delivery mode (similar to SMS) and with store-and-forward queuing (similar to email). In the future, MMS is expected to become the preferred messaging protocol for mobile users, since there are virtually no limits on what can be sent using it. An MMS message can contain graphics, data, animation, images, audio clips, voice transmissions, and video sequences. That said, it's important to point out that MMS messages are much larger than SMS messages, so MMS will not find widespread use until wireless bandwidth increases.

Outside of the SMS world, browser alerts and notifications are also used for messaging. They provide an alternative means of sending information to wireless Internet users via the microbrowser. In WAP 1.2, the concept of WAP Push was introduced. WAP Push enables a content server to push content to a WAP 1.2-enabled handset by sending a specially formatted XML document to the Push Proxy Gateway, which in turn forwards it to the handset.

Another form of quick-delivery text messaging is instant messaging (IM). Instant messaging from companies such as AOL, Microsoft, and Yahoo! is common on desktop clients, and is quickly moving into the wireless world. One difference between instant messaging and SMS is that IM uses the Internet to communicate. One of the challenges of using wireless IM is that no standard is in place, so IM vendors have their own proprietary interfaces in their IM servers. Though this situation sounds similar to that of SMSCs, with all the challenges that entails, actually it is quite different. With SMSCs you have to integrate with proprietary interfaces, over proprietary networks, whereas with IM, you are communicating over the Internet, meaning you do not have to have a fixed connection to the IM servers. The only level of enterprise integration that is required is communication with the various servers.

NOTE For the purpose of this chapter, we focus on the text-messaging capabilities in thin client applications. Although more advanced messaging with multimedia is possible, it is not commonly used at this time due to the current state of wireless networks and mobile devices. For a more complete look at messaging as it relates to all forms of mobile and wireless computing, see Chapter 5, "Mobile and Wireless Messaging."

Processing a Wireless Request

With an understanding of the various parts of the wireless Internet application architecture, we can now examine what is involved in each stage of a wireless Internet request. Figure 11.5 shows each step that a request goes through as it is processed.

The following is a detailed explanation of the steps of the process illustrated in Figure 11.5 (the numbers here correspond with the numbers shown in the illustration):

1. *Establish a wireless session.* If the device is not already connected to a wireless network, a connection has to be made. Most packet-data networks such as Mobitex and GPRS allow devices to be always connected, so this step may not be required. On networks that are not packet-based, such as GSM, the user will have to be authenticated to establish a connection.

2. *Submit a request.* The process starts with the user submitting a request through the client browser. The server uses this request object to obtain information about the user. It contains the URL of the page being requested, as well as other information about the device and browser being used. In addition, it specifies whether it is a GET or a POST request. This request is often encoded in a binary format to reduce the data sent over the bandwidth-limited wireless network. The data is transmitted wirelessly to the network tower, or base station, at which time it is transmitted over a wireline connection to the wireless gateway.

3. *Translate a request.* The wireless gateway decodes the request object from the binary format to a text format and forwards it to the enterprise wireless application server. This request is converted from the wireless protocol (for example, WAP) to HTTP and transported over a wireline IP-based network connection. In order to convert from the wireless protocol to HTTP, the data has to be decrypted. After the conversion, it can once again be encrypted, most commonly using Secure Sockets Layer (SSL). This decryption and reencryption process is often called the WAP gap. Because of this moment of weakened security, many vendors allow the enterprise to host the wireless gateway on-premises, so that data conversion can occur within the corporate firewall, adding additional security.

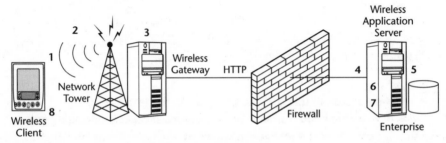

Figure 11.5 Stages of a wireless Internet request.

NOTE The wireless gateway is not always required. For wireless applications using an HTTP-based client, it is possible to go directly from the client browser over the Internet to the wireless application server. WAP 2.x allows for this type of architecture, removing the requirement for a wireless gateway.

4. *Receive a request.* The request is received by the wireless application server. In most cases, the wireless interface to the enterprise application is by HTTP (or HTTPS) using Java servlets/JSPs, ASPs, or other Web interfaces. The wireless application server/Web server can receive the request and perform the appropriate transformation based on the client device/browser combination. The server also performs many other tasks as described earlier in this chapter. If the corporation does not want to maintain this server in-house, many vendors offer hosting services: They host the wireless server and connect to the enterprise only for data access.

5. *Identify the wireless client.* Using the HTTP request object, the wireless server can determine which device and microbrowser is making the request. The request object may also be used to determine other information, such as the image types supported, and preferred language. Using this information, the server can ensure that the appropriate content is sent back to the client. At this point, users may be authenticated to the enterprise system to make sure they have access to the data being requested.

6. *Process the request.* Once the user is authenticated, the server can process the request. The URL will specify the information that is required. If it is a static file, the server will simply return the file to the wireless gateway. If the user requires dynamic content, the data can be personalized based on user information. This will involve accessing one or more enterprise data sources to obtain the dynamic data.

7. *Transform content for device.* At this point, the enterprise information has to be transformed to the appropriate format for the client that made the request. This job may involve changing the markup language, the image types, as well as the richness of the user interface. In most cases this is accomplished on the wireless application server located in the enterprise, although in some situations the wireless gateway performs this translation. Many servers on the market use XML as the base data format and transform it using XSL stylesheets for specific devices and browsers.

8. *Return the content.* Once the content is formatted appropriately, it is sent back to the client. It will once again pass through the wireless gateway where it will be encoded to the format that the browser understands. Any information that the server wants to communicate back to the client browser is contained in the response object.

Wireless Application Protocol (WAP) Overview

The Wireless Application Protocol (WAP) is a worldwide standard for the delivery and presentation of wireless information to mobile phones and other wireless devices. The idea behind WAP is simple: simplify the delivery of Internet content to wireless devices by delivering a comprehensive, Internet-based, wireless specification. The WAP Forum released the first version of WAP in 1998. Since then, it has been widely adopted by wireless phone manufacturers, wireless carriers, and application developers worldwide. Many industry analysts estimate that 90 percent of mobile phones sold over the next few years will be WAP-enabled.

The driving force behind WAP is the WAP Forum component of the Open Mobile Alliance. The WAP Forum was founded in 1997 by Ericsson, Motorola, Nokia, and Openwave Systems (the latter known as Unwired Planet at the time) with the goal of making wireless Internet applications more mainstream by delivering a development specification and framework to accelerate the delivery of wireless applications. Since then, more than 300 corporations have joined the forum, making WAP the de facto standard for wireless Internet applications. In June 2002, the WAP Forum, the Location Interoperability Forum, SyncML Initiative, MMS Interoperability Group, and Wireless Village consolidated under the name Open Mobile Alliance to create a governing body that will be at the center of all mobile application standardization work.

The WAP architecture is composed of various protocols and an XML-based markup language called the Wireless Markup Language (WML), which is the successor to the Handheld Device Markup Language (HDML) as defined by Openwave Systems. WAP 2.x contains a new version of WML, commonly referred to as WML2; it is based on the eXtensible HyperText Markup Language (XHTML), signaling part of WAP's move toward using common Internet specifications such as HTTP and TCP/IP.

In the remainder of this section we will take a look at the WAP programming model and the various components that comprise the WAP architecture. Where it is applicable, we will supply information on both the WAP 1.x and 2.x specifications. (More information on the leading markup languages used in wireless Internet applications is provided in Chapter 13.)

WAP Programming Model

The WAP programming model is very similar to the Internet programming model. It typically uses the *pull* approach for requesting content, meaning the client makes the request for content from the server. However, WAP also supports the ability to *push* content from the server to the client using the Wireless Telephony Application Specification (WTA), which provides the ability to access telephony functions on the client device.

Content can be delivered to a wireless device using WAP in two ways: with or without a WAP gateway. Whether a gateway is used depends on the features required and the version of WAP being implemented. WAP 1.x requires the use of a WAP gateway as an intermediary between the client and the wireless application server, as depicted in Figure 11.6. This gateway is responsible for the following:

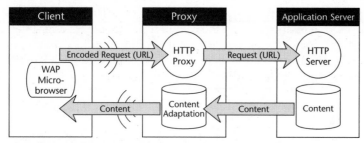

Figure 11.6 WAP Programming model using a wireless gateway (or proxy).

- Translating requests from the WAP protocol to the protocols used over the World Wide Web, such as HTTP and TCP/IP.

- Encoding and decoding regular Web content into compact formats that are more appropriate for wireless communication.

- Allowing use of standard HTTP-based Web servers for the generation and delivery of wireless content. This may involve transforming the content to make it appropriate for wireless consumption.

- Implementing push functionality using WTA.

NOTE The WAP gateway is often called the *WAP proxy* in the WAP 2.x documents available from the OMA. In this chapter we continute to refer to it as the WAP gateway; just be aware that both terms are used to refer to the same technology.

When developing WAP 2.x applications, you no longer are required to use a WAP gateway. WAP 2.x allows HTTP communication between the client and the origin server, so there is no need for conversion. This is not to say, however, that a WAP gateway is not beneficial. Using a WAP gateway will allow you to optimize the communication process and facilitate other wireless service features such as location, privacy, and WAP Push. Figure 11.7 shows the WAP programming model without a WAP gateway: Note that removing it makes the wireless Internet application architecture nearly identical to that used for standard Web applications.

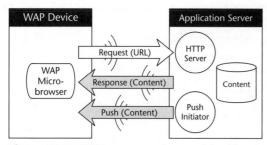

Figure 11.7 WAP programming model without gateway.

Both WAP programming models require the same core set of steps to process a wireless Internet request. These steps are based on the common pull model used for Internet applications; that is, a request/response method for communication. If you are interested in more details about how a wireless request is processed, refer back to the previous section of this chapter entitled *Processing a Wireless Request*.

WAP Components

The WAP architecture comprises several components, each serving a specific function. These components include a wireless application environment, session and transaction support, security, and data transfer. The exact protocols used depend on which version of WAP you are implementing. WAP 2.x is based mainly on common Internet protocols such as HTTP and TCP/IP, while WAP 1.x uses proprietary protocols developed as part of the WAP specification. We will investigate each component and its related function.

To begin, we will look at how WAP conforms to the Open Systems Interconnection (OSI) model as defined by the International Standards Organization (ISO). The OSI model consists of seven distinct layers, six of which are depicted in Figure 11.8 as they relate to the WAP architecture. The physical layer is not shown; it sits below the network layer and defines the physical aspects such as the hardware and the raw bitstream. For each of the other six layers, WAP has a corresponding layer, which will now be described in more depth.

Wireless Application Environment (WAE)

The Wireless Application Environment (WAE) is the application layer of the OSI model. It provides the required elements for interaction between Web applications and wireless clients using a WAP microbrowser. These elements are as follows:

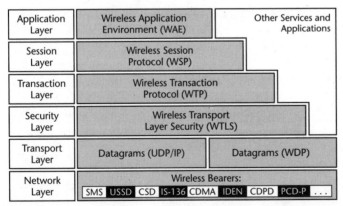

Figure 11.8   WAP architecture and its relationship to the OSI model.

- A specification for a microbrowser that controls the user interface and interprets WML and WMLScript.

- The foundation for the microbrowser in the form of the Wireless Markup Language (WML). WML has been designed to accommodate the unique characteristics of wireless devices, by incorporating a user interface model that is suitable for small form-factor devices that do not have a QWERTY keyboard.

- A complete scripting language called WMLScript that extends the functionality of WML, enabling more capabilities on the client for business and presentation logic.

- Support for other content types such as wireless bitmap images (WBMP), vCard, and vCalendar.

WAP 2.x extends WAE by adding the following elements:

- A new markup language specification called WML2 that is based on XHTML-Basic. Backward compatibility with WML1 has been maintained.

- Support for stylesheets to enhance presentation capabilities. Stylesheet support is based on the Mobile Profile of Cascading Style Sheets (CSS) from the W3C, and supports both inline and external style sheets.

NOTE WAP 2.x WAE has backward compatibility to WML1. This is accomplished either via built-in support for both languages or by translating WML1 into WML2 using eXtensible Stylesheet Language Transformation (XSLT). The method used depends on the implementation by the device manufacturer.

WAP Protocol Stack

The WAP protocol stack has undergone significant change from WAP 1.x to WAP 2.x. The basis for the change is the support for Internet Protocols (IPs) when IP connectivity is supported by the mobile device and network. As with other parts of WAP, the WAP 2.x protocol stack is backward-compatible. Support for the legacy WAP 1.x stack has been maintained for non-IP and low-bandwidth IP networks that can benefit from the optimizations in the WAP 1.x protocol stack.

We will take a look at both WAP 1.x and WAP 2.x, with a focus on the technologies used in each version of the specification.

WAP 1.x

The protocols in the WAP 1.x protocol stack have been optimized for low-bandwidth, high-latency networks, which are prevalent in pre-3G wireless networks. The protocols are as follows:

Wireless Session Protocol (WSP). WSP provides capabilities similar to HTTP/1.1 while incorporating features designed for low-bandwidth, high-latency wireless networks such as long-lived sessions and session suspend/resume. This is particularly important, as it makes it possible to suspend a session while not in use, to free up network resources or preserve battery power. The communication from a WAP gateway to the microbrowser client is over WSP.

Wireless Transaction Protocol (WTP). WTP provides a reliable transport mechanism for the WAP datagram service. It offers similar reliability as Transmission Control Protocol/Internet Protocol (TCP/IP), but it removes characteristics that make TCP/IP unsuitable for wireless communication, such as the extra handshakes and additional information for handling out-of-order packets. Since the communication is directly from a handset to a server, this information is not required. The result is that WTP requires less than half of the number of packets of a standard HTTP-TCP/IP request. In addition, using WTP means that a TCP stack is not required on the wireless device, reducing the processing power and memory required.

Wireless Transport Layer Security (WTLS). WTLS is the wireless version of the Transport Security Layer (TLS), which was formerly known as Secure Sockets Layer (SSL). It provides privacy, data integrity, and authentication between the client and the wireless server. Using WTLS, WAP gateways can automatically provide wireless security for Web applications that use TLS. In addition, like the other wireless protocols, WTLS incorporates features designed for wireless networks, such as datagram support, optimized handshakes, and dynamic key refreshing.

Wireless Datagram Protocol (WDP). WDP is a datagram service that brings a common interface to wireless transportation bearers. It can provide this consistent layer by using a set of adapters designed for specific features of these bearers. It supports CDPD, GSM, CDMA, TDMA, SMS, FLEX (a wireless technology developed by Motorola), and Integrated Digital Enhanced Network (iDEN) protocols.

WAP 2.x

One of the main new features in WAP 2.x is the use of Internet protocols in the WAP protocol stack. This change was precipitated by the rollout of 2.5G and 3G networks that provide IP support directly to wireless devices. To accommodate this change, WAP 2.x has the following new protocol layers:

Wireless Profiled HTTP (WP-HTTP). WP-HTTP is a profile of HTTP designed for the wireless environment. It is fully interoperable with HTTP/1.1 and allows the usage of the HTTP request/response model for interaction between the wireless device and the wireless server.

Transport Layer Security (TLS). WAP 2.0 includes a wireless profile of TLS, which allows secure transactions. The TLS profile includes cipher suites, certificate formats, signing algorithms, and the use of session resume, providing robust wireless security. There is also support for TLS tunneling, providing end-to-end security at the transport level. The support for TLS removes the WAP security gap that was present in WAP 1.x.

Wireless Profiled TCP (WP-TCP). WP-TCP is fully interoperable with standard Internet-based TCP implementations, while being optimized for wireless environments. These optimizations result in lower overhead for the communication stream.

NOTE Wireless devices can support both the WAP 1.x and WAP 2.x protocol stacks. In this scenario, they would need to operate independently of each other, since WAP 2.x provides support for both stacks.

Other WAP 2.x Services

In addition to a new protocol stack, WAP 2.x introduced many other new features and services. These new features expand the capabilities of wireless devices and allow developers to create more useful applications and services. The following is a summary of the features of interest:

WAP Push. WAP Push enables enterprises to initiate the sending of information on the server using a push proxy. This capability was introduced in WAP 1.2, but has been enhanced in WAP 2.x. Applications that require updates based on external information are particularly suited for using WAP Push. Examples include various forms of messaging applications, stock updates, airline departure and arrival updates, and traffic information. Before WAP Push was introduced, the wireless user was required to poll the server for updated information, wasting both time and bandwidth.

User Agent Profile (UAProf). The UAProf enables a server to obtain information about the client making the request. In WAP 2.x, it is based on the Composite Capabilities/Preference Profiles (CC/PP) specification as defined by the W3C. It works by sending information in the request object, allowing wireless servers to adapt the information being sent according to the client device making the request.

External Functionality Interface (EFI). This allows the WAP applications within the WAE to communicate with external applications, enabling other applications to extend the capabilities of WAP applications, similar to plug-ins for desktop browsers.

Wireless Telephony Application (WTA). The WTA allows WAP applications to control various telephony applications, such as making calls, answering calls, putting calls on hold, or forwarding them. It allows WAP WTA-enabled cell phones to have integrated voice and data services.

Persistent storage interface. WAP 2.x introduces a new storage service with a well-defined interface to store data locally on the device. The interface defines ways to organize, access, store, and retrieve data.

Data synchronization. For data synchronization, WAP 2.x has adopted the SyncML solution. As outlined in Chapter 10, "Enterprise Integration through Synchronization," SyncML provides an XML-based protocol for synchronizing data over both WSP and HTTP.

Multimedia Messaging Service (MMS). MMS is the framework for rich-content messaging. Going beyond what is possible for SMS, MMS can be used to transmit multimedia content such as pictures and videos. In addition, it can work with WAP Push and UAProf to send messages adapted specifically for the target client device.

WAP Benefits

The WAP specification is continually changing to meet the growing demands of wireless applications. The majority of wireless carriers and handset manufacturers support WAP and continue to invest in the new capabilities it offers. Over the years WAP has evolved from using proprietary protocols in WAP 1.x to using standard Internet protocols in WAP 2.x, making it more approachable for Web developers. The following are some of the key benefits that WAP provides:

- WAP supports legacy WAP 1.x protocols that encode and optimize content for low-bandwidth, high-latency networks while communicating with the enterprise servers using HTTP.

- WAP supports wireless profiles of Internet protocols for interoperability with Internet applications. This allows WAP clients to communicate with enterprise servers, without requiring a WAP gateway.

- WAP allows end users to access a broad range of content over multiple wireless networks using a common user interface, the WAP browser. Because the WAP specification defines the markup language and microbrowser, users can be assured that wireless content will be suitable for their WAP-enabled device.

- WAP uses XML as the base language for both WML and WML2 (which uses XHTML), making it easy for application developers to learn and build wireless Internet applications. It also makes content transformation easier by incorporating support for XSL stylesheets to transform XML content. Once an application is developed using WML or WML2, any device that is WAP-compliant can access it.

- WAP has support for WTA. This allows applications to communicate with the device and network telephony functions. This permits the development of truly integrated voice and data applications.

- Using UAProf, the information delivered to each device can be highly customized. (Chapter 13 provides more details on how this information can be used to deliver user-specific content.)

- WAP works with all of the main wireless bearers, including CDPD, GSM, CDMA, TDMA, FLEX, and iDEN protocols. This interoperability allows developers to focus on creating their applications, without having to worry about the underlying network that will be used.

At present, all major wireless carriers support the WAP specification. This universal support is expected to continue as WAP evolves, providing a robust, intuitive way to extend Web content to wireless devices.

Summary

Thin client applications do not require any software on the device other than a micro-browser. The architecture is very similar to desktop Web-based applications, with the content generation, business logic, and enterprise integration all located on the server platform. The capabilities of wireless Internet applications are often dictated by the microbrowser that is being used. The features available for an application depend on many factors, including the markup language, graphics support, and display size on the device.

WAP is the de facto standard for building wireless Internet applications. It provides a specification for each layer of the wireless architecture, allowing carriers, device manufacturers, and developers to create applications that will work across many target platforms. In addition to WAP, other technologies can be used to display wireless Internet content, including HTML, cHTHL, and HDML.

In the next chapter we will continue to learn how to build thin client applications; there we investigate the application development process for thin client applications.

Helpful Links

Use these links to find out more information on the topics discussed in this chapter:

J2EE information	http://java.sun.com/j2ee
Java Community Process	www.jcp.org
Microsoft .NET	www.microsoft.com/net
Open Mobile Alliance	www.openmobilealliance.org
International Standards Organization	www.iso.ch

CHAPTER 12

Thin Client Development

Developing wireless Internet enterprise applications requires you to overcome numerous challenges. In addition to penetration and coverage issues currently inherent to wireless networks, you also have to be concerned with the myriad of microbrowsers and markup languages being used, not to mention enterprise data integration issues. Bringing together a complete, effective solution requires a solid understanding of how the application is going to be used, and under what circumstances access to the application is required. All that said, with the proper needs analysis and application design, you can meet these challenges and create very effective wireless Internet applications.

In this chapter we are going to discuss how to begin the development of a wireless Internet application. We will step through the various stages of the development process, starting with the needs analysis phase and finishing with deployment options. As we move through this process, helpful hints will be provided for avoiding pitfalls commonly encountered when developing this type of application. A section at the end of the chapter describes the common thin client application models, and outlines their target audiences, technical challenges, and types of solutions available for each application type.

NOTE This chapter does not discuss the general software development cycle; rather, it focuses on the development process in respect to developing thin client wireless applications.

The Development Process

Developing a wireless Internet application is very similar to and at the same time, very different from, developing a desktop Internet application. It is similar in that they share a common architecture and often use the same technologies for content generation and enterprise integration. It is different in that the client interface and navigation, along with the hardware on which they are being executed, are not at all the same. When you add in the issues introduced by the wireless network, such as low bandwidth, high latency, and sporadic coverage, you end up with an entirely new set of technical and physical challenges that have to be addressed.

Complicating the process is the fact that you are often working with new technology and standards that are not well understood. This is especially true when it comes to the wireless communication protocols, markup languages, device types, and development environments. When creating wireless Internet applications, you will do most of the development and testing on your desktop, often using device emulators to simulate how the application will actually perform when deployed. Once you are comfortable with the simulated application, you will then need to establish the required network connectivity to test it on real devices, over the wireless networks on which you are going to deploy.

All of these steps require careful analysis and planning. In order to create a successful wireless application, you first will need a good understanding of what the application is supposed to do, then design it in such a way that it meets those requirements without being cumbersome for the end user. In this section, we will discuss the stages of the wireless Internet development process, which is depicted in Figure 12.1.

Needs Analysis Phase

The needs analysis phase may be the most important of the entire development process. This is where you determine the goals of the application. Based on the information gathered here, you can then move on to designing and implementing the application itself.

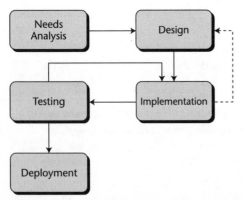

Figure 12.1 Wireless Internet application development cycle.

Questions to Ask When Researching User Requirements

Chapter 4, "Mobile Application Architectures," discussed the two main types of mobile and wireless applications: smart client and thin client. If you have not already determined the type of application required for your application, this is the time to do so. Based on this decision, you will be presented with different architectures, technologies, and development and deployment options. In order to make an informed decision, you will want to answer some key questions, namely:

- Who are the end users of this application?
- Do the target users have specific device requirements, or can the application provider dictate those?
- What is the overall goal of this application?
- What data integration is required? Does the user require data access at all times?
- Does this application require wireless connectivity? If so, what type of wireless access does it require, in which geographies?
- What are the primary usage scenarios for this application?

Finding the answers to these questions, along with others that will emerge, will allow you to determine the type of application that you should develop. Some of the key indicators that will help determine whether a thin client application is required are: the wireless data access, target users, and the typical usage scenarios.

Things to Consider During Needs Analysis

During the needs analysis phase of the development process, we are going to focus on topics related to developing thin client applications. If you are interested in the development process for smart client application, see Chapter 8, "Smart Client Development."

Once we gather information about the following items, we can make an educated decision as to the solution that is most appropriate, including the device type(s), wireless network provider(s), and development platform to use.

Wireless Access

Because thin client applications do not store any data on the client device, and because the content is downloaded dynamically, wireless connectivity is required to use the application. This should be one of the first items you investigate when developing thin client applications. You will need to determine which wireless carriers provide coverage in the area in which the application is being deployed. For example, in urban centers, you will most likely find that you will have several options for wireless access; but when deploying in some of the more remote, rural locations, you may have only one option or, in many cases, no options. In the single-option case, you will have to decide if it is acceptable; and if it is not, a smart client architecture may be more suitable for your application.

In addition to wireless coverage, wireless penetration may also be a concern. Often, in locations within areas that have coverage, the network cannot penetrate the physical surroundings. Examples include confined areas such as subways and office buildings, as well other places such as airports and train stations.

Finally, bandwidth of the network can become an issue. As discussed in Chapter 1, "Welcome to Mobile and Wireless," and Chapter 3, "Wireless Networks," pre-3G wireless networks have very limited bandwidth, restricting the amount of data that can be downloaded in a reasonable amount of time. When designing an application, you will need to keep this in mind, and keep the optional content, such as graphics, to a minimum.

End User

Who is the target audience for this application? Is the application aimed at the consumer market, where you will have limited control over the devices and wireless networks being used; or is it a corporate application, where the devices and connectivity are determined and provisioned by a central source? The target audience of the application will influence many decisions that you have to make. If you have some level of control over the technology the end user will be utilizing, it will make the development and testing process much easier. Trying to build a solution that will work on all of the device/network combinations is a daunting task, although there are many commercial solutions available to make this job easier, as covered in Chapter 14, "Wireless Internet Technology and Vendors."

If you are building an application for an audience for which you can influence the device being used, you will want to consider the users' level of technical background and comfort level with wireless applications. If the audience is not comfortable with, say, pen-based user input, you will want to target devices that have QWERTY keyboards or other data input mechanisms. Similarly, the geographies in which this application is going to be used will dictate which wireless network provider is best suited for proper coverage.

As stated in the introduction in Chapter 1, this book is primarily aimed at those developing m-business corporate applications, although much of the information also directly applies to the consumer market. In either case, keeping the requirements of the end user in mind will allow your applications to reach their full potential.

Application Goals

Determining the goal of the application will provide you with information on where you should focus your efforts. Most successful wireless Internet applications have a very specific goal. They are not often used for generic Web-surfing or information-gathering purposes. Determining what the desired functionality is, and producing an application that meets it, are important objectives. If the goal is to provide timely financial information, for example, the application should be designed to enable the user to access this information quickly and efficiently, without having to navigate through multiple pages. At the same time, if the application is designed for remote information gathering, a device should be chosen that allows for rapid data entry. In this scenario, a cellular phone most likely would not be adequate, whereas a Pocket PC or Palm device might be.

Beyond the user interface, other factors may come into play. The data being requested may reside in a variety of enterprise systems. At this point, the means of enterprise integration and security concerns will be of paramount importance. For each application being developed, you should have a clear set of goals against which the success of the application can be judged.

Usage Scenarios

Understanding how and when the application will be used will have a substantial impact on the device chosen and the application design. If access to the application and the data it provides is crucial to job performance, as in the case of a field service worker, then researching the options for wireless connectivity and running a pilot of the application are recommended. If the application is going to be used in an environment where it may get knocked around a lot, or where it can quickly get dirty, it may make sense to choose a ruggedized device with a suitable data input mechanism. In this case, a tablet-based device that does not have small keys would work well. Similar consideration should be given to the display of the data. If large amounts of data are going to be retrieved and displayed, a device with a larger form factor, such as a PDA or Tablet PC, will most likely be better than a WAP phone.

Apply the same type of logic to the level of wireless connectivity required. If the application is going to be used in a fixed location, implementing a wireless LAN solution would be more manageable and cost-effective than relying on a wide area wireless network.

Content Source

The primary goal of most wireless applications is to extend the reach of enterprise information to remote workers. For wireless Internet applications, this enterprise information is often in the form of an existing Web application. It is unrealistic to expect to be able to display an entire Web application on a wireless device; instead, the most useful parts of the application must be utilized. This is not a trivial task. Internet applications aimed at desktop users typically use HTML, as well as client scripting languages such as VBScript or JavaScript. Because most wireless microbrowsers do not support these languages, they have to be transformed to the appropriate language for the wireless client.

Existing Web content is not the only data being used for thin client applications. For many corporate applications, the wireless Internet application is developed with direct integration to the same business logic used by the Web application. By integrating at this level, the content for the wireless application can be customized specifically for wireless users, thereby avoiding the challenges of transforming desktop applications. Experience has shown that this type of development—that is, creating the wireless applications in parallel with the desktop applications—leads to a better experience for the wireless user and fewer headaches for developers.

Design Phase

Once you have completed the needs analysis, you will have a clear idea of the requirements of the ultimate solution. This does not mean that you will know exactly which device or network will meet your needs, but at least you will know what you are looking for. Ideally, you will have limited the choices to a few wireless devices, wireless networks, and markup languages that will meet your criteria. Predicting which technology will be best for your application will be very difficult if you have not had any experience using it. So at this point, you should keep an open mind when making the final decision as to which technology to implement. Typically, a developer may try a couple of possible solutions before feeling comfortable making a final decision. This approach makes sense. If you do not have experience developing a WML-based WAP application, for example, you will have difficulty knowing whether it is better suited than an HTML-based application. (To learn more about the leading markup languages used in wireless Internet applications, see Chapter 13, "Wireless Languages and Content Generation Technologies." There you'll find an overview of the leading wireless markup languages, as well as the content delivery technologies that are available.)

During the design phase of the development cycle, you have a chance to design each layer of your wireless application. You can focus on the issues that are often present in building these applications, such as creating an effective user interface for targeting multiple devices and microbrowsers, developing the server-side business logic, and integrating with enterprise data sources.

User Interface

The user interface is the most scrutinized part of a wireless application. Comments such as "WAP is dead" have led people to believe that the technology behind the wireless Internet is not sufficient for creating effective wireless applications. This viewpoint does not reflect reality. In most cases, wireless Internet applications have not succeeded as expected because of application design. True, limited devices capabilities, low bandwidth, and high latency of wireless networks have contributed to the slow adoption of wireless Internet applications, but even if these aspects were improved, the wireless Internet might not thrive if certain fundamental technical design issues are not addressed.

Latency

Anyone who has used a wireless Internet application has been frustrated by the wait time involved with downloading applications. Very often, this delay is incorrectly blamed on the low bandwidth of wireless networks; in fact, most often, network and application latency causes the delay. The amount of content it takes to fill the screen of most wireless devices is quite small, meaning that even slow wireless networks, with download speeds hovering around 14.4 Kbps can get the content within a few seconds. This amount of time seems reasonable. Unfortunately, the actual time that users wait for a given Web page may end up being in the tens of seconds, if not minutes, because of network latency.

In order to combat this delay, your application design will have to minimize round-trip interactions with the enterprise server. Each request ends up traveling through the public Internet, hopping through random routes, before it reaches its destination. This adds latency. So keeping the number of requests low allows the user to interact with the application without having to constantly wait for new data to download.

Human Interaction

The best wireless thin client applications have very few screens to navigate, and can be operated with one hand. This ease of use allows users to operate the application while on the move. While one-handed operation is possible on WAP phones, it is usually not possible on PDAs, since the input is typically pen-based. In either case, one of the main design goals of your wireless application should be simplicity, especially for the user interface. Keeping the focus of the application on the content, rather than on images and graphics, usually helps to accomplish this goal. Also, make it easy for a new user to navigate through the application with minimal effort. In most cases, the user is using the application to solve a problem or accomplish a goal, not to browse. This is especially true for WAP applications. When the screen only has six lines of display, the human interaction with the application has to be streamlined. The point is to keep the application as simple as possible, allowing the user to perform the desired task without having to navigate through unnecessary pages or enter large amounts of text. If this goal can be attained, user satisfaction, hence adoption of your application, will increase dramatically.

When working with a device for which you have not created applications in the past, running a pilot program early in the development cycle can provide some invaluable insight into the usability of the application. And you will gain that knowledge at a stage when it can still be incorporated into the application design without too much additional cost.

NOTE If you are interested in learning more about the usability and design guidelines for WAP applications, visit some of the Web sites listed in the *Helpful Links* section at the end of this chapter. Many of the device and browser vendors, as well as the Open Mobile Alliance, provide excellent information on this topic.

Device Independence

If you are deploying an application aimed at a broad audience, where you are looking for widespread adoption, supporting a large number of devices is a definite plus. This is especially true for applications that may be used in various modes by the same user. For example, if an airline deploys an application that lets users check flight arrival and departure schedules, it will want to make sure that people can access this data from the full range of wireless devices as well as desktop browsers. In this way, customers can check their flight times before they leave home, and again on their way to the airport, each time from a different device, using a different network for connectivity.

For corporate applications, such device independence may not be required. In this scenario, you may only be deploying to one or two targeted devices; nevertheless, you should design your application so that it can work with other devices that may be deemed necessary in the future. With wireless technology evolving at a rapid pace, the device of choice in the future could be much different from what you choose today.

Personalization

One way to minimize the content being downloaded, while still generating useful information, is to personalize the data. Personalization levels can be set at an individual user level or at a category of user level. In either case you end up with more specialized content that is of interest to the current application user. When a person logs in to the application, everything that is displayed to him or her from that point on has been predetermined to be of interest to him or her. All of the content can be generated dynamically, based on user specifications. Ideally, there will be a personalization application that can be accessed using a desktop PC application, allowing the user to set up specific requirements.

Other forms of personalization happen at the application level. Rather than creating one complex application for a variety of users with different roles, you can create separate, targeted applications for each function that needs to be addressed. In this way, you will be able to keep the user interface simple and minimize the number of screens the application requires.

Another application type that works very well with personalized content is messaging applications. A messaging application can provide a one-way stream of information, notifying the user when a particular event occurs. Using the flight schedule example, if a user is scheduled on a specific airline flight, he or she could be notified if the schedule changes, rather than having to navigate through an application to find the same information. Based on the notification, you can send relevant information directly to the user, such as a screen to rebook the flight, if necessary.

Enterprise Integration

Enterprise integration is a term used to describe any communication to systems that are not on the device. It encapsulates integration with enterprise databases, business applications, XML data, Web content, and legacy data, among other things. For the purpose of designing your wireless solution, you will need to determine to which enterprise systems you require access. All of the logic to integrate with these enterprise systems will be located on the server. The client device contains no business logic and, therefore, no enterprise integration capabilities.

Because wireless Internet applications are based on a distributed architecture, similar to desktop Internet applications, the enterprise integration layer should be very similar as well. At this phase in the development cycle, you will want to consider which systems need to be communicated with and how that communication is going to take place. For example, you may have an enterprise relational database and a corporate business application, such as SAP, to which you require access. You will then have to determine how that access will be accomplished. In the case of the relational database, you may want to use a native driver, ODBC or JDBC. When it comes to SAP integration, you have the choice to use a commercial adapter for SAP or to create your own adapter based on the APIs that SAP exposes. These decisions may not directly influence the wireless client, but they will affect the overall performance and capabilities of the application.

Once you have access to the data source in question, you are then required to adapt that data to a format appropriate for the wireless microbrowser. This adaptation can be accomplished in a number of ways, all of which are discussed in Chapter 13.

One other form of enterprise integration common for Web applications is Web services. This integration is accomplished using a defined XML interface, allowing you to incorporate a wide variety functionality and information into your application. The logic for Web services resides on a server, which can either be hosted in-house or be supplied by third-party vendors. (A summary of Web services and their role in mobile computing is provided in Chapter 18, "Other Useful Technologies.")

Business Logic

The business logic for wireless Internet applications is executed on the enterprise server (which is often a wireless application server). This logic is responsible for many wireless-specific features such as content generation, session management, and messaging. Each of these features is implemented with wireless applications in mind; however, not all of the logic has to be developed that way. If you have general logic that is responsible for data manipulation, enterprise integration, or processing other specific business rules, it should be shared across all applications within the enterprise. Avoid duplicating business logic, if at all possible.

In many corporations, separate teams are responsible for wireless Internet applications and for desktop Internet applications. This separation of responsibility often leads to duplication of efforts. Instead, because wireless Internet applications share an architecture similar to desktop Internet applications, it is more effective to design them to share the same business logic and enterprise integration modules.

Implementation and Testing Phase

It is during the implementation and testing phase of the development process when developers start to get excited, because this is when they start to see tangible progress. This is not to say that the needs analysis and design phases are not valuable; in fact, the implementation of the application will go much more smoothly if enough effort is given to the application design. Likewise, user-satisfaction levels will be much higher if the needs analysis phase is completed properly.

As most software developers know, creating an enterprise application does not happen in one step. Even at the implementation phase of the development process, many incremental steps have to be executed, and often repeated, before an application is complete. This is true for desktop Internet applications, and is definitely true for wireless Internet applications. The first step is often the development of a prototype. This allows the developers to obtain feedback on the application design and features early in the development cycle, when there is still an opportunity to make changes to the application. If you wait until the application is nearly finished before field testing, required changes will be very time-consuming, leading to additional costs and a later release date, not to mention frustration for the developers.

HELPFUL TIP The wireless client for Web content does not have to be developed separately from the desktop client. Designing and developing these applications together often results in a better-integrated solution that shares business logic and enterprise integration logic, reducing complexity and cost. Whenever possible, think about the wireless aspect of an Internet application at the design phase.

Prototypes

Developing a prototype of your application will help you determine if the decisions you made during the design phase of your application are accurate. By field testing the prototype, you will acquire valuable information early on, saving you headaches and, more importantly, additional costs later in the development cycle.

During the prototype phase, you do not have to perform all of the testing using a real device on a wireless network. Much of the user interface, business logic, and enterprise integration can be tested using emulators. Only when it comes to determining the wireless coverage and penetration of a given network, or when evaluating the application's responsiveness and performance, do you actually require wireless connectivity.

The following is a list of questions you will want to answer during the prototype stage, before moving on to the final implementation. Some you will answer by using emulators; others will require the use of physical devices on wireless networks:

- Which device is most suitable for your application? Take into consideration the means of wireless access, the microbrowsers available, the device form factor, means of data input, and, of course, cost.

- Which wireless network is appropriate for your application, and does it perform as expected? Look at coverage for the geographies in which your application will be used. You may also want to factor in the time and effort required to get a connection to the carrier being evaluated.

- Does the user interface provide the most efficient way for the user to operate the application? Does it match the device characteristics?

- Does the application work properly on the device/network/browser combinations that may be used? Is there an optimal environment that should be recommended for most users?

- Is the appropriate data available in the application? Is it hard to navigate to the core functionality that the application provides?

- Does the enterprise integration layer work? Is it scalable to meet the needs for your application?

- Have security concerns been addressed? Are there holes where corporate data is left unprotected? Depending on your security requirements, an on-premise gateway solution may be required.

- Does the application provide an upgrade path for new features? Will it be adaptable for new wireless networks as they arrive?

The preceding questions are some of the major ones for which you will want to have answers during the prototype and initial testing stages of the application. Unfortunately, prototyping is often omitted from the software development cycle to save time and money. In reality, neither savings is achieved. The information gathered during this prototype phase will allow you to make modifications to your application before it is too late.

Keep in mind that a prototype does not have to be an entire application. You can simply put together a couple of screens that will represent the larger application, and execute a sample of the data source integration. Even this level of effort will result in huge dividends when you move on to the development and deployment of your application.

Development Tools and Emulators

Choosing a development tool for your wireless Internet applications is not an easy task. As you will see in the remainder of this section, there are many development tools available, all with various strengths and weaknesses. It is worthwhile to take the time to find a tool that fits your development needs, as well as personal preferences. You will probably spend a considerable amount of time using it during development, so you should make sure you like it.

Until recently, most development tools for wireless devices were provided by device or browser manufacturers, often free of charge as part of a software development kit (SDK). This situation is changing as leading Internet development tool vendors continue to add support for wireless markup languages and emulation environments to their offerings. Usually, vendors charge a licensing fee to use these tools, but the expense is often worthwhile, because of productivity gains achieved from the cross-device and browser capabilities of these tools. A third choice for development tools comes from wireless platform vendors. These vendors typically provide a tool that is tightly integrated with their platform, often taking advantage of proprietary extensions that have been developed within their frameworks. In most cases, you will be interested in using the tool only if you are using the other parts of the vendor's platform as well. For this reason, this discussion does not address tools from wireless platform vendors. Chapter 14, "Wireless Internet Technology and Vendors," covers this topic along with vendors that provide wireless software infrastructure.

In addition to the development environments, the tools also provide emulators to simulate how the application will look and feel without having to deploy it to a physical device. Simulation is often necessary, because, in all likelihood, you will not have every device/network combination available to you for testing. Even if you do, testing the application of every possible deployment combination is not practical or cost-effective.

The following sections provide information on the leading wireless device types and their corresponding development tools.

NOTE If you are interested in content-generation techniques, such as using Java servlets, JSPs, or ASPs, you can find information on these technologies in Chapter 13.

Wireless Application Protocol (WAP)

There are more development tools and SDKs for WAP than for any other platform. All of the leading device manufacturers such as Nokia, Ericsson, and Motorola, provide development kits, as does Openwave Systems, which develops WAP browsers and gateways. The nice thing about these tools is that they provide emulation environments for their devices. Just make sure you test using the emulator for each device you are targeting, because the WAP implementation varies from device to device, and for browser vendors, from version to version of their products.

Table 12.1 lists the major WAP infrastructure providers, along with a URL where you can obtain the related development kit. In most cases, you will find that you can download the SDK free of charge.

Figure 12.2 depicts the Openwave 4.1 and 5.0 emulators. It is a good idea to use the emulator that most closely resembles the physical device characteristics you are deploying on. For most of the current Web-enabled phones, the version 4.1 emulator is suitable.

Table 12.1 WAP SDKs

COMPANY	URL FOR DOWNLOAD	DESCRIPTION
Ericsson	www.ericsson.com /mobilityworld	The WapIDE 3.2.1 Software Development Kit enables you to develop WAP applications and test how they behave on different Ericsson phones.
Motorola	http://developers .motorola.com /developers	The Mobile Application Development Kit (ADK) provides an environment for writing, editing, and testing WAP applications for Motorola phones. It is also the legacy toolkit for development of VoiceXML/VoxML applications.
Nokia	www.nokia.com/wap/ development.html	The Nokia Mobile Internet Toolkit provides developers with a PC-based testing and simulation environment in which they can create WAP and other mobile Internet applications, including those based on the eXtensible Hypertext Markup Language (XHTML), Cascading Style Sheets (CSS), and Multimedia Message Service (MMS).

Table 12.1   *(Continued)*

COMPANY	URL FOR DOWNLOAD	DESCRIPTION
Openwave Systems	www.openwave.com /products/developer _products/sdk/	The Openwave SDK provides an integrated development environment and Openwave Mobile Browser emulator to enable developers to test their mobile applications on a desktop PC. This is one of the most commonly used emulators for WAP development.

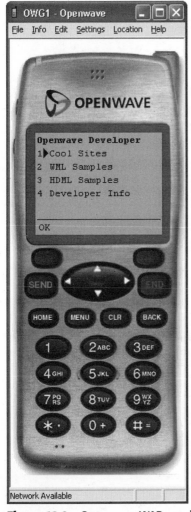
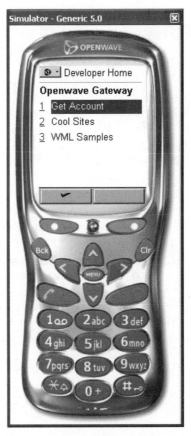

Figure 12.2   Openwave WAP emulators.

Table 12.2 Microsoft Windows CE Development Kits

URL FOR DOWNLOAD	DESCRIPTION
www.microsoft.com/mobile /developer/default.asp	Information on both Pocket PC and the Microsoft Smartphone 2002 is available on this Web site. You can also download development and emulation environments that allow you to test and debug your applications without using a physical device.
msdn.microsoft.com/vstudio /device/mobilecontrols/default.asp	ASP.NET mobile controls consists of a set of server-side Web form controls that allow a wireless developer to author cross-device user interfaces. These controls extend the functionality of ASP.NET. Microsoft has integrated ASP.NET mobile controls directly into the Visual Studio .NET environment.

Windows CE

Microsoft provides the leading microbrowser and development tools for Windows CE-based devices, including both Pocket PC and Microsoft Smartphone 2002. One nice feature of these devices is that they use Pocket Internet Explorer (PocketIE), which is a very capable Web browser that supports HTML, WML, cHTML, and client-side scripting. As these devices grow in popularity, expect to see more development platforms that target these devices.

The development and emulation environment is free to download from Microsoft's developer site. In addition, Microsoft supplies ASP.NET mobile controls (formerly the Mobile Internet Toolkit), which allows for cross-platform wireless development from a single environment. Table 12.2 lists the URLs where you can obtain more information on these products.

Palm OS

A variety of browsers and development kits are available for the Palm OS, provided primarily by the leading device manufacturers, including Palm and Handspring. In addition, browsers are provided by wireless Internet service providers (WISPs), including GoAmerica, Neomar, and AvantGO. More information on these browsers is provided in Chapter 14, where we discuss wireless technology and infrastructure vendors.

The standard browser provided by Palm is based on the Web-clipping model, using HTML as the markup language. WML browsers, as well as non-Web-clipping HTML browsers are also available. See Table 12.3 for a listing of development kits and microbrowsers available for the Palm OS.

Many current Palm OS-based devices have integrated wireless modems, or make it possible to add one using an expansion slot. These developments have made Palm devices more suitable for wireless Internet applications than they have been in the past, when wireless connectivity was cumbersome to obtain.

Symbian OS

The best place to get technical information for the Symbian OS is on the Symbian Developers Network, as listed in Table 12.4. Symbian is a joint venture between Ericsson, Nokia, Motorola, and Psion, so as you would expect, you can also find information regarding Symbian OS on these manufacturers' developer sites as well.

Table 12.3   Palm OS Development Kits

COMPANY	DOWNLOAD URL	DESCRIPTION
GoAmerica	www.goamerica.com /partners/developers/	The Developer Zone gives you the tools you need to create a wireless Web-based application that can be accessed via the Go.Web browser.
Handspring	www.handspring.com /developers/	The Development Kit for Handspring handheld computers is a collection of development software, documentation, and associated files.
Neomar	www.neomar.com /developers/index.html	Neomar enables the extension of applications and services to wireless intelligent devices such as RIM, Palm, and Pocket PC via open wireless standards (currently, WAP).
Palm	www.palmos.com/dev /tech/webclipping/	The Web-clipping architecture includes client-side applications that run on a Palm OS-powered handheld, proxy servers for handling translation between the Web-clipping application format and HTML, and content servers. The client-side application is called a Web-clipping application (WCA). It is constructed in HTML and translated into the Web-clipping application format.

Table 12.4 Symbian OS Development Kits

COMPANY	DOWNLOAD URL	DESCRIPTION
Symbian	www.symbian.com /developer/index.html	Complete SDKs for developing Symbian OS-based applications, including wireless applications using C++, Java, and WAP.
Ericsson	www.ericsson.com /mobilityworld	See Table 12.1.
Motorola	http://developers .motorola.com/developers	See Table 12.1.
Nokia	www.nokia.com/wap	See Table 12.1.

Research In Motion (RIM) BlackBerry

The BlackBerry devices from Research in Motion are primarily aimed at providing wireless access to corporate email; and because they have an integrated modem, decent display capabilities, and an effective data input mechanism, they are used for wireless Internet applications as well. The most common microbrowsers for the RIM devices are provided by wireless ISPs that supply wireless connectivity for RIM users. These browsers typically use WML as the markup language, making the application development similar to WAP applications.

Table 12.5 RIM Blackberry OS Development Kits

COMPANY	DOWNLOAD URL	DESCRIPTION
GoAmerica	www.goamerica.com /partners/developers/	See Table 12.3.
Neomar	www.neomar.com /developers/index.html	See Table 12.3.
Research In Motion (RIM)	www.blackberry.net /developers/index.shtml	Applications can be created for RIM handhelds using the free SDKs available on the RIM developer site.

General Web Tools

In addition to all of the device- and browser-specific SDKs, many of the leading Web development tools have added support for wireless application development. Very often, this support is accomplished using XSL stylesheets for transforming XML data into markup-language-specific content such as WML, cHTML, VoiceXML, or HTML, or by using XHTML. While having the robust capabilities of these tools is definitely a plus when creating wireless applications, we are not going to focus on these vendors since wireless is only a small part of their overall offerings.

That said, if you are planning on creating a Web solution in conjunction with your wireless offering, these tools may be exactly what you are looking for. Table 12.6 gives you the names of the leading vendors, along with their URLs, where you can find more information.

Table 12.6 General Development Tools for Wireless Internet Development

COMPANY	DOWNLOAD URL	PRODUCT AND DESCRIPTION
Macromedia	www.macromedia.com /software/dreamweaver/	*Dreamweaver*: A complete, easy-to-use environment for creating very sophisticated Web applications. Many wireless platform vendors have plug-ins for Dreamweaver for their wireless Internet development.
Macromedia	www.macromedia.com /software/homesite/	*HomeSite*: A tool geared for Web developers who prefer hand-coding their applications. Homesite provides tools for creating fast, effective Web sites, as well as the ability to manage the development process. It also has support for XHTML, which is the base language of WAP 2.0.
Microsoft	www.microsoft.com /frontpage/default.htm	*FrontPage*: A general-purpose Web development tool with support for XML, which allows you to create content geared for wireless access.

Physical Devices

Moving from an emulation environment to a deployment environment introduces two new variables to your application: the physical device and the wireless network. This is different from smart client applications, which can be tested on a physical device without requiring wireless connectivity. Since all of the business logic and content-generation technology resides on the server, a wireless connection is required to enable application access. Let us take a look at each new variable that real devices introduce.

First we have the device itself. You will need to obtain the actual devices that you plan on using during deployment. You will also want to make sure you test the application with the browser that you are planning on using. This is important because each microbrowser might display the content differently, and each device will have different display and navigation features. Some will have 12-digit numeric keypads; others will have alphanumeric keyboards; while still others will have pen-based input. If you are deploying your application for an audience that may use any number of devices, it is unlikely that you will have each of these devices for testing. As an alternative, try to get a good representation of the target devices and browsers, and test on those. For example, you might want to have one Palm device, one Pocket PC device, and one WAP device to test the major form factors and markup languages. If you do not have these devices available to you, at least test on the most widely used devices, since that will satisfy the majority of your target audience.

Testing with a physical device satisfies only half of the testing sequence. The wireless network is the other half. Wireless connectivity can introduce a lot of unknowns into the overall usability of your application. As discussed in Chapter 3, many different wireless networks may be used for deployment, including wireless wide area networks and wireless local area networks. Depending on the deployment requirements, you may need to test your application on one or both types of networks. Wireless LANs provide high-speed connectivity in a controlled environment. Public domain wireless networks offer coverage over a wide range of geographies, with less predictability. They may also introduce performance issues, due to the high latency and low bandwidth, and accessibility issues, due to coverage and penetration. Testing on actual devices over the targeted wireless networks is the only way to get an accurate assessment of the application's performance and usability.

Deployment Phase

Over the past several years, many corporations have moved away from traditional client/server architectures to distributed Web architectures. Many factors have influenced this migration, but the one that stands out is the ease of deploying Internet applications. With client/server applications, such as smart client mobile applications, a client component has to be deployed to each device that is going to use the application. Even with the software management and deployment solutions available, this still is a time-consuming and high-maintenance task. All of this can be avoided by moving to the Internet model, where only a Web browser is required on the client. All of the data storage and business logic is kept on the server, and is accessed in real time over a network. This same model applies to wireless Internet applications, although some additional work is required.

The first step in deploying the wireless Internet application is to set up the deployment environment. This can be accomplished in one of two ways: on-premise or hosted. Most companies run the required servers on-premise, so we will look at that scenario first. "On-premise" means that the entire suite of server-side software is going to be maintained by the corporation (as opposed to a third party), within the corporate firewalls. This includes the wireless application servers, Web servers, and enterprise data sources, as well as the hardware it will run on. For many corporations, the deployment configuration will mirror the configuration that was used for testing. For large-scale deployments, or those that are mission-critical, staging servers are set up, so that application developers can test the application in an environment with the same characteristics as the one being used for deployment. Once they are satisfied with the quality of the application, it is moved to the deployment servers.

If the effort required to set up and maintain the required enterprise servers is too large or too costly, or if a decision has been made to outsource this process, a hosted solution may be the answer. In a hosted setup, a third-party organization can host your application server and/or enterprise data sources for you. These companies usually charge a monthly fee for this service, which is dependent on the number of servers required and connections to the outside world.

In addition to setting up the application itself, wireless connectivity has to be addressed. As discussed in Chapter 11, "Thin Client Overview," this may require a wireless gateway to provide access to the wireless carrier. Fortunately, the wireless carrier provides this gateway service, converting the wireless protocol being used to HTTP, which is directly supported by wireless application servers. The only thing required of the enterprise is to have a listener on a port using HTTP or HTTPS, and make sure that port is open in your corporate firewall.

In some cases, when multiple wireless networks are being used, or when additional wireless optimization or security is required, a wireless ISP is used. Wireless ISPs can provide you with access to multiple wireless carriers from a single location, called a network operations center (NOC). The NOC then communicates with your enterprise over a secure HTTP connection. The benefit of using a wireless ISP is that you can communicate over multiple wireless networks and have all of the technical support, provisioning, and billing provided by a single company. In most cases, wireless ISPs also provide you with browser software for the devices they support, often with optimized wireless communication capabilities. (Information on the companies that provide this service can be found in Chapter 14.)

NOTE In some cases, for security reasons, you will also have the option to install the wireless gateway on-premise, within your corporate firewalls. This allows any decryption and reencryption to happen behind the firewall, protecting you against the WAP gap. More information on this issue and other related issues is provided in Chapter 6, "Mobile and Wireless Security."

At this point, most of the tasks associated with deployment are taken care of. Once your servers are all configured, and the wireless network access is initiated, you simply let your users know the URL of your wireless Internet application. They then enter this URL into their microbrowser to gain access to your application.

Thin Client Application Models

Thin client wireless applications can be created using a number of different application models. One is based on existing Web content that is transformed for different client devices that require access. The other involves creating new wireless applications from the ground up. Both of these solutions are applicable to various situations.

Extending Existing Web-Based Applications

Many corporations use both internal and external Web sites to disseminate information and to allow users to interact with pertinent data. These Web sites are usually developed using HTML in conjunction with client scripting languages such as JavaScript or VBScript. They have rich graphics, multimedia, and various document formats, all of which makes them unsuitable for consumption by wireless clients which have much more limited capabilities. This creates a need for a way to transform these Web sites into a format appropriate for wireless devices.

Most wireless devices use HDML, WML, or a scaled-down version of HTML. They are unable to handle regular HTML and rarely have support for advanced layouts or graphics. So to display the information contained on existing HTML-based Web sites, essentially two options are available:

- Rewrite the Web site in the appropriate markup languages for the wireless devices, such as WML, HDML, or cHTML
- Use technology to transform the existing Web content into content appropriate for wireless devices

Which option is best suited for your situation depends on the complexity of the existing Web site, what functionality is required on the wireless device, and your Web architecture. If your goal is to make the entire Web site available to wireless users, as may be required in a wireless portal, rewriting the entire Web site is not practical. Why not? Because once you start re-creating the content in new formats, you also create a divide between the main Web site content and the wireless content. That means that every time the content needs to be updated, work will have to be done on both the wireline and the wireless clients. This is clearly unacceptable.

In this scenario, using transcoding technology is probably the path you want to take. This technology gives you the ability to extract existing Web content and transform it into different markup languages appropriate for wireless devices. This goal can be accomplished in many ways, depending on the vendor providing the solution. The following are some of the technologies that may be involved in this type of solution:

- HTML parsing
- HTML reduction to suitable formats for wireless devices
- HTML reformatting to WML, HDML, XHTML, VoiceXML, or other markup languages
- HTML-to-XML conversion

- XML-to-other XML variants
- Image reduction and reformatting
- Business document reformatting

As you can see, many different types of conversion are required to automatically convert existing Web pages to formats appropriate for wireless devices. Furthermore, due to the complexity of existing Web applications and the limitations of wireless Internet applications, this type of automatic transcoding rarely ends up meeting expectations. As you can imagine, reworking a Web site designed for desktop users into one fit for viewing on a wireless device is not easy. Even if you manage to reformat the content, the navigation is vastly different on wireless devices, adding new complexity to the situation. Figures 12.3 and 12.4 illustrate the differences between desktop, Pocket PC, and WAP devices.

To help solve this problem, some vendors provide more of a programmatic way of extending existing Web applications. This approach allows developers to interact with various content sources, as well as the existing business logic, to create a wireless client for the existing Web application. In this scenario, the goal is usually to re-create only parts of the Web site that provide true value to the wireless user, as opposed to re-creating the entire Web site for wireless access. Rewriting only some sections of the Web site for wireless users is not as problematic as rewriting the entire site. Many of the technologies listed earlier are also required for this type of transcoding; the difference is that you have more control over which sections of the Web application are used and how they are transformed.

Figure 12.3 Web site viewed using Internet Explorer.

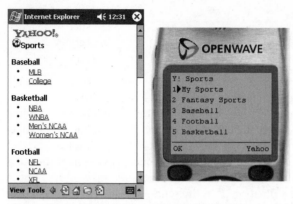

Figure 12.4   Web site viewed with PocketIE and a WAP browser.

Of course, for both of these solutions, the wireless device being targeted plays a significant role in the level of transformation required. WAP devices usually have very limited screen sizes, therefore dramatic changes are required to make the content useful. On the other hand, Pocket PC and Palm devices often have HTML browsers and can handle more complex layouts and graphics. Figure 12.3 shows the Yahoo! sports Web site on desktop Internet Explorer. Figure 12.4 shows what the same Web site looks like on WAP and Pocket PC devices after it has been transformed. You can see that the WAP device requires major design and navigation changes to the Web site, while Pocket PC is somewhere in between. (In Chapter 14, the features offered by the leading wireless application server and transcoding technology vendors are covered in more depth.)

Creating New m-Business Applications

For many corporate wireless applications, simply extending an existing Web site is not the optimal solution. Instead, new applications are created with specific goals in mind. These applications usually provide access to information, allowing users to have real-time access to critical data, so that they can make better-informed decisions. These applications can also allow the user to insert new data or update the data presented. Creating this application from the ground up gives you flexibility that is not possible when extending an existing Web site. You have the opportunity to direct the design of your application to meet the needs of wireless users—of course keeping the screen size and bandwidth constraints in mind. The most successful wireless applications are simple and to the point, so that the user can get at the application's core value with minimum effort. In order to accomplish this goal, key design and architecture decisions should be made at an early phase of development.

In Chapter 11, we covered the architectural layers in a wireless Internet application. Recall that this included the client presentation layer, the middleware business logic layer, and the data integration layer. When creating new applications from the ground up, you will have to address the requirements at each layer to build a successful solution. This is your chance to go through the steps covered earlier in this chapter to create a well-thought-out solution that will meet the needs of the target audience. The following are the main layers that you will have to develop for your new application:

Support for multiple clients. You will most likely have to support client devices using different browsers and requiring multiple markup languages. Ideally, this will involve creating the presentation content once and then adapting it for each client device. This may involve some level of transcoding capability. Many techniques are available for creating this content; they are discussed in depth in Chapter 13.

Reusable business logic. Because you may potentially have several wireless clients, and possibly traditional desktop clients, you will want to encapsulate your business logic in reusable components. This will save you from having multiple sets of the same logic. Some of the common object models being used are Enterprise JavaBeans (EJBs), COM or .NET-Managed objects from Microsoft, or just plain Java or C++ classes. Most wireless application server vendors will support at least one of these component models for creating your business logic.

Data Integration. At some point, your application will be required to access a form of enterprise data. This may be a relational database, business applications, or Web services. In any case, you will be required to have an enterprise integration layer that provides access to these data sources.

When you start down the path of building this type of application, choosing a wireless platform vendor can make the task easier. In most cases, the vendor will provide you with tools, frameworks, and data connectivity components that remove the low-level complexities, allowing you to focus on developing the application itself. For client development, such tools and features may include device-detection capabilities, session management, and content repurposing. The vendor will usually also provide an engine for executing your business logic, and possibly tools to help create it as well. And for enterprise integration, you can minimally expect the ability to communicate with relational databases using ODBC or JDBC, as well as some XML integration support. When you add all of this together, the added capabilities and support will get you well on your way in the development of your application. When putting together a new application for your wireless users, you would be wise to consider the other client types for which, in the future, may require access to the same business logic and enterprise data. Chances are that desktop Web access will be required not long after the wireless interface is deployed, simply because users are not always going to want to access this application from a wireless device. At times they will be using a desktop Web browser, from which they will prefer to access the application for the simple reasons that, by doing so, they will save communication costs and most likely have a more productive, higher-speed interface to the data being presented. This is just one example of a situation in which other types of client access may be required. The point is, design an open architecture that is scalable for adding new clients.

Offline Web Support

One of the growing trends in wireless Internet applications is to provide offline Web support. In its simplest form, this support gives the user access to key Web data, even when a wireless connection is not available. The concept is similar to smart client applications, but usually implemented somewhat differently. Rather than creating client-side applications that contain the presentation layer, business logic, and persistent data storage, offline Web applications make it possible to store Web pages and their related data in a persistent store, either custom developed or sometimes using the cache capabilities of the client browser. Then, when a user requests a Web page when he or she does not have wireless coverage, the page will be taken from the client storage location, thereby providing offline access.

This capability is usually of interest to corporations that want to create wireless applications with a browser interface but are interested in obtaining some of the benefits that smart client applications provide. The goal is to get real-time content when a connection is available and cached content when it is not. When working with the cached content, any changes that are made on the client need to be queued up and sent to the server when connectivity is restored. While this seems like a great solution, it does introduce some restrictions.

For thin client applications, all of the business logic is executed on the server, meaning that when a connection to the server hosting this logic is unavailable, so is the ability to run the logic. This situation poses a problem: Even if you have cached the entire set of Web pages for a particular application, chances are that it will need to access business logic and data in order to be effective. Because the logic and data are not stored on the client, the usefulness of the application without connectivity is severely limited.

This situation is clearly a problem for corporate business applications, though it may not be as serious for some consumer Web sites. For example, if you are interested in reading the latest news or weather information, you may be perfectly happy with Web pages that have been cached for a few hours. In this case, you do not need to be able to interact with the application. It is in these scenarios where this application type has experienced some level of success.

For business application developers, there may still be hope. A new form of hybrid smart client/thin client technology is being made available by several vendors. It is usually based on J2ME, which gives you the ability to store actual business and data objects on the device while still having the ability to receive real-time data. Currently, it is really a scaled-down form of a smart client application, but there are promising signs that J2ME technology will provide us with new, more flexible forms of wireless applications. Additionally, increased support for client scripting languages such as JavaScript allows offline Web applications to offer some business logic on the client for more sophisticated applications.

Summary

The many unique characteristics of wireless Internet applications add complexity to the traditional Internet application development process: You have to worry about a variety of microbrowsers, multiple markup languages, and communicating over a variety of wireless protocols. Therefore, take extra care to design applications that can operate effectively on devices with small screen sizes and low bandwidth.

A variety of development tools and SDKs are available to help with the development of thin client applications. The majority of these products come with device emulation environments. Emulators allow application developers to build and test the applications on the same physical machine; they are not, however, intended to replace testing on physical devices over wireless networks.

In the next chapter we explore the various wireless languages and technologies available for generating wireless content.

Helpful Links

Follow these useful links to find out more information on the topics discussed in this chapter:

Ericsson Developer Information	www.ericsson.com/mobilityworld/sub/open/index.html
Microsoft Developer Network (MSDN)	http://msdn.microsoft.com
Nokia Developer Information	www.nokia.com/wap/index.html
Open Mobile Alliance	www.openmobilealliance.org
Wireless Developer Network WAP Channel	www.wirelessdevnet.com/channels/wap

CHAPTER 13

Wireless Languages and Content-Generation Technologies

In the previous two chapters we covered the various technologies involved in developing wireless Internet applications. Now it is time to take a closer look at the wireless markup languages used to create these applications, as well as the technologies available for creating dynamic, data-driven content. As you know by now, there are many options available for building wireless Internet applications. If you are fortunate, you may have the option to select the markup language to use for your development project, but in all likelihood, you will not have the luxury of making this decision; instead you probably will have to support a variety of languages to address a wider audience.

When it comes to creating dynamic wireless content, most organizations have existing Web technology that they want to leverage. We will take a look at how the leading content-generation technologies work and how they apply to wireless application development.

> **NOTE** The sample code in this chapter has been tested with the Openwave SDKs and Pocket Internet Explorer as discussed in Chapter 12, "Thin Client Development."

Wireless Content Types

The creation of wireless Internet applications involves many types of content, including markup languages, images, multimedia, and business oriented files. For most phone-based applications, such as those using WAP, you will only have to be concerned with the wireless markup language and some basic images. But if you are targeting higher-end devices, such as those based on Windows CE or Symbian OS, you will have the option of incorporating multimedia content and business files, such as Microsoft Word or Excel, into your applications. The content types that are supported will be dictated by a combination of the microbrowser and the mobile operating system being used.

In this section we are going to focus on wireless markup languages. The markup language controls the majority of the application, including the user interface, navigation, and content.

We are not going to focus on the other content types, as they vary significantly among devices and microbrowsers.

Markup Languages

When developing desktop Internet applications, HTML is the markup language of choice. If a developer is comfortable with programming HTML, he or she will be well prepared to create content for all of the leading browsers, including Internet Explorer, Netscape, and Mozilla. Unfortunately, this is not the case in the wireless world, where HDML, WML, HTML, cHTML, and XHTML are all used for various browsers. The factors that contribute to which markup language will be required in a given development effort include geographic location, device type, and the microbrowser being used.

This section provides an overview of each of these markup languages to familiarize you with the main concepts of each language and where each is commonly used. To help demonstrate these, we will also look at some sample code.

> **NOTE** The goal of this chapter is to give you a feel for language concepts in general, not to teach the in-depth workings of each language. The sample code is the first screen of an inventory application, where users can select how they would like to perform an inventory search. It is included for illustrative purposes only; these samples are not intended as complete applications.

Later in this chapter, we will take a look at the various content-generation technologies that can be used to create data-driven wireless thin client applications.

HDML

For those developing applications exclusively for markets outside North America, the Handheld Device Markup Language (HDML) is not important. Device manufacturers outside of North America did not ship HDML-based devices, but instead moved directly to WML devices. If, however, you are going to be deploying to users in North America, and you will not have control over the devices being used, you will want to read about HDML.

In 1997, Unwired Planet (now Openwave, and Phone.com in between the two) developed HDML, along with an HDML microbrowser called the UP.Browser. This took place before WAP came into existence, meaning that WML did not exist either. The idea behind HDML was to create a markup language that was suitable for use on cellular phones. (At the time, XML was quite new, so Unwired Planet did not base HDML on XML, thereby imposing limitations it could not foresee.) For the most part, the structure of HDML is loosely based on HTML, but it does not have anywhere near the capabilities that HTML offers. HDML is really a transition technology. As just stated, it was used before the advent of WML, which now has much wider industry support, and is deployed on a much larger scale.

> **NOTE** The full specification for HDML can be found on the W3C Web site, www.w3.org. It is still in working draft format, and has not been enhanced since its inception. This is a clear indication as to what the future holds for HDML.

Another limitation of HDML that is often considered an impediment to development is its lack of scripting capabilities. It does not support JavaScript or, like WML applications, WMLScript, making HDML quite restricted for executing logic locally on the device. Tasks such as validating user input, generating messages and dialogs, and other device-specific tasks can often be accomplished using HDML, but they have to be driven from the server, rather than the client. This increases network traffic, thereby reducing performance.

That said, HDML is still used in the U.S. and Canadian markets, primarily on CDMA- and CDPD-based devices. All North American devices that shipped before the end of 2000 used HDML-based browsers; so, as you can imagine, millions of these devices are still in use, and most likely will be for a few years to come. Fortunately, many of the WAP gateways will convert HDML to WML for you, making HDML content viewable on WAP-enabled devices. In addition, converting HDML to WML manually is not a difficult task, so you should be able to reuse many of your HDML applications once they are no longer required.

HDML Example

Let's take a look at some HDML code to get a feel for what it looks like. In Listing 13.1, you can see some straightforward HDML code that produces a list of choices when presented in an HDML browser, as shown in Figure 13.1.

The first elements to point out are the mandatory opening <HDML> and closing </HDML> tags in lines 1 and 16. Notice that this is similar to how HTML code is written, with the exception of the required version number, in this case version 3.0. Inside these tags, the HDML code is broken down into decks and cards. Each segment of code that is downloaded is a deck; hence, the entire set of code displayed in Listing 13.1 is one deck. Within each deck, you can have one or more cards. In this example, three cards are shown. The first starts on line 2, with the <CHOICE> tag, the second and third start on line 8 and 12, respectively, with the <ENTRY> tags. Each card acts as a single page that can be displayed on the wireless device. Within the HDML code, you have the option of moving to another card within the current deck, or to download another deck and link to it.

In this example, you can see both types of references in between the <CHOICE> </CHOICE> tags. These tags provide the user with a selection of choices. This is the preferred way to obtain input from a user. It is much more effective than having the user enter data using the keypad, which is cumbersome and slow. The choice card lets the user select any of the choices presented, and will then move the user to the destination specified by the DEST attribute. If, for example, the user selects Search by Name or Search by SKU on lines 4 or 5, he or she will be directed to the other cards within this deck. For example, selecting the Search by Name will move the user to the card starting on line 8, which has the matching NAME="ProductSearch". However, if the user selects the Inventory List choice on line 6, he or she will download the HDML code that is generated by inventorylist.jsp Java ServerPage.

```
1.  <HDML VERSION="3.0">
2.    <CHOICE>
3.      <CENTER><b>Inventory Search</b>
4.      <CE TASK="GO" DEST="#ProductSearch">Search by Name
5.      <CE TASK="GO" DEST="#SKUSearch">Search by SKU
6.      <CE TASK="GOSUB" DEST=inventorylist.jsp>Inventory List
7.    </CHOICE>
8.    <ENTRY NAME="ProductSearch" KEY="ProductName">
9.      <ACTION TYPE="ACCEPT" TASK="GO"
DEST="ProductSearch.jsp?Product=$ProductName">
10.     Enter Product Name:
11.   </ENTRY>
12.   <ENTRY NAME="SKUSearch" KEY="SKU">
13.     <ACTION TYPE="ACCEPT" TASK="GO" DEST="SKUSearch.jsp?SKU=$SKU">
14.     Enter SKU:
15.   </ENTRY>
16.</HDML>
```

Listing 13.1   HDML code for inventory sample.

If you look at the second card starting on line 8, you can see that we are obtaining user input in another way, via the <ENTRY> tag. In this case, the user will input the product name, then pass this value to the ProductSearch.jsp page, as shown on line 9. The third card works in a similar manner, except the SKU is entered and the value is passed to the SKUSearch.jsp page. In this way, you can create HDML applications that retrieve dynamic content based on the values entered by the application user.

One last item to note in this listing is that because HDML is not based on XML, quotes are not required around the attribute values. Notice that the destination specified by the DEST attribute in line 5 has quotes around it, while the DEST in line 6 does not. Both of these are valid for HDML applications. That said, it is a good idea to use quotes, since they are required in the other more common wireless markup languages such as WML and XHTML.

Figure 13.1 Openwave HDML Emulator showing output from sample code in Listing 13.1.

HELPFUL RESOURCE For more information on creating HDML applications, the HDML style guide provided by Openwave is a great resource. It can be downloaded from http://demo.openwave.com/pdf/styleguides /hdml_style.pdf.

WML

The Wireless Markup Language (WML) is a part of the Wireless Application Environment (WAE) as defined in the Wireless Application Protocol (WAP). Unlike HDML, WML is based on XML, making the syntax requirements somewhat stricter. For example, if a tag is not closed, the WML deck will not work—plain and simple. This is just one case; there are many other differences between WML and HDML (and HTML, for that matter) that warrant mentioning. They will be discussed later in this section after we take a look at some basic aspects of WML.

WML is a successor to HDML, developed by Openwave, Nokia, Motorola, and others as represented by the WAP Forum (now the Open Mobile Alliance). It is targeted directly at mobile information devices such as cellular phones, PDAs, two-way pagers, and smart phones. Since these devices do not typically have large displays or support for advanced graphics and layouts, the markup language has to be lightweight and designed for the unique requirements of wireless connectivity. This is where WML fits into the picture. Because it was developed by a consortium of companies, all of which have a vested interest in the success of wireless Internet applications, WML has garnered tremendous support throughout the wireless industry. It is in widescale deployment throughout Europe, North America, South America, and parts of Asia. Most of the handsets being shipped in these regions support WAP and include WML browsers. The most common versions of WML are 1.1 and 1.2.1, although WML2 is gaining momentum.

In general, when creating Internet applications for wireless phones, WML is the markup language of choice. This may change with the release of WAP 2.0, which uses XHTML as the markup language, although its widespread adoption is still some time off. Just as HDML is still commonly used in North America even several years after the arrival of WAP, WML will continue to be used for many years to come as device manufacturers slowly move to XHTML. Fortunately, both WML and XHTML are XML-based, making the transition between the two straightforward. Actually, today, many companies are developing their applications in XHTML, and use conversion tools and transcoding technology to translate the XHTML into other markup languages such as WML. Taking this approach makes future markup language transitions much more manageable.

For many Web developers, the move to WML can be bittersweet, simply because many of the screens are 12 characters wide and 5 lines high; if you are used to creating sophisticated Web pages, with much of the emphasis on presentation and graphics, WML applications present a significant change. In most cases, wireless Web applications are focused on accessing data, with a specific task in mind. They are not geared at generic Web surfing. For example, if you want to check the inventory level of a particular product, a wireless Internet application would work quite well. You could simply do a search based on a product name or, better yet, an SKU, and receive the response you are looking for. It would not, however, be effective as a product catalog, where many details and, often, images are required. The focus of successful WML applications is on making sure that, one, all of the relevant data is there, and, two, that the navigation to that data is not too cumbersome.

Another major difference between desktop Web development and wireless development are the browsers. For wireless applications, the microbrowsers are quite simple. Unlike the immensely complex desktop browsers, such as Internet Explorer and Netscape, the wireless microbrowsers are often limited to very simplistic commands such as Submit, Back, Link, and More. Interaction with the browser is accomplished using soft keys, which can be programmed within your WML code. (You can find more information on wireless Internet technology, including microbrowsers, in Chapter 14, "Wireless Internet Technology and Vendors.")

WML Example

The best way to learn more about WML programming is by looking at an example. Listing 13.2 shows WML code that could be used for the first screen of an inventory search application. Figure 13.2 shows how this code would look in the Openwave WAP emulator.

If you are familiar with XML, and have done some Web page development, then programming WML applications will not be difficult. Similar to HTML and HDML pages, all of the content is contained within opening and closing tags, in this case <wml> and </wml>. The only exceptions are two lines at the beginning that are there present for XML compliance:

- The first line of every XML document is <?xml version="1.0">.
- The second line is the Document Type Definition (DTD) to which it conforms. In this case, we are using the WapForum DTD for WML 1.1. This is not a requirement for all XML documents.

Since WML is XML-based, it has to adhere to the rules of XML. The following are some of the major differences that you will notice in WML versus HDML and other non-XML-based languages:

- All WML elements must be closed.
- WML elements must be properly nested.
- WML documents must be well-formed.
- WML tag names must be lowercase.

These rules may seem quite strict at first, especially if you have been doing a lot of HTML development. Do not worry; after developing an application or two, you will get used to it, which is important because other XML-based languages such as XHTML are becoming increasingly popular. We will take a closer look at XHTML later in this chapter. Now we will continue to investigate the WML code shown in Listing 13.2.

WML is divided into decks and cards, similar to HDML. This example contains one deck containing three cards. Each WML deck contains at least one card and has a maximum size of 1492 bytes (note, however, that smaller decks, ideally under 800 bytes, download quicker and are more manageable by the device's memory cache). The cards in this example start on lines 4, 26, and 33. Each card will be displayed on the phone individually, equivalent to an HTML page for desktop browsers. Figure 13.2 shows the contents of the first card, lines 4–25 on an Openwave emulator.

```
1.   <?xml version="1.0" encoding="UTF-8"?>
2.   <!DOCTYPE wml PUBLIC "-//WAPFORUM//DTD WML 1.1//EN"
"http://www.wapforum.org/DTD/wml_1.1.xml">
3.   <wml>
4.     <card id="card1">
5.       <p align="center"><i>Inventory Search</i></p>
6.       <p align="left">
7.       <select>
8.         <option>Search by Name
9.           <onevent type="onpick">
10.            <go href="#ProductSearch"></go>
11.          </onevent>
12.        </option>
13.        <option>Search by SKU
14.          <onevent type="onpick">
15.            <go href="#SKUSearch"></go>
16.          </onevent>
17.        </option>
18.        <option>View Inventory List
19.          <onevent type="onpick">
20.            <go href="Inventorylist.wml"></go>
21.          </onevent>
22.        </option>
23.      </select>
24.      </p>
```

Listing 13.2   WML code for inventory sample. *(continued)*

```
25.   </card>
26.   <card id="ProductSearch">
27.   <!--WML code here for Product Search-->
28.     <p>
29.       Enter Product Name:
30.         <input name="product" emptyok="false"></input>
31.     </p>
32.   </card>
33.   <card id="SKUSearch">
34.   <!--WML code here for SKU Search-->
35.     <p>
36.       Enter SKU:
37.         <input name="sku" emptyok="false"></input>
38.     </p>
39.   </card>
40. </wml>
```

Listing 13.2   *(continued)*

Line 5 demonstrates the use of tags, in this case, the paragraph tag, <p>, as well as an attribute with a value, align="center". All text within a card should be enclosed within <p> </p> tags; and, remember, they have to be closed, and cannot be nested. If you want to add a line break, you can use the familiar <br> tag. Similar to the other WML tags, it has to be closed using either <br/> or
</br>. Also of interest are the quotes around the attribute value. All WML attribute values have to be enclosed in double quotes.

As we move to the more functional areas of the example, we have the <select> tag in line 7. This presents the user with the options specified inside it. In our example we have three options:

- Search by Name.
- Search by SKU.
- View Inventory List.

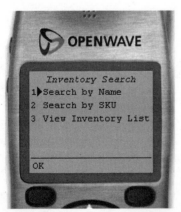

Figure 13.2   Openwave WML Emulator showing output from sample code in Listing 13.2.

This is a better technique for user input than requiring them to type on the device's keypad. Within the <select> tag, the <option> tag denotes each option.

WML has several options for moving between cards and decks. The example uses the <onevent> tag in conjunction with a <go> tag. When an item is selected, the user is moved to the URL argument specified in the <go> tag. In the example, lines 10 and 15 will take the user to other cards within the same deck, in this case to cards with id="ProductSearch" and id="SKUSearch", respectively. The link on line 20 is different in that it links to another deck, which will then be downloaded and displayed. If you need to send information back to the server, this can be accomplished by using the <postfield> element within the <go> tag. Other linking methods can also be used in WML. Some common tags are <do> and other syntaxes of the <option> tag, such as using the onpick attribute:

```
<option onpick="#ProductSearch">Search by Name</option>
```

Most applications are not considered complete unless they enable user input. In WML, this is accomplished using the <input> tag. In lines 30 and 37, the user can enter input and have it stored in the related variable. For the Product Name input, we store the value entered into the product variable as specified in the name="product" attribute. There is also an option to set the type attribute to either "text" or "password", depending on whether the text should be displayed or concealed with asterisks. If you want to have default text displayed, possibly to reduce the amount of user input required, you could set the value attribute.

HELPFUL RESOURCE For more information on creating WML applications, the WML style guide provided by Openwave is a great resource. It can be downloaded from http://demo.openwave.com/pdf/styleguides/wml_style.pdf.

WMLScript

WMLScript provides a way to add more advanced capabilities to WML applications. Just as JavaScript of VBScript may be used in conjunction with HTML, WMLScript is used with WML to provide client-side logic. This is an important concept, because each request to the server can add a significant amount of wait time for the user. The following are some of the most common uses of WMLScript for WAP applications:

- Validation of user input
- Access device facilities, such as adding numbers to an address book
- Creation of dialogs and messages, avoiding round-trips to the server

If you want to use WMLScript in your applications, you will create the script in files with the extension .wmls. Rather than including the WMLScript in the WML file, as JavaScript would be in an HTML file, WMLScript is contained in a separate file and invoked from the WML card just as any other URL. The following is an example:

```
<a href="validateuser.wmls#foo($(user))>Validate User</a>
```

You will also have to ensure that both your microbrowser and the WAP gateway have WMLScript support. In addition, you may have to register the MIME types with your HTTP server.

Within your WMLScript pages, you can create both For and While loops; create functions, and libraries of functions; display dialog boxes; and have advanced interaction with the microbrowser.

> **NOTE** It is beyond the scope of this book to go into depth about WMLScript's capabilities, so if you are interested in learning more about WMLScript, visit http://demo.openwave.com/pdf/50/wmls_dev_guide.pdf to view the Openwave WMLScript Developer's Guide.

HTML

Most developers have some experience with HTML; and if they do not, an abundance of material is available on the subject. For these reasons, this section is brief, its purpose only to provide some background information and details regarding the flavors of HTML that are being used for wireless applications.

HTML is the de facto language of the World Wide Web. Web developers worldwide use various versions of HTML to create rich, sophisticated Web sites, primarily viewed with either Internet Explorer or Netscape browsers. Over time, the complexity of HTML has increased dramatically. In 1995, HTML version 2.0 was defined. At that time, Web sites primarily displayed static data only, providing a means for companies to display brochures and similar content, and for individuals to display various types of personal information. Before long, more complex (and interesting) sites started appearing online, many of them based on HTML 3.2, which was released in 1997. It was around this time that we started seeing dynamic content, often generated using Common Gateway Interface (CGI) executables. Soon afterward, HTML 4.0 was introduced, adding a new level of functionality to Web development. Interactive, transaction-based, Web applications are now the norm. It is not unusual to perform bank transactions, make purchases, or book entire vacations online. In many cases, this is the preferred way to interact with many institutions.

In order to support these advanced applications, server-side technology had to evolve dramatically. Businesses now have several options for creating dynamic content, including Java servlets and JavaServer Pages (JSPs), Active Server Pages (ASPs), as well ISAPI, NSAPI, and CGI executables. Some of these technologies are discussed later in this chapter in the section called *Content-Generation Technologies*.

Not to be left behind, the Web browsers had to evolve as well. Netscape and Internet Explorer continued to have new features, making it easier for developers to create Web applications. As the applications evolved, some of the rules of the language were relaxed, to the point where even poorly written HTML code displayed without errors. Developers started noticing that they could get away with these "shortcuts," and HTML code became increasingly sloppy. Today it is estimated that close to 95 percent of HTML pages are not programmed according to the specification. The most common offenses are unclosed tags and incorrect nesting of elements.

Enter wireless Internet applications. It only made sense that the language of the wireless Internet would be the same as the language of the World Wide Web. This introduced a problem, however. HTML was much too complex, requiring too much processing power for parsing, to be suitable for wireless devices. The solution was to introduce subsets of HTML that could be used for wireless devices. These subsets include Compact HTML, and Web Clippings. Both of these languages are subsets of HTML, with the more complex aspects of the specification removed.

The HTML syntax supported by wireless devices often resembles HTML versions 2.0 and 3.2. In Listing 13.3 is an example of HTML code that can be used for wireless Internet applications. Many of the HTML browsers for the Palm OS, Pocket PC, and Symbian OS have support for a decent subset of HTML 3.2, on which this sample is based.

```
1.    <!DOCTYPE HTML PUBLIC "-//W3C//DTD HTML 3.2 Final//EN">
2.    <html>
3.    <head>
4.      <title>Inventory List</title>
5.    </head>
6.    <body>
7.    <P align=left><FONT size=4><STRONG>Inventory Product
List</STRONG></FONT>
8.    <br>
9.    <br>
10.   <TABLE cellPadding=3 border=1>
11.     <TR>
12.        <TD><STRONG>Product </STRONG></TD>
13.        <TD><STRONG>Quantity</STRONG></TD>
14.        <TD><STRONG>Price($)</STRONG></TD>
15.     </TR>
16.     <TR>
17.        <TD>Sony TRV30
18.        <TD>17
19.        <TD>1699.99
20.     </TR>
21.     <TR>
22.        <TD>Hitachi VMD875L
23.        <TD>24
24.        <TD>599.99
25.     </TR>
26.     <TR>
27.        <TD>Sony DCR-IP7BT
28.        <TD>11
29.        <TD>2199.99
30.     </TR>
31.     <TR>
32.        <TD>JVC GR-DV2000
33.        <TD>4
34.        <TD>1599.99
35.     </TR>
36.   </TABLE>
37. </BODY>
38. </HTML>
```

Listing 13.3   HTML code for inventory sample.

Figure 13.3   Mobile Internet Explorer showing output from sample HTML code in Listing 13.3.

One of the downsides of basing the wireless markup language on HTML is that you inherit much of the sloppy programming that goes with it. Look at the example in Listing 13.3 to see the common tags found in HTML documents, such as the opening <html>, <head>, and <body> tags in lines 2, 3, and 6, respectively. In line 7, is the first occurrence of a missing closing tag. The paragraph tag, <p>, is not closed with a corresponding </p> tag. This type of omission is common in HTML programming, and generally does not cause any errors in the browser. Move down the code and you will see similar omissions in the table entries. The <TD> tags for each product entry do not have a closing </TD> tag in lines 17–20, 22–24, 27–29, and 32–34. In addition, you can also see that the tags are not case-sensitive. For example, the opening <body> tag in line 6 is lowercase, while its corresponding closing </BODY> tag in line 37 is all uppercase. Again, this does not create an error in HTML browsers.

The output of the code in Listing 13.3 is shown on the Pocket Internet Explorer browser on the Pocket PC emulator in Figure 13.3. This browser supports most of the HTML 3.2 specification, omitting only very advanced features.

cHTML

The most popular subset of HTML is Compact HTML, or cHTML. Its popularity is based largely on the fact that cHTML is the markup language used by the phenomenally successful i-Mode service produced by NTT DoCoMo in Japan. i-Mode is fundamentally a product offering that allows users to access Internet content from wireless devices. It is really a wireless Internet brand, similar to AOL on the wired Internet. This type of approach has worked extremely well for NTT DoCoMo. Within the first year of its introduction, NTT DoCoMo had close to 10 million subscribers for the i-Mode service. That number easily surpassed the number of wireless Internet users in North

America and Europe combined. While i-Mode is not the only factor contributing to the success of the wireless Internet in Japan, it definitely is a major contributor.

cHTML was submitted as a Note to the W3C in 1998, and is available for discussion purposes only. The W3C has not endorsed cHTML nor made any contribution to the language in any way. The original specification for cHTML from 1998 has remained unchanged, much as the specification for HDML. In many ways, the role of these two languages is similar, as cHTML is also a transition technology. Just as HDML has slowly moved aside for WML, cHTML is being replaced by XHTML in the i-Mode service. This is a very welcome move, allowing XHTML to become the language of the wireless Internet, in use by the two most prevalent offerings, WAP 2.x and i-Mode.

Like others in this category, the goal of cHTML is to create a markup language that is suitable for small information devices such as smart phones, communicators, and PDAs. Interestingly, cHTML is based on all three HTML specifications—2.0, 3.2, and 4.0—so it can take advantage of the HTML-based content resources, software tools, and documents. It has taken aspects from each of these HTML versions to create a complete language for limited-resource devices. The creators of cHTML have taken the following requirements into account:

- Hardware restrictions of the devices, including limited memory, low power central processing unit (CPU), and small displays.
- The navigation restrictions of the device: Moving between sets of information has to be possible with a minimum number of operations.
- Wireless network agnostic: The markup language does not depend on the underlying wireless protocol being used.

The design of cHTML is based on four main principles:

- It is completely based on current HTML W3C recommendations.
- It is a light specification, allowing it to run on mobile devices.
- It can be viewed on a mono-screen display.
- It is easy for users to operate.

In order to accomplish the goals of the language, certain features of HTML have to be excluded. Some of these include the following:

- Support for frames and tables
- JPEG image support
- Image maps
- Multiple-character fonts and styles
- Background color and images
- Stylesheets

Though it benefits by being based on HTML, cHTML also loses by not being based on XML. You will, for instance, see the same sloppy programming in cHTML that is commonly found in HTML. Fortunately, the move to XHTML provides us with a language that is based on both HTML and XML.

HELPFUL RESOURCE If you are interested in seeing the complete cHTML Note submitted to the W3C, which contains a complete list of the tags supported by cHTML, visit www.w3.org/TR/1998/NOTE-compactHTML-19980209.

XHTML

This brings us to the eXtensible HyperText Markup Language, or XHTML, an XML-compliant version of HTML. To reiterate, XHTML is the direction in which wireless markup languages are moving. The convergence of markup languages on XHTML is expected to increase the rate of development for m-commerce and m-business applications. Companies building wireless Internet applications can now focus on one markup language, so they can concentrate their efforts on the applications they are building, rather than on the technical challenge of developing with multiple markup languages simultaneously.

XHTML 1.0 became an official W3C recommendation in early 2000, making it a Web standard. It is almost identical to HTML 4.0.1, with a few rules added for XML compliancy.

NOTE For the same reasons that HTML was not examined in depth, this section on XHTML will be kept brief, focusing only on the areas where XHTML differs from HTML, and on the profiles being used for wireless development, namely XHTML Mobile Profile (XHTML-MP) and XHTML Basic. In the not-too-distant future, comparing HTML and XHTML will no longer be required as XHTML is intended to be a replacement for HTML. Most of the new desktop Web browsers will support XHTML natively, resulting in a higher level of compatibility among browsers.

Why XHTML?

As noted earlier, desktop Web browsers allow for very poorly formatted HTML Web pages, making it difficult to have cross-platform, or even cross-browser, compatibility. Though much of the HTML code on the Web does not follow the rules of HTML, it will execute problem-free on a desktop browser. But browsers in other locations, such as on PDAs, mobile phones, and intelligent appliances, do not have enough computing power to compensate for poor programming. This is where XML comes into play.

XML requires all code to be programmed correctly, making the job of the interpreter or, in this case, the browser, much easier. XML itself is a way to describe data, whereas HTML is a way to present data. When we bring these two languages together, the result is a markup language that can be used now and that is adaptable to future change.

How XHTML Differs from HTML

Knowing the differences between XHTML and HTML will help you to start creating better HTML code now, so fewer changes will be required later when the browsers and other software programs start to require XHTML. The following are the most important differences:

- XHTML documents must be well formed. Because XHTML is XML-based, documents must conform to XML syntax rules.

- XHTML elements must be properly nested.

- Tags and attributes must be lowercase.

- All XHTML elements must be closed. You can use a closing slash with any empty element such as <br/> or <a id="page1"/>.

- All attribute values must be enclosed in quotation marks.

- A DOCTYPE declaration is required.

Let's look at an example to demonstrate some of these differences. Listing 13.4 contains the XHTML version of the HTML code shown earlier in Listing 13.3. Note that the changes made in converting the document from HTML to XHTML do not affect how it appears in a browser; that is, the code shown in Listing 13.4 will display as the HTML code did, shown previously in Figure 13.3.

```
1.   <?xml version="1.0"?>
2.   <!DOCTYPE html PUBLIC "-//OPENWAVE//DTD XHTML Mobile 1.0//EN"
"http://www.openwave.com/dtd/xhtml-mobile10.dtd">
3.   <html xmlns="http://www.w3.org/1999/xhtml"xml:lang="en">
4.   <head>
5.     <title>Inventory List</title>
6.   </head>
7.   <body>
8.   <p align="left"><font size="4"><strong>Inventory Product
List</strong></font></p>
9.   <br/>
10.  <br/>
11.  <table cellPadding="3" border="1">
12.    <tr>
13.      <th>Product</th>
14.      <th>Quantity</th>
15.      <th>Price($)</th>
16.    </tr>
17.    <tr>
18.      <td>Sony TRV30</td>
```

Listing 13.4 XHTML code for inventory sample. *(continued)*

```
19.     <td>17</td>
20.     <td>1699.99</td>
21.   </tr>
22.   <tr>
23.     <td>Hitachi VMD875L</td>
24.     <td>24</td>
25.     <td>599.99</td>
26.   </tr>
27.   <tr>
28.     <td>Sony DCR-IP7BT</td>
29.     <td>11</td>
30.     <td>2199.99</td>
31.   </tr>
32.   <tr>
33.     <td>JVC GR-DV2000</td>
34.     <td>4</td>
35.     <td>1599.99</td>
36.   </tr>
37. </table>
38. </body>
39. </html>
```

Listing 13.4   *(continued)*

We start to see the differences between the HTML and XHTML documents immediately. On line 1 we have the <?xml version="1.0"?> declaration, as required by all XML documents. On line 2 we have the required DOCTYPE declaration, in this case using the XHTML Mobile Profile. Notice that the <DOCTYPE> does not have a closing element. This is because it is not part of the XHTML document itself, and therefore does not require a closing tag. This declaration tells the browser which XHTML document type the document conforms to.

In addition to these differences, we also can see that this document is well formed: All the tags are in lowercase and are properly closed. For example, all of the table cells <td> have closing tags </td>, which were not present in the HTML document in Listing 13.3. Also, the attribute values are enclosed in quotation marks, as seen in lines 8 and 11.

In our simple example, the required changes were quite straightforward. Even so, the HTML document did require a significant number of changes to be proper XHTML. For developers who manage hundreds, or possibly thousands, of HTML pages, updating them all to XHTML can be a daunting proposition. If you find yourself in this position, you may want to look into the HTML Tidy or JTidy tools, which will do the HTML-to-XHTML conversion for you. (More information on these tools can be found at the Source Forge Web site at http://tidy.sourceforge.net.)

XHTML Basic and XHTML-MP

The markup language for the WAP Application Environment 2.0 (WAE) is XHTML Mobile Profile (XHTML-MP). This profile is a superset of XHTML Basic, as defined by the W3C. By taking advantage of XHTML modularization, XHTML-MP is very extensible, permitting the addition of new elements as required. In addition, documents based on the Basic profile of XHTML will be completely operable in XHTML-MP browsers.

XHTML Basic defines a document type that is rich enough for advanced content authoring yet can be used across a variety of clients, including desktops, PDAs, TV, and mobile phones. It is the mobile version of XHTML 1.0, supporting everything except features such as frames that are not appropriate for devices with small screens.

Similar to cHTML, XHTML Basic will be able to take advantage of the large number of HTML software programs and resources available. This means that developers will be able to test their applications in a desktop browser and have confidence that the layout will be an accurate representation of the browser on the mobile device. In addition, the information on XHTML given earlier in this chapter also applies to XHTML Basic.

Other than being XML-based, the other major difference between XHTML Basic and cHTML is XHTML Basic's support for cascading style sheets (CSS). CSS are executed on the browser, describing the screen presentation of documents without sacrificing device independence or requiring the addition of new markup language tags. Using CSS, an application developer can specify the presentation of the Web application in one place. If changes are required, they can be made to the style sheet and will be automatically reflected in all pages on the site. Residing on the client, CSS also help to alleviate the overhead on the server, where transcoding techniques are often used to transform documents to different markup languages. (Information on transcoding can be found in the *Thin Client Application Models* section of Chapter 12, "Thin Client Development.")

With support from both the Open Mobile Alliance and W3C, XHTML Basic is assured to have widespread acceptance now and in the future. (More information on XHTML Basic can be found on the W3C Web site at www.w3.org/TR/xhtml-basic. You can also download the full XHTML Mobile Profile reference from Openwave at http://developer.openwave.com.)

VoiceXML

One other markup language warrants mentioning in this chapter: the Voice eXtensible Markup Language (VoiceXML). VoiceXML is an XML-based markup language that allows access to Web applications and content via a telephone rather than a Web browser. Just as a Web browser interprets markup languages such as WML or XHTML, a voice browser interprets VoiceXML, then uses Text-to-Speech (TTS) to relay the information to the user.

VoiceXML was developed by the VoiceXML Forum, which was founded by AT&T, Lucent, Motorola, and IBM. In addition to the founding members, the VoiceXML Forum has an additional 400 members that support the growth of VoiceXML for voice-based applications. The future looks promising for VoiceXML. In 2000, VoiceXML 1.0 was accepted by the W3C as the basis for developing a dialog-based markup language. In 2002, VoiceXML 2.0 became an official W3C recommendation.

If you are interested in creating voice-based applications, you will want to read Chapter 15, "Voice Applications with VoiceXML," which is dedicated to the VoiceXML markup language and related VoiceXML technologies.

Content-Generation Technologies

Now that we have looked at the common wireless markup languages, it is time to look at the technologies most commonly used to create dynamic wireless applications. Each technology will provide you with a mechanism to interact with enterprise data, as well as server-side business logic so you can create interactive, data-driven wireless applications. Each technology will also allow you to create content specific to the device making the request. So, for example, you could have HDML, WML, and XHTML content all being driven from the same application on the server. This type of capability is of particular interest if you are deploying your application to a broad audience or to an audience on different continents.

Deciding which solution is best is not an easy task. The solution that is ideal for one application may not be suitable for another. The decision often comes down to the developer's skill set, in combination with the technology currently being used in the organization. In most cases you will find that the technology used to generate wireless content is the same that is used for creating HTML-based desktop applications.

We are going to look at five different technologies: CGI with Perl, Java servlets, JavaServer Pages, Active Server Pages, and XML with XSL stylesheets. These technologies can be divided into two categories: code-driven and page-driven. Both Perl and Java servlets are code-driven, with the markup language embedded into source code. These technologies are usually best suited for applications that require integration with server-side application logic, but do not require sophisticated user interfaces. The page-driven technologies embed source code into the markup language. Both JavaServer Pages and Active Server Pages fall into this category. They are good choices for applications in which there is a significant amount of client code, but perhaps not as much usage of server-side business logic. Finally we have XML with XSL stylesheets, which can be driven either by a code-centric or page-centric approach.

Keep in mind that these observations are only guidelines, and that each technology can be adapted for a variety of applications. As mentioned, the final decision often will be based on the skills of the developer, or development team, responsible for implementing the solution.

Common Gateway Interface (CGI) with Perl

The Common Gateway Interface (CGI) is one of the most widely used server-side technologies for Web development; it is supported on almost every Web server. CGI comprises a set of commands for communicating between a Web server and a program that processes information from a Web page. It was the first technology available for making Web sites dynamic. Unlike static content that is served up in a text format, CGI applications are executed on the server in real time, so that content can be generated on the fly. Many of the early dynamic Web sites used CGI to obtain user input, access enterprise databases, and return some result back to the browser. In order to communicate with other applications, there had to be a form of business logic associated with the application. CGI enables this program intelligence to be written in almost any language, though the more popular ones are C/C++, FORTRAN, Java, and Perl. Out of these and many others, Perl is the language most often used for creating CGI applications.

Now in its sixth version, Perl has a quite a history. Its first version was released at the end of 1987, long before people were thinking of building Internet applications. Perl is based on the C programming language, and is developed and maintained by thousands of developers, along with its creator, Larry Wall. Perl is an interpreted language, meaning that it does not have to be compiled into native bytecodes to execute. This has many benefits over compiled languages like C or FORTRAN, making it easier to debug, modify, and maintain.

Perl has grown in popularity for Web programming because it makes it possible to create quick and effective interactive applications. Listing 13.5 shows an example of a Perl application that outputs a WML deck. The code in this example is quite simple: It outputs Hello Wireless World! to the screen of the wireless device; nevertheless, it serves to demonstrate how the language is constructed. The first line of all Perl programs has to be the location of the Perl interpreter. In our example, it is located in the user/bin/perl directory. The second line specifies the content type. For most Web applications, this would be specified as text/html, but for WML applications, it is text/vnd.wap.wml. After that is taken care of, we are left to create our WML code for the application. Notice that, similar to C, each line ends with a semicolon. In this example, the code is not very dynamic at all. Each line uses a print statement to output some WML code. Lines 5 to 8 in the example create a WML deck with a single card that outputs "Hello Wireless World!"

There are other ways that this can be accomplished. One is to use a string variable to hold the code, then flush it to the screen all at once. Another is to create functions that would piece the WML together in logical units, inserting dynamic data where required. The specifics of Perl are beyond the scope of this overview, but the point is that Perl is indeed a complete programming language, capable of creating very advanced, dynamic applications.

```
1.  #!/usr/bin/perl
2.  print "Content-type: text/vnd.wap.wml\n\n";
3.  print "<?xml version=\"1.0\" encoding=\"iso-8859-1\"?>\n";
4.  print "<!DOCTYPE wml PUBLIC \"-//WAPFORUM//DTD WML 1.1//EN\"
    \"http://www.wapforum.org/DTD/wml_1.1.xml\">\n";
5.  print "<wml>\n";
6.  print "   <card id='card1'>\n";
7.  print "      <p>Hello Wireless World!</p>\n";
8.  print "   </card>\n";
9.  print "</wml>\n";
```

Listing 13.5 Perl CGI application.

The major disadvantage to using CGI for dynamic content generation is its scalability. Each request received by the Web server spawns a new process, creating its own set of environment variables, a separate instance of any required runtime environment, a copy of the program, and a block of memory for the program to use. In the case of Perl CGI programs, the Perl interpreter is invoked, requiring additional resources. For sites that do not get much traffic, this may not be a concern, but for sites that get thousands of hits a day, this can lead to significant performance problems and, potentially, system failure.

In response to the poor performance of CGI applications, Microsoft and Netscape each developed a proprietary set of APIs that could be used to write server applications. These APIs, known as Internet Server API (ISAPI) and Netscape Server API (NSAPI) did help with performance, but they also introduced new problems. The first problem is that these technologies are proprietary. When an application is created, it can only run on the platform for which it was developed. Moving the programs to a different environment is a costly task. The second problem is that the logic runs in the same process as the Web server. If the program causes a memory access violation, there is a chance it could crash the entire Web server, rather than just the one application.

The scalability problems of CGI and the proprietary nature of ISAPI and NSAPI led to the development of new technologies that are better suited for cross-platform server-side development.

Java Servlets

In 1998, Sun Microsystems introduced a technology called Java Servlets that addressed both the performance and the cross-platform obstacles that plagued earlier server-side technology. Written in Java, the servlet code is compiled into bytecodes that are interpreted by a Java Virtual Machine (JVM) on the Web server. This allows the servlet code to be moved to any server that has a JVM. Also, when a request is issued to a servlet, it creates a new thread to process the request. When the servlet has done its job, it is reused by other clients, meaning it does not have to be destroyed and re-created for each client request. This has a positive impact on both performance and scalability.

Servlets use a code-driven approach to create wireless content. This is similar to the CGI Perl approach in that each line of the WML or other markup language is output using a print statement, or to be precise, an out.println() statement. The code around the lines containing the output control the program flow, often making calls to other Java classes or Enterprise JavaBeans to execute logic or access enterprise data. This approach works very well for applications with simple user interfaces and for developers who prefer to program in Java instead of a wireless markup language.

The Java Servlet API is part of the Java 2 Platform, Enterprise Edition (J2EE). This gives servlets the capability to easily access other J2EE APIs, including Enterprise JavaBeans for component logic, JDBC for database access, and JMS for messaging integration. All of these technologies are commonly executed using application server technology. Most of the leading application servers on the market have full J2EE support, making them very capable platforms for creating Internet applications.

Over the past few years, the concept of a wireless application server has also emerged. These are usually application servers that provide specific class libraries, or frameworks, for creating wireless Internet applications. We are now going to look at an example of a Java servlet that generates XHTML code. (Chapter 14 has more information about the capabilities that are provided in wireless focused products.)

In Listing 13.6 is the complete Java code for a servlet that takes an HTTP GET request from the client and returns a simple XHTML form. The top five lines contain the import statements required for the class to compile. Line 6 contains the class definition. This class is called HelloWorld, and it extends HttpServlet. By extending HttpServlet, the class automatically gets servlet functionality without having to implement the entire Serlvet interface. HttpServlet implements the Servlet interface for us, saving us extra coding time. Within our HelloWorld class, we then move directly into the doGet(...) method, which is called every time a GET request is made to the servlet. If we wanted to do some initialization, we could have included an init() method as well. The servlet engine calls this method implicitly when the servlet is invoked. For our example, we did not require any initialization, so there is no init() method. If we were accessing a database or other data source, we might have included the connection to the data source in the init() method.

On line 8 you can see the doGet method takes two parameters, HttpServletRequest and HttpServletResponse. HttpServletRequest is used to get information from the client, such as the MIME type. The HttpServletResponse is used to send information back to the client; in this case, we are setting the content type to text/html, as shown on line 10, then outputting the XHTML content. We also use the HttpServletResponse object to get the PrintWriter for our output stream.

```
1.   // Import the required Java libraries
2.   import java.io.*;
3.   // Import the required Java Servlet libraries
4.   import javax.servlet.*;
5.   import javax.servlet.http.*;
6.   public class HelloWorld extends HttpServlet
7.   {
8.     public void doGet(HttpServletRequest req, HttpServletResponse res)
throws ServletException, IOException
9.     {
10.      res.setContentType("text/html");
11.      PrintWriter out = res.getWriter();
12.      out.println("<!DOCTYPE html PUBLIC \"-//OPENWAVE//DTD XHTML
Mobile 1.0//EN\" \"http://www.openwave.com/dtd/xhtml-mobile10.dtd\">");
13.      out.println("<html
xmlns=\"http://www.w3.org/1999/xhtml\"xml:lang=\"en\">");
14.      out.println("<head>");
15.      out.println("<title>XHTML Servlet</title>");
16.      out.println("</head>");
17.      out.println("<body>");
18.      out.println("<p align=\"left\"><b>Hello XHTML Wireless
World!</b></p>");
19.      out.println("</body>");
20.      out.println("</html>");
21.    }
22. }
```

Listing 13.6   Java servlet that outputs XHTML.

On lines 12 through 20 we output the XHTML code. You can see that each line of XHTML is output using the out.println(...) method. The XHTHML code generated by this servlet is quite simple. It would display "Hello XHTML Wireless World!" on the XHTML browser. Unfortunately, even for such straightforward code, the Java servlet is somewhat difficult to read.

The quotation marks in the XHTML code can confuse the Java compiler, so each quotation mark has to be escaped using a backslash. Line 12 demonstrates how confusing the XHTML code can become once it is embedded into the Java code and the escape characters are added. This demonstrates one of the drawbacks of using Java servlets for creating wireless content. Even a very simple application requires developers to have Java programming skills, and when they do, the resulting code is still quite difficult to develop and maintain. Each time the markup language is changed, the servlet has to be edited and recompiled. This could adversely affect other parts of the application.

Using Java servlets does, however, offer many advantages, namely a cross-platform, high-performance way to extend the capabilities of a Web server to provide dynamic wireless content. With wide platform support, servlets are a very attractive technology for those who are comfortable programming in Java.

JavaServer Pages

JavaServer Pages (JSPs) are ideal for developers who want the cross-platform and performance benefits of Java servlets but do not want to program using Java. JSPs are a page-driven technology that allows Java code to be embedded in the markup language, rather than the markup language being embedded in Java as with servlets. This allows Web developers to rapidly create information rich, dynamic Web pages that are easy to maintain. Unlike Java servlets, JavaServer Pages separate the presentation logic from the business logic, allowing changes to the user interface without affecting the program logic. This allows Web designers to create the user interface with their favorite Web development tools, while Java programmers create the business logic.

The contents of a JSP page can be divided into two main categories: elements and template data. The elements are XML-like tags that are used to control program flow and encapsulate Java logic. There are five types of elements that you may use: directives, declarations, scriptlets, expressions, and actions. All of these are processed on the server, which then performs the desired task. The template data is everything other than the elements, usually your markup language. The templates are ignored by the engine that processes the JSPs since they do not provide any server-side logic. In this way, you can create sophisticated Web applications with embedded logic to perform more advanced tasks. Let's look at a sample JSP to get an idea of what this means.

Listing 13.7 contains the source code for a complete JSP that outputs HDML code. As you can see, the majority of the code is an HDML template into which we can insert various JSP elements. Let's look at some of the more interesting segments of this example.

One of the most common problems of getting HDML (or WML, for that matter) to work with a JSP is related to the MIME type. The first line of your JSP file has to be the JSP page directive setting the contentType attribute. If this line is not present, then the microbrowser will not be able to display the content. The other attribute that needs to be set for the page directive is the language. Java is the only language supported in the current specification, so that is what we set it to. There are many other attributes, such as import, extends, session, errorPage, and buffer, that you may want to look into when creating your own JSPs.

On line 3 is a comment. You can add comments throughout your JSP to help a reader understand the code. Keep in mind that these are not Java comments and therefore are not removed on the server. So do not put anything in this type of comment that you do not want to be seen by the client. In this case, the comment is helpful for the JSP editor, but will not make any sense to those viewing it in a browser, since they will not see the code in lines 4 to 6. These lines contain JSP declarations. Declarations start with <%! and end with %>. In between, you can enter Java code to set variable values and other declarations. In our example, we are setting three strings with default values. Each line of Java code in a declaration ends with a semicolon.

Now we get into the HDML code. Lines 8 to 33 comprise the HDML template. The HDML code in this example consists of four cards. The first three display inventory information; the fourth uses Java code to output a personalized message. Let's look at the first card, which is on lines 9 to 13. After setting up the name of the card in line 9, we define the actions for the two soft keys on the device. On line 10, the accept key is set to move the user to the second card in the HDML deck, with the name item2. On line 11, the other soft key, defined as soft1, links to another JSP called details.jsp, passing a parameter on the URL. The parameter name is product_id, and the value is defined using a JSP expression. Expressions are very good for embedding values within your markup language. In this example, we are using the expression <%= item1_id %> to set the value to 101, as defined earlier in the declaration on line 4. Once the actions for the card have been set, we output the data that will be seen on the device. For the first card, the data is Sony-TRV30 Digital Video Camcorder. In Figure 13.4, you can see the output of the first card on the Openwave HDML simulator. The JSP code and the resulting output for the second card (lines 14 to 18) and third card (lines 19 to 23) produce outputs similar to the first card.

```
1.   <%@ page contentType="text/x-hdml"%>
2.   <%@ page language="java"%>
3.   <!-- string declaration -->
4.   <%! String item1_id="101"; %>
5.   <%! String item2_id="102"; %>
6.   <%! String item3_id="103"; %>
7.   <!-- HDML code to display Inventory list -->
8.   <HDML VERSION="3.0">
9.     <display name="item1">
10.      <action type="accept" task="go" dest="#item2" label="Skip">
11.      <action type="soft1" task="go"
dest="details.jsp?product_id=<%=item1_id %>" label="Details">
12.      Sony-TRV30 Digital Video Camcorder
13.    </display>
14.    <display name="item2">
15.      <action type="accept" task="go" dest="#item3" label="Skip">
```

Listing 13.7 Java Server Page that outputs HDML code. *(continued)*

```
16.      <action type="soft1" task="go"
dest="details.jsp?product_id=<%=item2_id %>" label="Details">
17.      Hitachi-VMD875L Digital 8 Camcorder
18.    </display>
19.    <display name="item3">
20.      <action type="accept" task="go" dest="#finish" label="finish">
21.      <action type="soft1" task="go"
dest="details.jsp?product_id=<%=item3_id %>" label="Details">
22.      Sony-DCR-IP7BT Micro MV Network Handycam
23.    </display>
24.    <display name="finish">
25.      <action type="accept" task="return" label="Done">
26.      <!-- Java scriptlet -->
28.      <%
29.          String username = request.getParameter("user");
30.          out.println("Thank-you for visiting " + username);
31.      %>
32.    </display>
33. </HDML>
```

Listing 13.7 *(continued)*

The final card in the deck is different from the first three. In this card, rather than displaying static data, we use a JSP scriptlet to embed Java code that produces the output. In a JSP, you can embed blocks of Java code between the <% and %> tags (this is a scriptlet). This code is parsed by the server and executed as-is in the resulting servlet. In our example, the Java scriptlet starts on line 28 and ends on line 31. This code uses the request object to get a parameter called user from the URL. It then uses an out.println() statement to send a thank-you message to the user.

Being able to program in Java throughout your JSP is a powerful tool. In addition to accessing implicit objects available to the JSP, you can also communicate with external classes to access databases and other business logic. A word of caution is in order here: With the power that Java brings, some programmers go too far, and embed much more Java code than they should. Once you have more than a few lines of embedded Java code, you may want to consider using a JavaBeans component for your logic. The JavaBeans component contains Java logic in a defined format. The logic in this bean can be reused in other parts of this application or in entirely separate applications without additional programming. JSPs have prebuilt tags for accessing JavaBeans components. The syntax follows:

```
<jsp:useBean id="inventoryBean" class="sample.InventoryData" />
```

After the bean is defined using this syntax, you can then call methods in the bean directly from your JSP or Java code. In the same way it is possible to create your own JSP tags. Any time that a set of logic will be required in several JSP pages, it may be worthwhile to create a JSP tag library. Some of the wireless application vendors have taken this approach and have created JSP tag libraries for wireless application development. (Information on these vendors is available in Chapter 14.)

Figure 13.4   Output from JSP shown in Listing 13.7.

Other than their development methodologies, JavaServer Pages and Java servlets are nearly the same. The JSP specification extends the Java servlet API. Both are part of the J2EE specification, allowing them to easily interact with one another and with other J2EE technologies. Actually, at runtime, JSPs are compiled into Java servlets! The first time a JSP is requested, it is parsed into Java code, then compiled into a Java servlet. This servlet is then executed on the servlet engine, and will return the resulting content. Subsequent requests do not require a recompile, so the JSP simply is executed in the same manner as any other servlet. Figure 13.5 shows the logic that the server uses to determine whether a JSP has to be recompiled. When a request comes in, the server determines whether the JSP code has been changed since the last time it was executed. If it has been changed, the JSP is parsed into the Java source and compiled into a servlet, which is then executed. If the JSP source has not changed, the resulting servlet can be executed immediately. By using this process, the server ensures that any changes made to the JSP are displayed, without introducing a performance penalty for JSPs that have not changed.

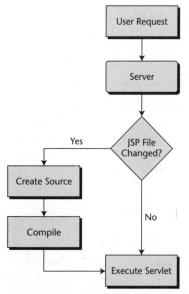

Figure 13.5   Server logic used for JSPs.

Both JavaServer Pages and Java servlets provide everything you need to create interactive, dynamic, high-performance wireless Internet applications. They can be executed on a variety of platforms from Solaris to Linux to Windows, giving you the widest range of platform flexibility of any of the technologies discussed. In addition, J2EE technology has a large following among vendors and developers alike, providing many options for developing and deploying your wireless applications. Fortunately, you do have a choice between using a code-centric servlet approach or a page-centric JSP approach to development. Since they are both executed as servlets at runtime, the decision to use JSPs or servlets can be based upon development preferences.

Active Server Pages

Active Server Pages (ASPs) provide similar capabilities to those in JSPs. They combine a markup language and scripting and server components into a single file called an Active Server Page. ASPs are almost as efficient as writing ISAPI applications, and much more efficient than CGI applications. They run as a service in Microsoft's Internet Information Server (IIS) and support multithreaded environments.

ASP technology has been around longer than both Java servlets and JSPs. ASPs were first introduced by Microsoft in 1997 as the server-side execution environment in IIS for running ActiveX scripts and ActiveX server components. In practice, ASPs are most commonly used with either JScript, Microsoft's version of JavaScript, or Visual Basic, Scripting Edition (VBScript). Both of these scripting languages are supported natively, making them much more practical than ActiveX scripts. The model most commonly used with ASPs is Microsoft's Component Object Model (COM), although other component models are also supported.

When we look at the sample ASP in Listing 13.8, we see that the syntax of ASPs and JSPs is quite similar. This listing shows an ASP that generates the WML equivalent of the HDML code generated by the JSP in Listing 13.7. Most of the basic elements are the same. All of the code between the <% and %> delimiters is processed on the server. Everything else is sent to the client-side browser.

Line 1 specifies the content type using the response object. In this case, we specify the content type to be text/vnd.wap.wml since we are creating WAP content. The second line specifies the scripting language. This line is optional. If it is not present, VBScript is the default value. Lines 4 to 6 specify values for the three variables that we use later in the WML code. (Notice that with VBScript you do not have to define a data type for the variables.) After these initial lines of script, we are left with our WML template and embedded VBScript expressions for the remainder of the ASP, in lines 7 to 48.

```
1.   <% response.ContentType = "text/vnd.wap.wml" %>
2.   <%@ Language=VBScript %>
3.   <!-- variable declaration -->
4.   <% Item1_id="101" %>
5.   <% Item2_id="102" %>
6.   <% Item3_id="103" %>
7.   <?xml version="1.0" encoding="UTF-8"?>
```

Listing 13.8   Active Server Page that outputs WML code.

```
8.  <!DOCTYPE wml PUBLIC "-//WAPFORUM//DTD WML 1.1//EN"
"http://www.wapforum.org/DTD/wml_1.1.xml">
9.  <wml>
10.   <card id="item1">
11.     <do type="accept" label="next">
12.       <go href="#item2"/>
13.     </do>
14.     <do type="cancel" label="details">
15.       <go href="details.asp?product_id=<%=Item1_id %>"/>
16.     </do>
17.     <p align="center"><b>Inventory Items</b></p>
18.     <p> Sony-TRV30 Digital Video Camcorder </p>
19.   </card>
20.   <card id="item2">
21.     <do type="accept" label="next">
22.       <go href="#item3"/>
23.     </do>
24.     <do type="cancel" label="details">
25.       <go href="details.asp?product_id=<%=Item2_id %>"/>
26.     </do>
27.     <p align="center"><b>Inventory Items</b></p>
28.     <p> Hitachi-VMD875L Digital 8 Camcorder </p>
29.   </card>
30.   <card id="item3">
31.     <do type="accept" label="next">
32.       <go href="#finish"/>
33.     </do>
34.     <do type="cancel" label="details">
35.       <go href="details.asp?product_id=<%=Item3_id %>"/>
36.     </do>
37.     <p align="center"><b>Inventory Items</b></p>
38.     <p> Sony-DCR-IP7BT Micro MV Network Handycam </p>
39.   </card>
40.   <card id="finish">
41.     <do type="accept" label="start over">
42.       <go href="#item1"/>
43.     </do>
44.     <!-- VBScript to get URL parameter -->
45.     <% userName = request.queryString("user") %>
46.     <p> Thank-you for visiting <%= userName %> </p>
47.   </card>
48. </wml>
```

Listing 13.8   *(continued)*

This application works much like the HDML example did. There are four cards. The first three display inventory information; the fourth shows a personalized thank-you message. On lines 15, 25, and 35, we embed the product_id in the WML code, using VBScript. The VBScript has been directly incorporated into the markup language using the <%=variable_name%> syntax. These lines show an href to another ASP page, passing a parameter with the URL. Finally, in lines 45 and 46 we have VBScript that uses the request object to obtain the username from the URL parameter called user. Once the username is obtained, it is then displayed with a thank-you message to the user. This example shows how you can send and obtain information to and from the ASP page.

As for the technologies previously discussed, we have only touched the surface of what ASPs can do. This is even more true in regard to the latest version, ASP.NET. The following is the definition of ASP.NET as stated by Microsoft on the ASP.NET developer site:(http://msdn.microsoft.com/library/default.asp?url=/nhp/Default.asp?contentid=28000440)

> *ASP.NET (formerly referred to as ASP+) is more than the next version of Active Server Pages (ASP); it is a unified Web development platform that provides the services necessary for developers to build enterprise-class Web applications. While ASP.NET is largely syntax-compatible with ASP, it also provides a new programming model and infrastructure that enables a powerful new class of applications. You can augment your existing ASP applications by incrementally adding ASP.NET functionality to them.*

Depending on your point of view, ASP technology is either very proprietary or very well integrated. It is a Microsoft technology that works great with other Microsoft technologies, as you would expect. As discussed, when creating an Active Server Page, you will use either VBScript or JScript as the scripting language, COM as the component model, and execute them in IIS, running on Microsoft server platforms. If you are currently using Microsoft technology for your development efforts, ASPs may be the ideal solution for creating your wireless Internet applications. If you do not typically use Microsoft technology, or use operating systems other than 32-bit Windows, you may want to look at the available Java technologies or an XML/XSL solution.

XML with XSL Stylesheets

The final approach that we will look at is converting eXtensible Markup Language (XML) data with eXtensible Stylesheet Language (XSL) stylesheets. This combination has generated much excitement in the industry due to its clean separation of data and presentation logic. Just as ASPs and JSPs improved on CGI and servlets in separating data from presentation, XML and XSL improve upon the ASP and JSP approach. Rather than using a page-based approach with embedded logic, XSL stylesheets can take raw XML data and format it to the desired markup language. This results in a separate XSL stylesheet for each markup language generated. For example, if you have one set of XML data that you want to present using XHTML, WML, and cHTML, you will end up with three separate XSL stylesheets for each page of generated content. If your application requires five decks for the WML application, and five pages for both the XHTML and cHTML applications, you would then require 15 XSL stylesheets to format the pages to the appropriate content. Essentially, each page you are developing requires its own XSL stylesheet.

We will go through an example of the entire process, starting with the XML document, then formatting it using an XSL stylesheet, and looking at the resulting WML code. Listing 13.9 is the XML document that contains the inventory data used for the application. It conforms to the XML syntax rules as discussed earlier in this chapter: All tags are case-sensitive; attribute values are enclosed in double quotes; all elements are properly nested; and all elements are well-formed. This is required for the XSL Transformation (XSLT) to take place.

XML Document

Listing 13.9 contains a complete XML document. The XML document is text, and can be stored in a text file, a database, or any other storage mechanism. Or it may be assembled by some of the technologies we have discussed earlier in the chapter, including CGI, Java servlets, JSPs, or ASPs. In any case, it is stored and transmitted in a text format and can be edited and viewed with any standard text editor.

If you are trying to do more than simply view the XML document, you will require an XML parser. The parser will divide the document into individual pieces that can be passed back to the application. If the XML is not well formed, the parser will return an error. It will not attempt to fix the problem in the file. If you choose to, you can go one step beyond having well formed documents. If, say, you want to define a set of precise rules regarding what is valid for the XML document, you will want to use a Document Type Definition (DTD). The DTD either can be created for a specific purpose or be an existing DTD that is appropriate for the task. If you decide to use a DTD, it can either be included in the XML document itself or the XML document can point to a Universal Resource Identifier (URI) where it is located. In order to make sure the XML document adheres to the DTD, you will require a validating parser. These parsers will not only check to make sure the document is well formed, but will also make sure it is valid, as defined by the DTD.

Many applications include XML parsers. These include Web browsers, such as Internet Explorer and Netscape; word processors, such as StarOffice; and many development tools, including Sybase's PowerBuilder. Actually, since XML is such a flexible format for formatted data, anytime data is encoded as text; an XML document can be used to define the document structure.

In our example, the XML document is well formed, but no DTD is specified. The root element for this document is <inventory>. This is the only element in our document that does not have a parent element. It does, however, have one child element, <product>. Each <product> element then has six child elements: <name>, <description>, <digitalstill>, <format>, <quantity>, and <price>. Even though <product> has six children, each child only has one parent. If we go one layer deeper, the <name> element has two children: <manufacturer> and <model>. In addition, each product also has one attribute defined, the id. Data can be held in either child elements or in attributes. The decision is up to the designer of the XML document.

In our document we have five products with associated data. Each product does not necessarily have the same set of child elements. These vary depending on the product itself.

NOTE In order to create effective XSL stylesheets, it is a good idea to know
the format of the XML data. We covered enough here to enable us to
understand the XSLT process; this section does not address the full extent of
the XML document structure.

```xml
<?xml version="1.0"?>
<inventory>

  <product id="101">
    <name>
      <manufacturer>Sony</manufacturer>
      <model>TRV30</model>
    </name>
    <description>Digital Video Camcorder</description>
    <digitalstill>1360 x 1020</digitalstill>
    <format>Mini DV</format>
    <quantity>17</quantity>
    <price>1699.00</price>
  </product>

  <product id="102">
    <name>
      <manufacturer>Hitachi</manufacturer>
      <model>VMD875L</model>
    </name>
    <description>Digital 8 Camcorder</description>
    <format>Digital8</format>
    <quantity>24</quantity>
    <price>599.00</price>
  </product>

  <product id="103">
    <name>
      <manufacturer>Sony</manufacturer>
      <model>DCR-IP7BT</model>
    </name>
    <description>Micro MV Network Handycam</description>
    <digitalstill>640 x 480</digitalstill>
    <format>Micro MV</format>
    <quantity>11</quantity>
    <price>2199.99</price>
  </product>
```

Listing 13.9　XML Inventory Data

```
<product id="104">
    <name>
        <manufacturer>JVC</manufacturer>
        <model>GR-DV2000</model>
    </name>
    <description>High-Band Digital Video Camcorder</description>
    <digitalstill>1600 x 1200</digitalstill>
    <format>Mini DV</format>
    <quantity>4</quantity>
    <price>1599.00</price>
</product>

<product id="105">
    <name>
        <manufacturer>Canon</manufacturer>
        <model>ES8200V</model>
    </name>
    <description>8 MM Camcorder</description>
    <format>HI8MM</format>
    <quantity>37</quantity>
    <price>399.00</price>
</product>

</inventory>
```

Listing 13.9   *(continued)*

XSL Stylesheet

Now that we understand the basic concepts about our XML document, we can examine how XSL stylesheets can be used to transform the data into various markup languages. There are many ways a stylesheet can be used to process XML data. One of the most common is to specify the stylesheet as line 1 of the XML document using the following syntax:

```
<?xml-stylesheet href="inventory.css" type="text/css"?>
```

This above syntax is often used with Cascading Style Sheets (CSS) that are processed by the Web browser. This is the method of transformation defined for the mobile profiles of XHTML. In our example, we are using an XSL stylesheet to do the transformation. XSL is the XML application that defines how one XML document is transformed into another. In our case, we will transform our XML data document into WML. In order for the translation to work, the XML document has to be well formed, but it does not necessarily have to be valid. Let's look at our XSL stylesheet in Listing 13.10.

Line 1 is the XML declaration. This is not required, but is good to include. Line 2 of the stylesheet is the root element of the document. It has to be either a stylesheet, as we have used, or a transform. Either one is equally valid and means the same thing to an XSLT processor. For our example, we have used the Apache XML Project's Xalan XSLT processor.

A template-based methodology is used for the XML transformation. To match an output with an input, you define templates in your XSL stylesheet. The templates are defined by the <xsl:template> element, as shown in line 4. The element has a match attribute that uses XPath to identify the XML input it matches. When a corresponding input is found, the contents of the template are executed. On line 4 we have a template looking for a match on inventory. This happens to be the root element of our XML document, so there will be exactly one match in the XML document. The output of the match is the <wml> tag, along with a card with id="inventory". Within the card is a select element, which then applies the product template, as shown on line 10.

By using the <xsl:apply-templates> element, we can change the order in which the stylesheet is applied. By default, the XSLT processor goes from top to bottom. When it sees the <xsl:apply-templates> element, it will know which element to process next. In our example, we move from line 10 to the product template defined on line 16. The output from the product template will then be embedded into the code in the inventory template.

The first thing we do in the product template is set the variable named product_id, as shown on line 17. Line 18 defines the value to which we set the variable; it is retrieved from the id attribute within each product element of our XML document. After the variable is set, we add the item (as defined in the product element) to an <option> element. For the value shown for each option, we go to another template called name. This gets the <manufacturer> and <model> data. The code for this is shown on line 30. After the name template is applied, we then add the href with a postfield element, as shown on lines 23 and 24. This process will happen five times, because we have five matches for the product template, one for each product in our XML document.

```
1.    <?xml version="1.0"?>
2.    <xsl:stylesheet version="1.0"
xmlns:xsl="http://www.w3.org/1999/XSL/Transform">
3.    <xsl:output method="xml" indent="yes" doctype-
system="http://www.wapforum.org/DTD/wml_1.1.xml" doctype-public="-
//WAPFORUM//DTD WML 1.1//EN" />
4.    <xsl:template match="inventory">
5.    <wml>
6.      <card id="inventory">
7.        <p align="center">Inventory Items</p>
8.        <p>
```

Listing 13.10 XSL Stylesheet to convert XML from Listing 13.9 into WML.

```
9.            <select name="productId" multiple="false">
10.              <xsl:apply-templates select="product"/>
11.            </select>
12.          </p>
13.        </card>
14.    </wml>
15.    </xsl:template>
16.    <xsl:template match="product">
17.        <xsl:variable name="product_id">
18.          <xsl:apply-templates select="@id" />
19.        </xsl:variable>
20.        <option value="{$product_id}">
21.          <xsl:apply-templates select="name"/>
22.          <onevent type="onpick">
23.            <go href="details.wml">
24.              <postfield name="product_id" value="{$product_id}" />
25.            </go>
26.          </onevent>
27.        </option>
28.    </xsl:template>
29.    <xsl:template match="name">
30.        <xsl:value-of select="manufacturer"/>-<xsl:value-of
select="model"/>
31.    </xsl:template>
32.    </xsl:stylesheet>
```

Listing 13.10   *(continued)*

> **NOTE** Internet Explorer 5.0 and 5.5 include an XSLT processor, but do not support XSLT 1.0. When testing your applications, do not depend on IE to give you accurate output.

The resulting WML code from applying the XSL stylesheet to our XML document can be seen in Listing 13.11. The output on the Openwave simulator is shown in Figure 13.6. This figure has a complete WML deck containing a card with a select list. The list contains the five products from our XML code. For each product the manufacturer and model is shown. If a particular product is selected, another WML document called details.wml is called, and the product_id is passed as a parameter on the URL. This was accomplished using the postfield element. All of this WML code was generated automatically by the XSL stylesheet. If we wanted to generate code for another markup language, we would have to create another stylesheet similar to the one in Listing 13.10, but with specific tags for the markup language being targeted.

```
<?xml version="1.0" encoding="UTF-8"?>
<!DOCTYPE wml PUBLIC "-//WAPFORUM//DTD WML 1.1//EN"
"http://www.wapforum.org/DTD/wml_1.1.xml">
<wml>
  <card id="inventory">
    <p align="center">Inventory Items</p>
    <p>
      <select multiple="false" name="productId">
        <option value="101">Sony-TRV30
          <onevent type="onpick">
            <go href="details.wml">
              <postfield value="101" name="product_id"/>
            </go>
          </onevent>
        </option>
        <option value="102">Hitachi-VMD875L
          <onevent type="onpick">
            <go href="details.wml">
              <postfield value="102" name="product_id"/>
            </go>
          </onevent>
        </option>
        <option value="103">Sony-DCR-IP7BT
          <onevent type="onpick">
            <go href="details.wml">
              <postfield value="103" name="product_id"/>
            </go>
          </onevent>
        </option>
        <option value="104">JVC-GR-DV2000
          <onevent type="onpick">
            <go href="details.wml">
              <postfield value="104" name="product_id"/>
            </go>
          </onevent>
        </option>
        <option value="105">Canon-ES8200V
          <onevent type="onpick">
            <go href="details.wml">
              <postfield value="105" name="product_id"/>
            </go>
          </onevent>
        </option>
      </select>
    </p>
  </card>
</wml>
```

Listing 13.11 Resulting WML code from XML in Listing 13.9 and XSL stylesheet in Listing 13.10.

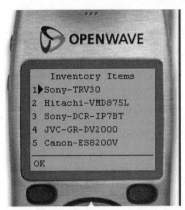

Figure 13.6 Openwave simulator showing the WML output from Listing 13.11.

While the XML/XSL approach is ideal at a conceptual level, it does impose some problems during delivery. The first issue is the sheer number of XSL stylesheets that have to be created. You will require 15 stylesheets for a typical application with five separate screens/decks that target three markup languages. The resources required to develop and test these may be prohibitive. This limitation may be less significant in the future as many of the protocols move to XHTML as the markup language and as better tools come on the market for XSL creation; but at this point in time it is something to think about.

Even if you do decide that developing the required XSL stylesheets is worth the effort, you may still come across one other problem: performance. Each time a device makes a request, the XSLT processor has to transform the XML to the appropriate markup language. This is a resource-intensive process. Even with the caching mechanisms available, the performance penalty associated with this process may lead to problems for applications that will have a large number of simultaneous users.

Summary

Many wireless markup languages are in use today, including HDML, WML, cHTML, HTML, and XHTML. Even though HDML is still supported by many North American devices, it is recommended that you use WML or XHTML for new development. This is because WAP 1.x is based on WML, while WAP 2.x, along with other wireless Internet standards, have moved to XHTML. For wireless applications, there are two specific forms of XHTML of interest: XHTML Basic and XHTML Mobile Profile. Several technologies can be used for dynamic content generation. Some of these, such as CGI with Perl and Java servlets, are code-driven; others, such as JSPs and ASPs are page-driven. In addition, a template approach can be taken using XML with XSL stylesheets.

The next chapter provides an overview of wireless Internet technologies and the vendor products that are available.

Helpful Links

Access these links for more in-depth coverage of the topics touched on in this chapter:

HDML and WML Style Guides from Openwave Systems	http://developer.openwave.com /support/techlib.html
Java Server Pages information from Sun Microsystems	http://java.sun.com/products/jsp
Java Servlets information from Sun Microsystems	http://java.sun.com/products/servlet/
Microsoft .NET homepage	www.microsoft.com/net/default.asp
Perl documentation and official FAQ	www.perldoc.com; www.perl.com
WAP 1.x and 2.x specifications	www.openmobilealliance.org
W3C recommendations and related information	www.w3.org
XHTML basic W3C recommendation	www.w3.org/TR/xhtml-basic/
XML and XSL information	http://xml.apache.org http://xml.oreilly.com/

Wireless Internet Technology and Vendors

Many technologies are available to aid with the development and deployment of wireless Internet applications. Using these technologies will make wireless application development much more approachable, increasing developer productivity. As demonstrated in the previous chapter, you can use common content-generation technologies to create wireless content. It is only when more sophisticated wireless Internet applications are required that existing Web technology is not adequate and problems start to arise; for example, applications that need to target multiple markup languages on multiple devices over multiple wireless networks. It is for these applications that products focused on wireless development and deployment become truly valuable.

In this chapter we are going to look at the technologies commonly used to implement wireless Internet applications. They are divided into four categories:

- Microbrowsers
- Wireless application servers
- Development tools
- Wireless service providers

For each category we will investigate the key technology features and summarize related vendor solutions. The goal is provide you with enough information so that you can make educated decisions as to which technology and vendors you want to evaluate.

Microbrowsers

The microbrowser is the most visible part of any wireless Internet application, as it is the only part of the solution that the end user sees. It is also the only part of the solution that is executed on the client device; the data, business logic, and content-generation routines all reside on the server. For these reasons, the microbrowser usually dictates the features that are available for any particular client device.

You have to choose a microbrowser that will meet the needs of your application. For mobile phones, where the browser is embedded on the device and cannot be changed, this means you will also be making a device decision. In turn, by choosing a particular device, you usually will also be tied to a certain wireless network. Fortunately, at this time, a single vendor, Openwave, provides the majority of browsers for this class of device.

For more capable devices, no such limitation exists. Several browsers are available for Palm, Pocket PC, and RIM devices. Many of the browser vendors are also wireless service providers that provide wireless network connectivity as well as hosting services. More information on these services is available later in this chapter. For now, we are going to focus on the technologies and vendors involved with the microbrowser itself.

Technology

The following paragraphs discuss some of the most common features found in wireless microbrowsers. This is not an exhaustive list, but it does provide a good starting point for comparing the features offered by different vendors.

Markup Languages

The markup language dictates the content types and capabilities that are available for wireless Internet applications. Most mobile phones support HDML, WML, or cHTML, while PDA browsers usually support WML or HTML. In some cases, the browser may support multiple markup languages. Microsoft's Pocket Internet Explorer, for example, supports HTML, WML, and cHTML. Unfortunately, not all browsers display the markup language the same. A WML site might work great in one browser but cause another browser to give a fatal exception. For this reason, content has to be catered to both the device and the browser on which it is going to be displayed.

In the future, we may not have to worry about the large number of markup languages, as many of the technologies, including WAP and i-Mode, are converging on XHTML as the markup language of choice.

> **NOTE** An in-depth look at markup languages and content-generation techniques is provided in Chapter 13, "Wireless Languages and Content-Generation Technologies."

Image Support

Not all browsers provide the same level of support for images. Many of the mobile phone browsers support black-and-white wireless bitmap (WBMP) images, while browsers on the more advanced devices may provide full JPEG and GIF support. WBMP files are bitmap images suitable for wireless devices, whereas JPEG and GIF files are common desktop image formats. You would be wise to limit the number of images on your wireless Web sites to minimize the download times over bandwidth-constrained networks.

Push Technologies

Support for push technologies is growing in importance for wireless Internet users. The most common technologies include HDML notifications and WAP Push. The Openwave HDML browser supports HDML notifications. Browsers that support WAP version 1.2 or later support WAP Push, although support on the WAP gateway is also required for this functionality. (More information on Push messaging can be found in Chapter 5, "Mobile and Wireless Messaging.")

Offline Support

The ability to access offline content is the most recent feature being added to many microbrowsers. Initially, this feature was in the form of a browser cache, but it has evolved to include access to persistent data stores and, on Pocket PC devices, local Web servers. This trend is expected to continue, as the rollout of 3G wireless networks is not providing the true pervasive access to wireless data they were expected to.

Web Clippings

Palm developed the Web-clipping architecture to provide support for advanced, feature-rich, HTML-based applications. It separates the static components of the Web site (such as images, HTML templates, etc.) from the dynamic content (data that changes frequently). The static information is then transformed into the Web-clipping application format (.pqa file) and deployed to the device in advance. Since the HTML templates and images are not downloaded at runtime, advanced user interfaces can be created without performance penalties. The Web-clipping format is supported on all wireless Palm OS-powered devices. GoAmerica offers a similar technology called MobileClips for Go.Web microbrowsers.

Security

The type of security supported depends on the device and wireless protocol being used. The most common form of security for wireless Internet applications is encryption of the communication stream between the microbrowser and the gateway or Web server. For WAP applications, this support is provided by WTLS, while for many HTML applications it is provided using SSL/TLS. (Detailed information on mobile and wireless security is provided in Chapter 6, "Mobile and Wireless Security.")

Device Support

Many device manufacturers either include a microbrowser or recommend a browser for use on their devices. These browsers usually are tested extensively and work well on the device with which they are included. If you are unhappy with the default browser, you usually have the option to replace it with one of your choice. You simply have to find a browser that supports the mobile operating system and chipset you are using, then download it to the device. This is true for all devices except mobile phones, which typically ship with an embedded microbrowser that cannot be replaced.

Microbrowser Vendors

This section provides an overview of the leading vendors and their products. We do not make recommendations for any vendors, as this is a decision that must be made in consideration of the features of the applications that are going to be deployed. Moreover, it is common that applications will be developed for, and deployed to, more than one type of device and more than one flavor of microbrowser.

Openwave Mobile Browser

URL	www.openwave.com/products/device_products/ mobile_browser/index.html/Markup
Markup Languages	HDML, WML, XHTML-MP, cHTML
Operating Systems	Usually embedded on mobile phone OS

Summary:

The Openwave Mobile Browser is by far the most common microbrowser being used today. By mid-2002, close to 50 device manufacturers had licensed the Openwave Mobile Browser for use on almost 200 mobile phones. There are now more than 70 million shipped devices that include this browser. A complete list of the devices using the Openwave Mobile Browser is provided at http://upmkt.openwave.com/dev _phones/phones.cfm. (Note: The Openwave Mobile Browser is commonly known as the UP.Browser. The UP stands for Unwired Planet, the former name of the company.)

Openwave is the creator of HDML, and the company has the only browser on the market that supports it. Version 3.x of the Openwave Mobile Browser is based on

HDML. As WAP grew in popularity, Openwave released the 4.x series of browsers with native WML support. These browsers also have HDML translation support for backward compatibility. The most recent release of the Openwave Mobile Browser is version 6.2. It has native support for WAP 2.0, XHTML Mobile Profile, WML 1.x, WAP Push, WML GUI for M-Services, MMS, and cHTML. A full summary of the browsers and their related SDKs is available at http://developer.openwave.com /resources/sdk.html.

Microsoft Pocket Internet Explorer

URL	www.microsoft.com
Markup Languages	HTML 3.2, WML, cHTML, XHTML
Operating Systems	Windows CE 3.0, Windows CE .NET (Pocket PC, Pocket PC 2002)

Summary:
Microsoft's Pocket Internet Explorer (PocketIE) is the most capable microbrowser available. It has complete support for HTML 3.2, with support for tables, forms, and frames. In addition, the latest versions have added support for WML and cHTML. PocketIE also supports Microsoft JScript, Bitmap, GIF, and JPEG image types. For security it supports SSL versions 2.0 and 3.0.

When using PocketIE, you can access many existing Internet sites without modification. It also has automatic state detection to determine if the device is connected to the Internet. If it is not, it diverts the browser to a cached version of the Web page, if available.

AvantGo Client

URL	www.avantgo.com
Markup Languages	HTML, JavaScript
Operating Systems	Pocket PC, Palm OS, RIM, Symbian OS

Summary:
The AvantGo client browser was built with offline support in mind. Content is delivered using the channel concept, where users can preselect the channels they are interested in and synchronize the corresponding content to their device. A channel represents a single Web site or application. Applications are written in HTML along with JavaScript. The AvantGo browser comes preinstalled on many mobile devices. It is a good choice for cross-platform deployments. It works best when used with the AvantGo M-Business Server. In December 2002, iAnywhere Solutions (www.iany where.com) made an offer to purchase AvantGo. By the time you read this, the acquisition should be complete.

GoAmerica Go.Web

URL	www.goamerica.com/goweb/
Markup Languages	HTML, WML
Operating Systems	Palm OS, Windows CE 2.11 and 3.0, RIM, 32-bit Windows

Summary:
The Go.Web browser comes with the Go.Web Internet service from GoAmerica. It offers support for both HTML and WML, allowing you to access most wireless Internet applications. Because it supports a broad range of mobile operating systems, it is a good choice for applications that need to be deployed to a variety of devices. The latest versions of Go.Web offer complete offline support with store-and-forward message capabilities. In addition, Go.Web supports push capabilities as defined in WAP 1.2.

NOTE The features supported for each version of Go.Web depend on the operating system on which it is being used.

Neomar Microbrowser

URL	www.neomar.com/products/index.html
Markup Languages	WML
Operating Systems	Palm OS, RIM, Windows CE, J2ME

Summary:
The Neomar Microbrowser is available for all leading PDA-style devices, allowing you to create your WML content once and view it on many devices. Neomar has put emphasis on creating a friendly user experience by incorporating data compression, support for GIF and WBMP images, and on-device caching. If you are looking for offline data or push capabilities, you can install a plug-in for the browser, called the Neomar Intelligent Client Engine (ICE).

For security, Neomar has implemented WPKI support, with support for WTLS 2.0 and HTTP authentication. Similar to Go.Web, the Neomar Microbrowser is often used with Neomar's wireless Internet service offering.

Palm Web Browser

URL	www.palmos.com/dev/tech/webclipping
Markup Languages	HTML, cHTML
Operating Systems	Palm OS

Summary:
The Palm Web Browser is a full-featured HTML browser. It enables access to any URL on the Internet, whether designed for the Palm OS or not. It will automatically adapt desktop Web content to the appropriate size of the Palm handheld screen. While

this does not work perfectly for every Web site, it does allow the user to access a whole new set of content that was previously unavailable without customization. In addition, it provides the ability to save information for offline viewing and caches a history of visited sites.

The browser maintains support for the Web-clipping architecture, allowing developers to create more advanced client applications without having to worry about the associated download times over wireless networks. The Palm Web Browser is available for devices running Palm OS 5.0 and above.

Opera Software

URL	www.opera.com/devices/
Markup Languages	HTML
Operating Systems	Symbian OS, Linux, QNX

Summary:

The Opera browser is best known for its cross-platform capabilities; it runs on seven different operating systems: BeOS, Symbian OS, Linux, Mac, OS/2, QNX, and Windows. It is a high-performance browser that has full support for HTML on mobile platforms. The Symbian reference design includes support for Opera. Other manufacturers such as Sharp and Psion include the Opera browser with their devices. Other features of the Opera browser include support for JavaScript, HTML 4.0.1, HTTP 1.1, XML, and SSL 2.0 and 3.0.

Device-Specific Browsers

In addition to the browser products covered so far, a number of proprietary browsers have been created by device manufacturers. These browsers all are embedded on mobile phones and are WML-based. They usually are similar to the Openwave Mobile Browser in appearance and feature set.

Wireless Application Servers

Internet applications employ a three-tier architecture, with one or more data sources, a client browser, and a Web server. Wireless Internet applications are no different, except that the Web server is often replaced with or enhanced by a wireless application server. These servers extend the capabilities offered by standard application servers by adding support for wireless application development and deployment, usually based on J2EE technology. In many cases, the technology provided by a wireless application server vendor is installed onto existing application servers, such as those provided by BEA Systems, IBM, or Sybase.

Comparing J2EE application servers is beyond the scope of this book, but it is safe to say that performance, security, scalability, business logic management, component support, and enterprise data connectivity are among the features that you should consider when selecting an application server.

Technology

Every wireless application server vendor has its own list of features that the company markets as critical for a successful solution. That said, these features do not always coincide with the features required by the typical wireless application developer. This section provides a summary of the core features that you can expect to find in any complete wireless application server implementation.

> **NOTE** Do not confuse these offerings with WAP gateways. A WAP gateway is typically deployed at the wireless carrier site and does not have to be implemented by individual enterprises. It is responsible for the communication from a wireless device to the gateway using WAP, and then from the gateway to the enterprise server, using HTTP. That said, many of the features described here may also be supported by WAP gateways.

Request-Handling Proxies

In most cases, the wireless application server has to be able to work in harmony with the enterprise Web server. Corporations already have Web sites, and these should not be adversely affected when a wireless solution is implemented. To prevent this, request-handling proxies, or redirectors, can be installed on your existing Web server. These proxies enable desktop Web requests to be handled by the existing Web server infrastructure and to forward requests from wireless devices to the wireless application server. These capabilities provide the following benefits:

- You can maintain the same URL for all Internet clients, wireless or otherwise.
- You can maintain the same Web server and firewall configurations.
- It can provide load-balancing capabilities for your wireless clients, improving scalability and performance.

Content Transformation

Content that looks great on a Pocket PC device may not be at all suitable for a WAP phone, and vice versa. With so many microbrowsers and devices on the market, it is important that content can be delivered in a customized format. Ideally, this can be accomplished without having to write a new version of the application for every browser/device combination you require. The concept behind content transformation (often called transcoding) is to be able to reuse existing content, such as that from an HTML Web site, for wireless clients. There are many ways to accomplish this, with varying degrees of effort and overall effectiveness. If you are going to be deploying to many wireless browsers, you will want to evaluate the content-transformation techniques that are provided by the vendor you select. Keep in mind that if it looks too good to be true, it probably is.

Device/Browser Identification

In order to provide customized content for specific microbrowsers, you need a way to identify them. The most common way is by viewing the UserAgent that is part of the HTTP request header. The UserAgent will typically identify the microbrowser in use, along with its version number. Once you know this information, you can obtain the properties of that specific browser and format your content accordingly. Some products also keep a repository of microbrowser information, so you do not have to do this research yourself.

In addition to the UserAgent, the HTTP header can also provide you with much more information about the device itself, including the preferred language (for example, English or Japanese), whether cookies are supported, the maximum packet size for the device, and supported MIME types.

Another way to determine a device's capabilities is by using Composite Capabilities/Preferences Profile (CC/PP). CC/PP is a collection of the capabilities and preferences of any client that accesses the World Wide Web. It contains information about the hardware platform, system software, and any applications used by the user. In addition, it can also contain information about individual user preferences, allowing for very customized content delivery. (For more information on CC/PP, visit the W3C site at www.w3.org/Mobile/CCPP.)

Dynamic Content Generation

To be effective, wireless Internet applications require dynamic content. This may include real-time weather information, stock quotes, or business-related data such as price lists and inventory levels. Wireless application servers will typically support Java servlets, JSPs, ASPs, or XML and XSL to provide this capability. The best wireless Internet applications provide dynamic data personalized for the current user and customized for the microbrowser and device they are using. (Chapter 13 covers the leading content-generation techniques, using code examples.)

Session and State Management

Most wireless microbrowsers on the market do not provide client-side session management capabilities. To understand the impact for Internet applications, turn off cookie support in your Netscape or Internet Explorer browser and visit your favorite Web sites. You will quickly see just how important this capability is. You will lose the ability to store valuable information on the client for user authentication and personalization. As well, information gathered on one page will not be available on another.

To overcome this limitation, wireless application servers maintain session information on the server. This is usually accomplished in one of two ways: URL rewriting or persistent cookies. URL rewriting works by appending a unique session ID to the end of the URL to identify users as they move from page to page. This approach should only be used where there is no client support for cookies. Persistent cookies allow for the server to maintain session information, such as username, preferences, and application variables in a global object on the server. This information is available until the user logs off or until the session times out. This provides a convenient way to maintain a session even when the user device is temporarily disconnected from the network. When users reconnect, they can reestablish their session and continue where they left off.

> **NOTE** Many wireless clients, such as WAP phones, do not store cookies on the device, but still allow for sessions to be managed using cookies. This is possible because the WAP gateway stores cookies on behalf of the client and adds the cookies to any HTTP request made for that client.

Enterprise Integration

The ability to access enterprise data is crucial for any m-business application. Wireless application server products should provide at least standard database access using JDBC, ODBC, or native database drivers. In addition, a way to interact with packaged enterprise resource planning (ERP), customer resource management (CRM), and supply chain applications is also very useful. Very often, this level of integration is provided using a prebuilt custom adapter that is created in the component model supported by the server.

Another level of enterprise integration that is growing in importance is the ability to integrate with Web services and to create your own Web services. All of the leading software vendors, including Microsoft, Oracle, Sybase, Sun Microsystems, and BEA Systems, are big backers of Web services. With this level of corporate support, Web services will soon be inherent to all forms of enterprise applications. (An overview of Web services is provided in Chapter 18, "Other Useful Technologies.")

Messaging Integration

Messaging capabilities are also of growing importance for wireless applications development. Technologies such as WAP Push, HDML notifications, Short Message Service (SMS), and instant messaging (IM) can help to improve the effectiveness of wireless Internet applications. Selecting a server that allows you to incorporate these protocols into your application will prove to be beneficial. (Complete information on mobile and wireless messaging systems is provided in Chapter 5.)

Security Services

Wireless devices typically connect to the enterprise using public carrier networks. This means that sensitive enterprise data could be traveling over inherently insecure lines. For this reason, it is important to have a complete suite of security capabilities that will prevent the data from being observed or altered by unauthorized users.

To make sure you know who is accessing the server applications, user authentication capabilities have to be included in the solution. This can range from a simple username and password combination to a more complex (and more secure) solution using digital signatures. Once a user is authenticated, authorization can be implemented using access control lists (ACLs). ACLs provide a way to control to which data and logic a user has access.

To prevent unauthorized access to your data, public key cryptography solutions such as SSL/TLS or WTLS are required. It is essential that the wireless application server you select have support for these technologies. Finally, the overall solution should be able to work within the firewall environment you have in place, ideally without having to add a new entry point into the system. (Complete information on mobile and wireless security is provided in Chapter 6, "Mobile and Wireless Security.")

Scalability

As more applications are deployed, and as more users start to take advantage of those applications, the scalability of the platform becomes increasingly important. To meet these growing needs, select a product that has strong performance and that can scale without having to adapt a new architecture. This often can be accomplished using clustering technology, wherein many servers work together as a single unit. When a request comes in to the cluster, it is allocated to a particular server. This capability is referred to as load balancing. Load balancing provides a way to evenly distribute the number of users across the number of servers in a cluster to maximize performance. Many possible algorithms can be used to distribute the load, ranging from round robin to weighted distributions that are based on each server's processing power.

Operating System Support

This category is self-explanatory. There are many server-side operating systems currently in use. Most vendors usually classify them into two broad categories: Windows and UNIX. On the Windows side, support for Windows NT, 2000, and XP is required. On the UNIX side, the most common OS supported is Solaris, although Linux support is spreading. In some cases, vendors will also provide support for HP-UX, AIX, among others. (The issue of client-side operating system support was covered earlier in this chapter in the section on microbrowsers.)

Development Tools

A development tool is not a core requirement for a wireless application server, but it will definitely improve productivity. This is especially true if the developer is new to the technologies being used. A number of features are useful to have in a development tool, including support for drag-and-drop client-side development, support for multiple markup languages, built-in device emulators, and application server integration. Not all of these features are required for every application, but they are useful to have if the need arises. The topic of development tools is covered in more depth later in this chapter.

Wireless Application Server Vendors

The list of vendors that offer wireless application servers is in a constant state of change. During the height of the wireless explosion in 2000, there were easily 50 vendors that could be listed in this category. Two years later, the list is considerably shorter, with approximately 10 leading vendors and a handful of challengers. It is expected that this list will continue to change as the industry matures and consolidation continues.

One of the most interesting things about this vendor list is the similarity—or perceived similarity—of the offerings. Commentary is ongoing regarding the similarities between the product data sheets and whitepapers produced by these vendors. When you go beyond the marketing, you will discover that each vendor has many unique features and unique opinions on how wireless Internet applications should be developed. To do these products justice, each one would require its own chapter in this book. Since that is not feasible, each product is listed in Table 14.1, with a URL where you can go to find more information. With the rapid change being experienced in the wireless industry, vendor Web sites are often the best source for up-to-date information.

For vendors that have mobile platform offerings, the wireless application server is only one component of the overall solution. When making a decision as to which vendor is best for your needs, it is recommended that you look at the entire platform, since many components of the platform will provide value to your mobile solution.

There is no clear leader in this market, so the companies are listed in alphabetic order.

Table 14.1   Other Vendors with Wireless Application Server Offerings

VENDOR	PRODUCT NAME	PRODUCT URL
Aether	Aether Fusion	www.aethersystems.com
Air2Web	Mobile Internet Platform	www.air2web.com
AvantGo	AvantGo M-Business Server	http://avantgo.com/products /mbus_server_app.html
Broadbeam	Broadbeam Mobile Solutions System	www.broadbeam.com/products /mobile_platform.asp
Covigo	Covigo Platform	www.covigo.com/products
Everypath	Everypath Server	www.everypath.com /products/server.shtml
Extended Systems	Mobile Solutions Platform	www.extendedsystems.com
iAnywhere Solutions	iAnywhere M-Business Studio Server and Message Anywhere Studio	www.sybase.com/products /mobilewireless
IBM	Websphere Transcoding Publisher	http://www-3.ibm.com /software/pervasive /products/

Table 14.1   *(Continued)*

VENDOR	PRODUCT NAME	PRODUCT URL
iConverse	Mobility Platform	www.iconverse.com /products/platform.asp
Microsoft	Mobile Information Server	www.microsoft.com /miserver/default.asp
Oracle	Oracle 9i Application Server	www.oracle.com/ip/deploy /ias/mobile/index.html
724 Solutions	724 Solutions	www.724solutions.com /products/platform.asp

Development Tools

There are two main types of development tools: those tied to a specific platform and those open for all platforms. For wireless Internet applications, most browser vendors and device manufacturers offer free software development kits (SDKs) that allow you to develop applications for their device using any server-side platform. These SDKs were described in the *Development Tools and Emulators* section of Chapter 12, "Thin Client Development," so this section focuses on the development tools that are geared to developing applications for multiple devices and microbrowsers. First, the common features in these tools are examined; then a summary is given of the vendors, along with their Web addresses.

Technology

The right development tool can dramatically increase developer productivity by removing low-level complexities, enabling the developer to focus on the application logic. Many of the tools for wireless application development allow for multidevice and multibrowser development from a single environment. These development tools are great for building the client interfaces, but they usually do not expedite the development of business logic to such an extent. Here are some of the key features to look for in a wireless Internet application development tool.

Rapid Application Development

One of the foundations of today's development tools is the concept of rapid application development (RAD). Timelines for creating advanced, multimodal applications are typically tight, so development tools have to provide prebuilt, drag-and-drop, "what you see is what you get" (WYSIWYG) component-based development. These features allow for rapid prototyping and accelerated development on wireless applications. Many of the early development tools for wireless applications were based on proprietary frameworks, essentially locking you into a specific vendor. Over the last year, however, there has been a migration to open technologies, such as Java technologies and XML, making the tools much less proprietary.

Multichannel Support

Mobile access to enterprise data can be established via multiple device channels, including wireless microbrowsers, voice, and messaging. Creating an effective wireless Internet application for one platform is difficult; making it work on multiple platforms, each with different client interfaces and modes of interaction, is an even more daunting task. Applications have to be able to work with any current or emerging network protocol, device, microbrowser, and markup language, and it is unrealistic to expect developers to have the knowledge and experience to create such a diverse range of applications without a development tool to aid them. The development tool should eliminate the complexities and idiosyncrasies of these mobile technologies, allowing developers to design for multiple clients in a single design environment.

Beyond helping in the application development process, the tool also should aid with overall application management and maintenance.

Built-in Emulators

During the development cycle, wireless Internet applications are usually tested using device emulators. Most microbrowser vendors provide emulators for their products, so that developers can test their applications without having to deploy them to a server that can be accessed using a wireless network. Having these emulators incorporated into the development tool makes testing and debugging an application much easier, as you can do it all from the same environment.

Extensibility

The wireless landscape is constantly changing. Devices that are popular today may not be used at all tomorrow. Therefore, it is important that the applications you develop today will be suitable for the devices and networks of the future. One way to ensure this is by separating the presentation layer from the business logic and data integration layers. This allows businesses to quickly adapt applications to new devices, data sources, and business processes. The development tool should support this layered approach to application development.

Support for Standards

Support for open standards is an important aspect of any mobile platform. Support for XML and J2EE allows for integration with existing J2EE-compliant application servers, so that you can choose best-of-breed solutions. Similarly, support for the standard markup languages (HDML, WML, cHTML, HTML, XHTML, and VoiceXML) lets you deploy your applications to a variety of client devices without modification. Finally, the mobile platform you choose should allow for development using standard networks and security protocols, including WAP, WTLS, HTTP, and HTTPS.

Server Integration

The development tool you select should integrate with the wireless application server you are using. Minimally, this means that it must use compatible technology. For example, if your development tool creates JSPs, make sure JSPs are supported in your server platform. You will find that when vendors include a development tool as part of their platform, these tools usually have very tight integration with the server, often providing automatic deployment and debugging capabilities. Unless you have a good reason for not doing so, it is usually recommended that you use the development tool that comes with the platform you have decided to use.

Development Tool Vendors

We can divide wireless application development tool vendors into two categories: open development tools and platform-specific tools. Most of the open tools are either software development kits (SDKs), provided by microbrowser vendors, or Internet tools that have added wireless capabilities. (These tools are covered in Chapter 12.) Platform-specific tools typically provide a more robust set of features, but, as their name signifies, they are created for a specific server platform. This means that by adopting the development tool, you are adopting the entire platform. This is often not ideal, as the cost of deployment for many of these platforms is beyond what many developers are willing or able to pay.

If you are planning to adopt a platform for your mobile application development, you will want to evaluate all aspects of the platform, including the wireless application server, development tool, support for smart client applications, and mobile messaging support.

Table 14.2 lists the several development tools that are available from mobile platform vendors. Due to the rapid ongoing changes in the wireless Internet industry, detailed information on each tool is not provided, as it would quickly become outdated. Instead, it is recommended that you research the current state of each vendor before making any development tool or platform decisions. The URL for each product is provided here to help aid in that research.

Table 14.2   Development Tool Vendors

VENDOR	PRODUCT NAME	PRODUCT URL
Air2Web	Nomad Publisher	www.air2web.com /solutions_mip.jsp
BroadBeam	Mobile Solutions System	http://www.broadbeam.com /products/mss.asp
Covigo	Covigo Studio	www.covigo.com/products
Everypath	Everypath Studio	www.everypath.com/products /standard-studio.shtml

(continues)

Table 14.2   Development Tool Vendors *(Continued)*

VENDOR	PRODUCT NAME	PRODUCT URL
Extended Systems	Mobile Solutions Platform	www.extendedsystems.com /ESI/default.htm
iConverse	iConverse Mobile Studio	www.iconverse.com /products/studio.asp
MagnetPoint	MagnetStudio	www.magnetpoint.com /Products.html
Mircrosoft	Mobile Information Toolkit	msdn.microsoft.com /vstudio/device/ mobilecontrols/default.asp
MySkyWeb	SMARTTookit	www.myskyweb.com /products/index.html

Wireless Internet Service Providers

Wireless Internet service for mobile devices can be obtained in two main ways:

- Set up a contract with a wireless carrier directly.
- Use a wireless service provider.

For individual users, either one of these solutions will be effective, as they both can offer service plans for the device for which you require wireless service. The decision for corporations is a bit more complex.

Wireless carriers typically provide support for one network protocol and for a set of devices that work using that protocol. If an organization supports a diverse set of devices, requiring access to a variety of networks, then working with several individual carriers is not ideal. This is because carriers often offer discounts based on volume. If the organization is forced to set up contracts with several carriers, it will not have the same volume as it would generate with a single carrier. Another drawback is that the organization will have to maintain relationships with a number of companies, all providing a similar service.

This is where wireless service providers come in. They provide wireless Internet service to a variety of devices, using a variety of networks. These networks are often rebranded using the service provider's name. For example, Palm.net rebrands other carriers' Mobitex service under the Palm.net network name. The same is true for GoAmerica, although with other network protocols. The result is that these providers can give an organization wireless Internet access for most, if not all, of the devices within the organization. The organization can work with a single vendor, allowing it to establish both a business and technical relationship that is mutually beneficial. These wireless service providers are often called mobile virtual network operators (MVNO) since they do not own their own spectrum or wireless network infrastructure.

Table 14.3   Wireless Service Providers

SERVICE PROVIDER	URL
Aether Systems	www.aethersystems.com
Earthlink	www.earthlink.net/mobile
GoAmerica	www.goamerica.com
Neomar	www.neomar.com
Palm.NET	www.palm.net
Research In Motion (RIM)	www.blackberry.net
Virgin Mobile	www.virginmobile.com

MVNOs typically offer a variety of service plans to meet the needs of their diverse customer base. For enterprise usage, the most common plans are those that offer unlimited data access. These plans make the cost predictable, which makes budgeting much easier. In addition to wireless Internet service, many service providers also offer a software platform that includes a microbrowser and a wireless gateway. The micro-browsers are usually supplied for all of the platforms on which the particular service provider operates, typically Palm, Pocket PC, and RIM devices. These browsers are optimized for the wireless gateways, to improve performance, and include additional features such as data compression. The wireless gateways often provide some basic content-conversion capabilities, making it easier for users to view existing Web content without modification.

Finally, some wireless service providers also offer device provisioning and application hosting services. This allows them to increase their value to an organization by supplying all of the wireless network capabilities that an organization may require. Table 14.3 lists some of the leading wireless service providers.

Summary

The most visible part of a thin client solution is the microbrowser. It defines the presentation layer, which includes the markup language used, image types, offline capabilities, and device support. Many different microbrowser solutions are on the market, all with different features. Some are only appropriate for a particular platform, such as PocketIE and Palm Web Browser, while others are targeted at a wide range of platforms. Within the corporate firewall, a wireless application server provides the core features that are required to develop and deploy wireless Internet applications. It improves developer productivity by supplying features aimed at mobile development. The server can also provide integration with other enterprise systems such as ERP, CRM, or relational databases. Finally, there are development tools that provide advanced features for wireless application development: Drag-and-drop programming, multichannel support, built-in emulators, and server integration are beneficial when creating enterprise applications. All of these components are required during the development and deployment of a complete wireless Internet solution.

The next chapter explains how voice applications can be created using VoiceXML.

Helpful Links

The vendor links given throughout the chapter should provide you with the majority of additional information you require. However, here are some additional resources that you may find helpful:

CC/PP Working Group	www.w3.org/Mobile/CCPP
Open Mobile Alliance Technical documents	www.wapforum.org/what/technical.htm
W3C Web Services	www.w3.org/2002/ws/

CHAPTER 15

Voice Applications with VoiceXML

When building mobile and wireless applications, most corporations do not give adequate consideration to the benefits that a voice interface provides. Voice applications make it possible to extend enterprise data to anyone with a telephone, wireless or otherwise. Unlike the proprietary, inflexible systems of the past, the new wave of voice applications provides an open, robust solution for a variety of business and consumer applications.

This chapter discusses building voice applications using the Voice eXtensible Markup Language, VoiceXML. Unlike other applications described in this book, VoiceXML provides a voice, rather than visual, interface to enterprise systems. The VoiceXML architecture is very similar to that of Internet applications, with the Web browser replaced by a voice browser, and the handheld device replaced with a telephone. Voice interfaces give true universal access to your applications.

After reviewing the history of VoiceXML, we will examine the VoiceXML architecture and then provide information on building VoiceXML applications. The goal is to give you enough information so that you can make educated decisions as to whether a voice interface is suitable for your enterprise applications. An overview of the VoiceXML language is also given, to enable you to get started building custom solutions.

NOTE Most of this chapter is based on the working draft of the VoiceXML 2.0 specification released by the W3C. The code samples have been developed and tested with TellMe Studio, a leading VoiceXML platform.

Why Voice?

Before looking at the specifics of VoiceXML technology, it's a good idea to first consider the reasons why voice applications are being deployed. Some of the reasons are due to limitations of other application types, while others result from core benefits that voice solutions provide. Let's start with one of the most compelling reasons, accessibility.

There are more than 1 billion telephones in the world. This far exceeds the number of computers with Internet access and dwarfs the number of Internet-enabled mobile devices. Voice applications allow each of those telephones to act as an interface to the enterprise. In addition to universal accessibility, the medium of interaction is also desirable. Voice is the most efficient and preferred user interface for many applications, especially where a visual interface is impractical, such as when driving an automobile.

Cost savings is another factor. Since the users of the application already have access to telephones, there is no additional cost associated with purchasing and maintaining devices. Voice applications are device-independent as well, meaning you do not have to be concerned with developing multiple markup languages for multiple wireless network protocols for multiple devices. Finally, training costs are minimal; users will not have to be trained to use a new device with an unfamiliar user interface.

Voice applications benefit from the characteristics of the telephone, especially mobile phones. Mobile phones are small, inexpensive, easy to operate, and have a long battery life. This makes the phone an ideal platform for portable applications. As capabilities of mobile phones evolve, such as with the addition of location-based services, voice applications will quickly be able to take advantage of this new functionality.

The limitations of current user interfaces also provide reasons for developing voice applications. While WAP is a technology with lots of potential, many early users of WAP-based applications have been frustrated because the display screens are small, and inputting data using a keypad is slow and cumbersome. For devices with more advanced capabilities, such as wireless PDAs, the support for graphics and other rich content types are important features. The ideal solution will involve multimodal access to data. For example, to get weather information, you might speak the city name into the device to have the weather information for that city displayed on-screen of the device.

Voice does not have to be the only means of data access. In most cases, voice is used in conjunction with other application types. You might, for example, already have a smart client application for which you want to provide an easy way to access real-time data without requiring a wireless connection to synchronize the data. Offering a voice solution will provide an effective way to extend your enterprise applications to both existing and new users alike.

Finally, we consider the area where voice applications are used most often. Call centers have been using Interactive Voice Response (IVR) systems for some time. When handling large volumes of calls from people seeking similar sets of information, a voice system can be very effective. An IVR system allows a company to provide better service while reducing costs. Rather than providing limited hours of availability, with long hold times, voice systems can allow for immediate access to data 24 hours a day. In most cases, users can find the information they need without requiring human interaction. Unfortunately, user satisfaction with traditional IVR systems has been low. Rigid menus and slow response times often cause users to hang up or to press 0 to get a human attendant, rather than use the IVR system to its full benefit. This is just one of the many application areas that can be improved by a more flexible voice solution in VoiceXML.

VoiceXML

Voice eXtensible Markup Language (VoiceXML) is an XML-based language for creating voice-enabled applications. It provides a standard way for developers to extend enterprise data and Web content to a new medium. Just as HTML describes the visual interface for a Web browser, VoiceXML describes the voice interface for a voice browser, allowing for audio input and output. VoiceXML leverages the Internet for development and delivery, making it easy for developers to add voice integration to existing systems.

The major goal of VoiceXML is to bring the advantages of Web-based development and content delivery to IVR systems. To do this, VoiceXML brings together many technologies, including speech recognition, keypad input, synthesized speech, digitized audio, and audio recordings. The result is an efficient and robust way to create user-friendly voice applications.

History of VoiceXML

Even though VoiceXML is a relatively new technology, it already has quite a history. In 1995, a group of researchers at AT&T Research were working to discover ways to use the Internet for telephony applications. The goal was to devise a system that could deliver Web content and services to ordinary phones. Over the next several years, the research continued, although in separate projects, in separate companies. AT&T, Lucent, and Motorola were all working on essentially the same thing: voice Internet.

By 1999, AT&T and Lucent each had its own version of the Phone Markup Language (PML), and Motorola had developed a technology called VoxML. At the same time, IBM was working on a similar technology called SpeechML. It quickly became clear that a joint solution was required if the voice Web market was going to succeed; consequently, these companies created an organization called the VoiceXML Forum (www.voicexml.org). Its members used the best features of each proprietary technology (with the majority of the syntax coming for Motorola's VoxML), along with some additions, to create the first version of VoiceXML, 0.9.

After VoiceXML 0.9 was published, the growing community of VoiceXML Forum members made huge improvements to the language, resulting in the release of VoiceXML 1.0 in March 2000. This release was well received, leading to several VoiceXML 1.0-compliant product offerings. With this initial success, the VoiceXML Forum submitted VoiceXML 1.0 to the W3C for consideration. With the future of the language in its hands, the W3C's Voice Browser Working Group put together version 2.0 of VoiceXML, which is now an official W3C recommendation. Even with all of the improvements to version 2.0, it is still very similar to version 1.0, making application upgrades trivial.

The VoiceXML Forum continues to flourish. Along with the founding members—AT&T, Lucent, Motorola, and IBM—there are close to 70 promoter members and almost 400 supporters of the technology! This is quite an accomplishment in just over three years. With such broad industry support, VoiceXML is destined to change the face of voice application development. It will not be long before we see a variety of consumer-oriented voice portals, as well as corporate applications, taking advantage of this flexible language.

Design Goals

The developers of VoiceXML had several goals, many centered on how VoiceXML relates to Internet architecture and development. Here are some of the top areas where VoiceXML benefits from the Internet:

- *VoiceXML is an XML-based language*. This allows it to take advantage of the powerful Web development tools on the market. It also allows developers to use existing skill sets to build voice-based applications.

- *VoiceXML applications are easy to deploy*. Unlike many of the proprietary Interactive Voice Response (IVR) systems, VoiceXML servers can be placed anywhere on the Internet, taking advantage of common Internet server-side technologies.

- *The server logic and presentation logic can be cleanly separated*. This allows VoiceXML applications to take advantage of existing business logic and enterprise integration. Using a common back end allows the development of different forms of presentation logic based on the requesting device.

- *VoiceXML applications are platform-independent*. Developers do not have to worry about making VoiceXML applications work on multiple browsers over multiple networks. When developing VoiceXML applications, the only concern is making sure it works with the VoiceXML browser being used. This leads to quicker development and less maintenance when compared to wireless Internet and desktop Web applications.

VoiceXML Architecture

VoiceXML uses an architecture similar to that of Internet applications. The main difference is the requirement for a VoiceXML gateway. Rather than having a Web browser on the mobile device, VoiceXML applications use a voice browser on the voice gateway. The voice browser interprets VoiceXML and then sends the output to the client on a telephone, eliminating the need for any software on the client device. Being based on Internet technology makes voice application development much more approachable than previous voice systems.

Figure 15.1 shows the architecture of a VoiceXML system and an Internet application. Showing both on the same diagram clearly shows the similarity between the two solutions. As you can see, the VoiceXML application does have some additional complexity when compared to the Internet application. Instead of using the standard request/response mechanism used in Internet applications, the VoiceXML application goes through additional steps on the voice gateway. Let's go through the steps of a sample voice interaction.

Just as Internet users enter a URL to access an application, VoiceXML users dial a telephone number. Once connected, the public switched telephone network (PSTN) or cellular network communicates with the voice gateway. The gateway then forwards the request over HTTP to a Web server that can service the request (Figure 15.1-1b). On the server, (Figure 15.1-2), standard server-side technologies such as JSP, ASP, or CGI can be used to generate the VoiceXML content, which is then returned to the voice gateway (Figure 15.1-3b). On the gateway, a voice browser interprets the VoiceXML code using a voice browser. The content is then spoken to the user over the telephone using prerecorded audio files or digitized speech. If user input is required at any point during the application cycle, it can be entered via either speech or tone input using Dual-Tone Multifrequency (DTMF). This entire process will occur many times during the use of a typical application.

As just stated, the main difference between Internet applications and VoiceXML applications is the use of a voice gateway. It is at this location where the voice browser resides, incorporating many important voice technologies, including Automatic Speech Recognition (ASR), telephony dialog control, DTMF, text-to-speech (TTS) and prerecorded audio playback. According to the VoiceXML 2.0 specification, a VoiceXML platform must support the following functions in order to be complete:

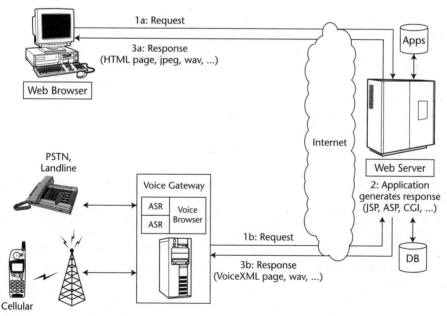

Figure 15.1   VoiceXML architecture.

Document acquisition. The voice gateway is responsible for acquiring VoiceXML documents for use within the voice browser. This can be accomplished within the context of another VoiceXML document or by external events, such as receiving a phone call. When issuing an HTTP request to a Web server, the gateway has to identify itself using the User-Agent variable in the HTTP header, providing both the browser name and the version number: "<name>/<version>".

Audio output. Two forms of audio output must be supported: text-to-speech (TTS) and prerecorded audio files. Text-to-speech has to be generated on the fly, based on the VoiceXML content. The resulting digitized speech often sounds robotic, making it difficult to comprehend. This is where prerecorded audio files come into play. Application developers can prerecord the application output to make the voice sound more natural. The audio files can reside on a Web server and be referred to by a universal resource identifier (URI).

Audio input. The voice gateway has to be able to recognize both character and spoken input. The most common form of character input is DTMF. This input type is best suited for entering passwords, such as PIN numbers, or responses to choice menus. Unfortunately, DTMF input is quite limited. It does not work well for entering data that is not numeric, leading to the requirement for speech recognition.

Transfer. The voice gateway has to be capable of making a connection to a third party. This most commonly happens through a communications network, such as the telephone network.

When it comes to speech recognition, the Automatic Speech Recognition (ASR) engine has to be capable of recognizing input dynamically. A user will often speak commands into the telephone, which have to be recognized and acted upon. The set of suitable inputs is called a grammar. This set of data can either be directly incorporated to the VoiceXML document or referenced to an external location by a URI. It is important to provide "tight" grammars so the speech recognition engine can provide accurate speech recognition in noisy environments, such as over a cell phone.

In addition to speech recognition, the gateway also has to be able to record audio input. This capability is useful for applications that require open dictation, such as notes associated with a completed work order.

NOTE Speech recognition is not the same as voice recognition. Speech recognition will work for nearly any voice, and does not have to be trained for individual users. It picks up speech patterns, rather than voice inflections. Voice recognition is more commonly used as a form of authentication to identify individual users.

Building VoiceXML Applications

Now that you have a solid understanding of the VoiceXML architecture, it is time to look more closely at the VoiceXML language itself. If you have experience with other XML-based markup languages, such as WML or XHTML, or even other tag-based languages, such as HTML or HDML, creating VoiceXML applications will not pose much difficulty for you. The main adjustment will be to the development of the user interface. Rather than sending content to a screen, it is spoken over a telephone. For this reason, it is very important to create clean, intuitive applications so users do not become frustrated and hang up. VoiceXML includes elements to help with this, as we will discuss later in this section.

NOTE We are not going to go into depth on every aspect of the VoiceXML language. This type of information can be readily found at many of the sites outlined in the Helpful Links section at the end of this chapter. Instead, we will look at the general language concepts of VoiceXML, then examine code in an example application.

Language Concepts

Before examining some VoiceXML code, let's take a quick look at the basic concepts behind a VoiceXML application.

Session

Once a user connects to the VoiceXML gateway, a session is started. This session is maintained as new VoiceXML documents are loaded and unloaded. The session ends only when requested by the user, the VoiceXML document, or the voice gateway. Each platform will have default session characteristics, many of which can be controlled by VoiceXML logic.

Dialogs

VoiceXML applications are constructed of one or many dialogs. Each dialog represents some form of conversational state with the user. After completing one dialog, you move on to another dialog, until the application is complete. There are two types of dialogs: forms and menus. A form collects user input, and a menu gives the user options to choose from. If at any point no dialog is specified, the VoiceXML application terminates automatically.

VoiceXML also has subdialogs. These are very similar to function calls, allowing the application to call out to a new dialog, then return to the original form. All of the variables, grammar, and state information is available upon returning to the calling document. Subdialogs can be used to create a set of components that may be used from several applications.

Applications

A VoiceXML application is a set of documents that share the same root document. This root document is automatically loaded any time a user interacts with any document in the application, and remains loaded until the user transitions to a document outside of the application. While it is loaded, the root document's variables are available to the other documents. It is also possible to specify grammars to be active for the duration of the application.

Grammars

Grammars make it possible to specify valid inputs from the user. Each dialog will contain at least one speech and/or DTMF grammar. In simple applications, only the dialog's grammars are active for that dialog. In the more complex, mixed-initiative applications, it is possible to have active grammars outside of the dialog being executed. Mixed-initiative refers to applications in which both the user and the gateway determine what will happen next.

Grammar creation is a very important aspect of designing intuitive, robust VoiceXML applications. It is essential to create grammars that accurately reflect the typical speech inputs from a user. If this is not achieved, either the prompts or the grammar will have to be changed. In some cases, the built-in grammars may be sufficient. VoiceXML 2.0 has the following built-in grammars: boolean, date, digits, currency, number, and time. In addition, many of the voice gateway vendors have proprietary grammars that you can use.

Events

When the normal execution of an application is interrupted, an event is thrown. In most cases, events are used when the user fails to respond to a prompt, or when the response is not suitable. They are also used when a user requests help or wants to exit the application. When an event is triggered, the <catch> element allows you to specify what the reaction should be. If there is no handler at the dialog level, the event can be caught at a higher level, since events follow an inheritance model.

Links

Links enable you to create mixed-initiative applications. They specify a grammar that is active when the user is within scope of the link. When the input matches the grammar, the user is redirected to the specified destination URI.

Scripting

If you require additional control over an application, which is not provided by standard VoiceXML elements, you can use scripting in the form of ECMAScript. This allows you to do such things as collect values of several fields in a single response.

Application Example

Listing 15.1 contains a relatively straightforward VoiceXML application. It demonstrates several of the concepts discussed in the previous section. This application allows a user to input data to a time-tracking system. Consultants who need to keep track of their hours as they work on customer projects would use such an application.

NOTE This code is for demonstration purposes only; it is by no means a complete application.

```
1.   <?xml version="1.0"?>
2.   <vxml version="2.0">
3.   <meta name="author" content="Martyn Mallick"/>
4.   <form>
5.     <block>
6.       Welcome to the voice time entry system.
7.       <goto next="#options"/>
8.     </block>
9.   </form>
10.  <!-- allow user to choose one of three options -->
11.  <menu id="options" dtmf="true">
12.    <prompt> What would you like to do? Say one of:
<enumerate/></prompt>
13.    <choice next="#entry">add entry</choice>
14.    <choice next="http://www.timeentry.example.com/vxml/delete.vxml">
delete entry</choice>
15.    <choice next="http://www.timeentry.example.com/vxml/list.vxml">
list entries </choice>
16.    <noinput count="1"> <reprompt/></noinput>
17.    <noinput count="2"> Please state what action you would like
<enumerate/></noinput>
18.  </menu>
19.  <!-- collect data for new time entry -->
20.  <form id="entry">

21.    <catch event="nomatch noinput" count="3">
22.      <prompt> Sorry, too many attempts.  Please try again later.
Goodbye.  </prompt>
23.      <throw event="telephone.disconnect.hangup"/>
```

Listing 15.1 Sample VoiceXML code for a time entry system. *(continued)*

```
24.    </catch>
25.    <field name="jobtype">
26.      <prompt>What is the job type for your entry? </prompt>
27.        <option>design</option>
28.        <option>development</option>
29.        <option>meeting</option>
30.        <option>travel</option>
31.        <option>vacation</option>
32.      <help>You must enter a valid job code to continue. Your options
are design, development, meeting, travel, and vacation.
<reprompt/></help>
33.    </field>
34.    <field name="hours" type="digits">
35.      <prompt> How many hours for job <value expr="jobtype"/>?
</prompt>
36.      <help> use the keypad to enter the number of hours worked
</help>
37.    </field>
38.    <field name="proceed" type="boolean">
39.    <prompt>Do you want to proceed with the entry for <value
expr="hours"/> hours for job type <value expr="jobtype"/>?  </prompt>
40.      <filled>
41.        <if cond="proceed">
42.          <prompt bargein="false">
43.            Your entry is being entered into the time system.
44.          </prompt>
45.          <!-- submit time entry to servlet for entry into database --
>
46.          <submit next="/servlet/entry" namelist="jobcode hours"/>
47.        </if>
48.        <clear namelist="jobcode hours proceed"/>
49.        <goto next="#options"/>
50.      </filled>
51.    </field>
52.  </form>
53.  </vxml>
```

Listing 15.1 *(continued)*

The first line of any document is the XML version number:

```
<?xml version="1.0"?>
```

This line is followed by the opening tag of the VoiceXML document, <vxml>, which includes the VoiceXML version number. The current version of VoiceXML is 2.0. The rest of the document is enclosed between the <vxml> and </vxml> tags. In our application, we have several form dialogs and one menu dialog. The first dialog in our application (lines 4 to 9) simply outputs a welcome message to the user, then transfers the application to dialog with the id="options". The greeting is contained within a

<block> element. In this case, the text is spoken using text-to-speech, although it is also possible to specify an audio file to be executed.

On line 10 is a comment. As with any application, properly commenting your code will make it much easier to understand for other developers. On line 11 starts the second type of dialog, a menu. The menu gives the user a choice of actions, which are listed on lines 13, 14, and 15. Each choice has a next attribute that specifies the location of the dialog or document that will be executed if that choice is selected. Also between the <choice> and </choice> tags is the text to which the user response will be compared. For example, if the user says "add entry," the entry form within this document will be executed. If the user says "delete entry," the delete.vxml document at the specified URI will be executed. One tag that is helpful is the <enumerate/> tag. Here it is contained within the <prompt> element on line 12. When it is reached, each of the options within the menu will be spoken to the user.

On lines 16 and 17 are <noinput> tags. These are executed if the user does not enter any input when prompted to in the menu. The first time this happens, the choice selection will be repeated, as specified on line 16 using the <reprompt/> tag. The second time there is no input, another prompt will be spoken to the user. Being able to change the prompts can prove helpful for rewording the request in case the user was unclear what was expected.

Lines 19 to 52 contain the entry form. If "add entry" was selected in the menu, this is the location that is executed. The <form> element contains an id with the name entry. This is used to move between forms. On lines 21 to 24 is a <catch> element that will be executed if a nomatch or noinput occurs three consecutive times. If executed, the <prompt> on line 22 will be spoken, followed by line 23 containing a throw of the predefined event that will hang up the telephone, thereby ending the session. Since events use an inheritance model, the <catch> element can be executed for any field within this form.

On lines 25 to 33 is the first <field></field> element. The purpose of these tags is to obtain user input. The name specified on line 25 is the variable that will contain the selected option. So, for this example, the variable jobtype will contain design, development, meeting, travel, or vacation, as specified in the <option> tags on lines 27 to 31. In this example, rather than defining a grammar with five entries, we have used the <option> tag. It is a suitable replacement for a grammar in cases where there are only a few choices. When there is a larger set of choices, a grammar may be more suitable. When using a grammar, you can choose to use an inline grammar or point to an external file that contains the grammar. Again, the decision often comes down to the size of the grammar itself.

Another tag that we see for the first time is the <help> tag. The contents within the <help> and </help> tags are executed at any time during the field when the user says the word "help." Using these tags is an abbreviation for the tag <catch event="help">.

Since the jobtype field did not specify an action, the next element in the dialog is executed. In this case, that is the <field> on lines 34 to 37. In addition to setting the name attribute, we also specify that we want to use the built-in grammar of type="digits". This allows the user to enter any sequence of digits, either using the keypad or speaking in response to the prompt. The <prompt> on line 35 asks the user to enter the number of hours for the jobtype that was specified in the previous field. We are able to access the jobtype variable using the <value expr="jobtype"/> tag.

The final field on this form asks the user to confirm his or her entry before it is sent to a servlet for input into a database. On line 38 the <field> again uses a built-in grammar, this time of type="boolean". This specifies that the input can either be yes or no (or a similar variant such as yeah or nah). The variable used to store the user input is named proceed.

We again use the <value> tag to access variable data in the <prompt>, allowing the user to confirm whether the speech recognition engine heard the inputs correctly. Unlike the previous fields, on line 40 we now use the <filled> tag to define the actions to take once the data is correctly entered. The <filled> element is commonly used to specify an action to perform when some combination of fields are filled by user input. In our case, it is when the user inputs either yes or no in response to the prompt.

If the response was yes, proceed is set to true, making the <if> statement from lines 41 to 47 execute. When the time entry is confirmed, a <prompt> tells the user that his or her entry is being entered to the time system. We specified the bargein="false" attribute in the <prompt> element. This ensures that the entire <prompt> is read before the user can interrupt it. This is useful when it is important that the user hears certain information, such as a licensing agreement or confirmation number. On line 46, the data entries are sent to a servlet for processing.

If the response to the <prompt> was no, then the <if> condition would not be met, so the next line of code executed would be line 48. On this line we clear all of the variables before going back to the options menu using the <goto> tag on line 49. From lines 50 to 53 the open tags are closed, ending with the </vxml> tag to complete our VoiceXML document.

> **NOTE** The best resource for detailed information on VoiceXML 2.0 is the VoiceXML 2.0 specification located at www.w3.org/TR/voicexml20/.

Now that we have examined the VoiceXML code, let's "listen" to a typical conversation using the code in Listing 15.1. In the following dialog, "Gateway" represents the output from the voice gateway and "User" represents the input by the user of the application.

Gateway: Welcome to the voice-time entry system.

Gateway: What would you like to do? Say one of "add entry, delete entry, list entries."

User: (no input)

Gateway: What would you like to do? Say one of "add entry, delete entry, list entries."

User: (no input)

Gateway: Please state what action you would like: add entry, delete entry, list entries.

User: Add entry (or presses 1 on keypad).

Gateway: What is the job type for your entry?

User: Help.

Gateway: You must enter a valid job type to continue. Your options are design, development, meeting, travel, and vacation.

Gateway: What is the job type for your entry?

User: Sleep.

Gateway: I'm sorry, I didn't get that. (this is a platform-specific response)

Gateway: What is the job type for your entry?

User: Design.

Gateway: How many hours for job design?

User: One seven (or enters 17 on keypad).

Gateway: Do you want to proceed with the entry for 17 hours for job type design?

User: Yes.

Gateway: Your entry is being entered into the time system.

Gateway: What would you like to do? Say one of "add entry, delete entry, list entry."

User: Goodbye (on most voice platforms exits the application).

Voice Vendors

Many vendors provide VoiceXML platforms and related products. Some supply the entire voice gateway, while others specialize in a particular technology such as speech recognition. If you are interested in learning more about the offerings from any particular vendor, check out the URLs in Table 15.1.

Table 15.1 VoiceXML Vendor List

VENDOR NAME AND URL	PRODUCT CATEGORY
AT&T www.naturalvoices.att.com	Advanced text-to-speech engine
IBM www.ibm.com/software/speech/	Complete VoiceXML platform, including speech recognition, text-to-speech, and development tools
Lucent www.lucent.com	VoiceXML gateway solution
Voxeo www.voxeo.com	Voice platform, including speech recognition, text-to-speech, and VoiceXML browser

(continues)

Table 15.1 VoiceXML Vendor List *(Continued)*

VENDOR NAME AND URL	PRODUCT CATEGORY
VoiceGenie www.voicegenie.com	VoiceXML gateway, development tools, and packaged voice applications
TellMe www.tellme.com	Complete VoiceXML platform with online development tools
BeVocal www.bevocal.com	Voice software aimed at wireless carriers
Nuance www.nuance.com	Speech recognition, voice authentication, and text-to-speech software
Speechworks www.speechworks.com	Automated speech recognition solutions

NOTE For a complete list of the nearly 400 companies that are part of the VoiceXML Forum, visit www.voicexml.org/.

Summary

Voice applications can help overcome some of the challenges commonly experienced when building wireless Internet applications, such as limited screen size, unreliable wireless coverage, and difficult data entry. More than 1 billion telephones in the world can take advantage of these applications.

The industry-leading technology for creating voice applications is VoiceXML. VoiceXML architecture is very similar to the Internet architecture, but instead of a client-side visual Web browser, VoiceXML employs a server-side voice browser. In doing so, VoiceXML brings together many technologies, including text-to-speech, automatic speech recognition, and DTMF input to create user-friendly voice applications. The specification for VoiceXML is managed by the W3C.

The market for voice applications is in its infancy. The number of VoiceXML applications will grow dramatically as corporations start to see and experience the benefits that voice integration can provide.

The next chapter brings us to Part IV of this book. There we will cover technologies for extending email and other personal information management applications, as well as mobile device management solutions.

Helpful Links

To help you keep track of this important technology, log on to these Web sites:

TellMe Studio	http://studio.tellme.com
VoiceXML 2.0 working draft	www.w3.org/TR/voicexml20/
VoiceXML Forum	www.voicexml.org
	www.voicexml.org/tutorials/index.html
Voice Channel on WirelessDevNet	www.wirelessdevnet.com /channels/voice/

Beyond Enterprise Data

So far we have focused on the creation of mobile and wireless applications that provide access to corporate data. Part Two covered the creation of smart client applications to enable offline access to enterprise data, while Part Three covered wireless Internet applications. In Part Four, we are going to look at technology that does not necessitate enterprise data access, although it still plays a role in the deployment of enterprise mobile solutions.

The first topic is mobile information management (MIM), which comprises both personal information management (PIM) and mobile device management capabilities. These technologies are becoming increasingly important as mobile devices proliferate and become the responsibility of enterprise IT staffs. The second topic is location-based services (LBS). Much of the hype surrounding LBS has revolved around the consumer market, but there are many corporate applications that can benefit from location information. Finally, we will take a look at four additional technologies that could have an impact on mobile development projects. In this part, one chapter is dedicated to each of these topics, as follows:

- Chapter 16, "Mobile Information Management"
- Chapter 17, "Location-Based Services"
- Chapter 18, "Other Useful Technologies"

Mobile Information Management

The term mobile information management (MIM) comprises two separate but related technologies: personal information management (PIM) and *mobile device management*. PIM applications include email, calendar, task lists, address books, and memo pads. Access to these applications is often the reason why consumers purchase mobile devices. In the corporate world, technology is available to integrate these PIM applications with corporate servers. The functionality these products provide is often referred to as PIM Sync.

As more employees use mobile devices, the task of managing them becomes increasingly difficult for IT staff. For small deployments of fewer than 30 to 40 users in a small number of locations, mobile device management software may not be a requirement, as it is possible to install and configure these devices in a reasonable amount of time. When there are more than 40 users, or when users are distributed in a vast number of geographic locations, mobile device management software can provide substantial benefits for both the deployment and management of software and devices.

Historically, these solutions were offered as separate packages, but this is changing as it becomes clear that many organizations are interested in both technologies. Many of the leading MIM vendors have incorporated device management capabilities into their PIM solutions, or PIM capabilities into their device management solutions. This trend is expected to continue as vendors look to enhance their MIM offerings.

In this chapter the PIM Sync and mobile device management technologies are addressed in separate sections in order to better focus on the capabilities that each solution provides. This also serves to point out that it is not mandatory to use both technologies together; many vendors still have strengths in one area, making it worthwhile to choose a best-of-breed solution for each technology.

PIM Sync

Before deploying custom applications, many organizations prefer to introduce their mobile users to applications to which they are accustomed: email, calendars, address books, and to-do lists. These applications are commonly used in a desktop environment and translate very well to a mobile environment. Because mobile users are already comfortable with these applications, it's only necessary to introduce the users to mobile devices and, possibly, a wireless network. And because users are generally interested in these applications, enterprises find it easy to use them to introduce mobility to their workforces. Once devices are used for PIM applications, adding custom enterprise applications becomes a much smoother process; one that meets less resistance from the application users.

What Is PIM?

Personal information management (PIM) describes the set of applications used for everyday personal organization and communication. These applications include email, calendars, to-do lists, address book, and memo pads. For many mobile users, these applications are the most important, hence are a deciding factor in which mobile device to purchase. In the consumer space, PIM applications are often the only software applications, other than a Web browser, used on a mobile device.

> **NOTE** Some companies do not include email in the definition of PIM, preferring to separate their solutions as email and PIM. For the purposes of this book, however, email is included in PIM.

For enterprise usage, PIM applications can provide tremendous productivity gains for mobile workers. Having the ability to receive and send email messages while away from the office allows for faster reaction times to internal and external events, often resulting in better response times to important issues. Remote access to calendars and address books enables employees to stay on top of important meetings and appointments while they are away from the office. For the corporate user, the mobile PIM solution must be able to integrate with the leading enterprise groupware solutions, including Microsoft Exchange and Lotus Domino.

> **NOTE** When referring to the groupware product by Lotus, the client is called Lotus Notes and the server is called Lotus Domino. For the remainder of the chapter, we will be referring to Lotus Domino since we are interested in server-side integration.

We will take a closer look at each of the applications incorporated into a PIM solution, focusing on the features that are important in a mobile environment.

Email

Email is the leading PIM application. Most mobile users want or require access to their corporate and personal email while away from the office. Companies such as Yahoo! and Hotmail provide access to their systems for mobile phone users, typically via a wireless Internet Web site that can be accessed from any device with a microbrowser. The user simply goes to the Web site, logs in, then sends and/or receives email messages. The overall experience is similar to using the desktop Web site—with some limitations. For example, on mobile phones, users can only view text and simple graphics. A typical wireless Internet email client is shown in Figure 16.1. As you can see, it is not possible to open or view attachments such as office documents or spreadsheets. This limits the usefulness for corporate users who must be able to perform tasks similar to those they would from their desktop systems. In addition, corporate users also require the ability to access their corporate email systems maintained behind corporate firewalls.

These limitations are addressed in many corporate mobile email solutions, which often include both client and server software. The PIM server resides within the corporate firewall to provide integration with the corporate PIM software, such as Microsoft Exchange or Lotus Domino. Most of the functionality for mobile PIM solutions is provided on the server. Depending on the solution being used, the client component may be either a Web client or a smart client. When a PIM smart client is provided, many vendors take advantage of the email client that comes with the mobile operating system.

In either case, it is important to look for some key features in a mobile email solution. These include the capabilities to do the following:

- Read new messages as they arrive, ideally with user notification when a message comes in.

- Reply to the sender in real time, without waiting for the data to be synchronized.

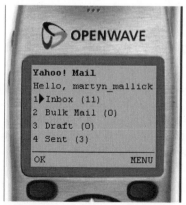

Figure 16.1 Wireless Internet email client.

- Forward messages in their original format. This is especially important for email messages that contain attachments.

- Compose new messages and send them from the mobile device. This should work the same way as from a desktop client, and include the ability to keep a copy of the sent message on the server for future retrieval.

- Delete messages from the server. A user does not want to have to delete the same message from both the mobile client and the desktop client.

- Apply read and unread message flags to the server data. Users do not want to read messages on their mobile device only to have them marked as unread when they view them from a desktop client, or vice versa.

Calendar

The calendar application is regarded as the second most important PIM application, after email. A calendar, essentially an electronic date book, is where employees keep track of and schedule meetings and appointments. Invitations to meetings are sent out over email, and once accepted, are automatically added to users' calendars.

For mobile applications, mobile users have to be able to access their corporate calendar information. This requires integration with enterprise PIM solutions, which typically is provided by the same server that is used for email integration. Similarly, many mobile PIM vendors take advantage of the calendar application that comes with mobile operating systems.

The useful features in a mobile calendar application include these capabilities:

- View existing meetings and appointments, and be able to retrieve their details such as location or dial-in number.

- Accept or decline meeting invitations.

- Create new calendar entries and send out invitations to other parties via email. The new entry should be reflected on the server for access from a desktop client.

- Change calendar entries and have the edits reflected on the server.

Contact Lists

Most mobile devices have the capability to maintain contact lists. In fact, this is a core feature of both mobile phones and PDAs. The contact lists maintained on mobile phones are often more limited in capabilities than those on PDAs. They make it possible to store phone numbers, email addresses, and other key contact information, but they do not provide the capability to store the same amount of information as that of a PDA. This is partially due to the memory limitations of the device and partially to the interface for viewing the information. On PDAs, the contact list can contain all of the core contact information and can usually be synchronized with a desktop contact list. This helpful feature means you do not have to enter every contact on the PDA, since changes made in one place will be reflected on the other.

The key features of a mobile contact application include these capabilities:

- Add, delete, or edit a contact, and have that change reflected on other clients that share the contact list.
- Search a global list of contacts.
- Send an email to a contact by launching the email client from the contact list.
- Autodial a contact on devices that have telephone capabilities.

Task Lists

When workers have remote access to task lists it helps them to stay focused and organized while on the move. Mobile task lists make it possible to view and update existing tasks, add new tasks to the list, assign due dates and statuses to tasks, and remove completed tasks from the list. Task lists are most useful when they reside on the device so the user does not have to establish a wireless connection to view upcoming tasks. It is also beneficial to have the task list integrated with the server so tasks can be synchronized to other clients, such as a desktop computer.

Memo Pad

The final application incorporated into PIM solutions is a memo pad. This application allows the remote user to jot down notes and other key pieces of information for future retrieval. Memo pads are useful only on devices that enable quick data entry. This effectively limits their usage to PDAs and other comparable devices.

PIM Architectures

There are two main architectures for PIM applications: wireless Internet and smart client. By now you are familiar with the differences between these two architectures, so we will not spend time explaining the differences here. Rather, we will examine the benefits of each and explore reasons why a mobile user might choose one over the other.

Wireless Internet

As with other wireless Internet applications, the main advantage to a wireless Internet PIM solution is real-time access to data. This may not be very important for contact lists, task lists, and the memo pad, but it definitely is a huge benefit for mobile email and calendars. The user will always be able to access the most recent email messages and calendar entries, thereby removing the need to periodically synchronize data back to the server. When data does not reside on the client, any changes made to the applications are automatically reflected on the PIM server. This type of solution is sometimes called an online PIM solution.

Another benefit of an online solution is universal access. Wireless Internet PIM applications can be accessed from any device that has a microbrowser. This allows PIM access from a wide variety of devices, including mobile phones, PDAs, and laptops. In a corporate environment, this means the IT staff does not have to install and configure the mobile devices either; all they have to be concerned with is the PIM server software and how it integrates with the enterprise PIM software.

Unfortunately, wireless Internet PIM solutions also have some serious drawbacks. These are the same as those covered in other parts of this book, but now impacting the PIM solution. Here are some of the most serious:

Wireless network access. In order to access any part of the PIM solution, users must have access to the PIM server. This means that they require access to a wireless network. If a wireless connection cannot be obtained, the data is not available. Some solutions attempt to address this by providing the ability to cache downloaded content, making some of the data available when disconnected.

Performance. Downloading PIM information over a wireless network can be slow. Users often want to quickly look up some contact information or check meeting details. The time it takes to get a wireless network connection and download this information can prevent the user from taking full advantage of the PIM solution.

Cost. Airtime costs can start to add up when a user constantly downloads information over a wireless network, for example, when a user has to download the same set of information many times, to access, say, his or her email or calendar entries.

Features. The client for the PIM applications is typically limited to a microbrowser, limiting the features of the solution. For example, if the microbrowser does not support the viewing of attachments, they will not be available. Additionally, if the client is accessible only from a mobile phone, data input can be a concern.

Smart Client

A smart client PIM solution involves client applications that reside locally on the mobile device. These applications provide offline access to PIM data with the ability to send and receive updates by synchronizing with the PIM server. This synchronization can happen in either a wired or a wireless environment. For example, users can synchronize the majority of the data using cradle synchronization, and use wireless communication only when it is required. This form of smart client-with-synchronization approach enables workers to have mobile PIM access without a wireless connection! Many of the early PDAs on the market used this approach for PIM access since wireless connectivity for PDAs was not available at the time. This does, however, limit the ability to communicate remotely, but it is still better than not having mobile PIM access at all.

There are many other benefits to a smart client PIM solution in addition to synchronization. The smart client PIM applications are more sophisticated than their wireless Internet counterparts. Because all the data is stored locally, performing searches or sorting the data is much more efficient. In addition, other applications on the device can be used to view and edit the attachments. Performance is not as much of a concern either, as many of these solutions take advantage of mobile database technology, meaning that large amounts of data can be stored and managed efficiently.

For devices that do have wireless connectivity, new data can be retrieved at any time. When a new email message is composed, or a new meeting invitation needs to be sent out, the remote user can send this information over the wireless network back to the PIM server, where it will be forwarded to the appropriate parties. At the same time, any new data for that user can be sent to his or her mobile device. In this way, users can use wireless connectivity for the most important information, and desktop cradle synchronization for the rest. One caveat here is that not all mobile users will have a companion PC to use for PIM synchronization, in which case, it is important to choose a solution that supports synchronization directly from the mobile device to the PIM server.

One of the main drawbacks of smart client PIM solutions is that they are suitable only for devices that allow for the installation of third-party software. This eliminates most mobile phones from consideration, leaving PDAs and smartphones as the target market. Another limitation is that, to access the PIM server, the client software has to be installed and configured on the mobile device. This typically means that remote workers require their own PDA for access; they cannot just use any device available, as is the case with wireless Internet solutions. To ease this situation, many smart client PIM vendors use the standard PIM client software that is installed on the PDA's operating system. To use the existing PIM clients, a piece of software is installed that takes care of the synchronization with the PIM server and the data management. This software works in conjunction with the standard PIM client as part of the overall smart client solution.

A nice feature is the incorporation of alerts and notifications with a smart client solution. The notification can include a segment of the email message, or possibly even a command to initiate synchronization. In this way, remote users can be notified when a new message arrives, and can decide whether they want to download the entire message immediately or wait until the next synchronization.

Mobile OS PIM Software

When users evaluate mobile devices, three of the main applications that they look for are email, calendar, and contact applications. In many cases, these are the only applications that a person may use on the device. In the corporate world, these are usually the first applications that are used. Not surprisingly then, each of the leading mobile operating systems comes with PIM client software. For the Palm OS, Pocket PC, and RIM OS, the operating system software provides all client components of a complete PIM solution. Symbian OS offers selected components, and allows for third-party vendors to provide the rest.

The PIM software that is included with these operating systems is targeted largely at the consumer market. Table 16.1 provides a quick summary of each operating system's PIM capabilities. This software typically allows for desktop synchronization, but not access to behind-the-firewall groupware servers. If you are looking for this level of functionality, you either have to upgrade to an enterprise-level software package from the operating system vendor or find another enterprise solution vendor that provides the required capabilities.

Table 16.1 Summary of Mobile OS PIM Software

MOBILE OPERATING SYSTEM	PIM COMPONENTS	FEATURES
Palm OS	Email, address book, date book, to-do list, memo pad	• Desktop synchronization using Hotsync • Software available to view common attachments • Categorization and sort features to quickly find what you are looking for
Pocket PC	PocketOutlook, contacts, calendar, notes, tasks	• Desktop synchronization using ActiveSync • View attachments with Pocket PC versions of Word and Excel • Direct integration with Outlook on a desktop PC • Handwriting recognition for taking notes
Symbian OS	Email, contacts, agenda	• SyncML-compliant client for building custom PIM applications • PC synchronization using infrared, Bluetooth, serial, and USB support for vCalendar and vCard in the application engine (discussed later in this chapter)
RIM OS (RIM BlackBerry devices)	Email, calendar, address book, task list, memo pad	• Provides "always-on" functionality with BlackBerry wireless devices • Desktop synchronization with leading groupware products • Built-in filtering software • Built-in security

Later in this chapter you'll find a list of vendors that can provide additional PIM functionality for these mobile operating systems. In many cases, these third-party solutions integrate with these existing client applications. This lets mobile workers use software they are familiar with while gaining access to corporate groupware products.

Core PIM Product Features

In addition to the specific PIM application features listed earlier in this chapter, there are also some overall product-level features that you will find helpful in a mobile PIM solution:

Device support. The PIM software has to support the main devices that will be used in your organization. This often includes Palm OS and Pocket PC PDAs and a range of mobile phones, but may also consist of laptops, two-way pagers, and tablets. Review your needs today as well as future requirements. Smart client PIM solutions should at least support Palm OS, Pocket PC, and possibly Symbian OS; and online solutions should have both WML and HTML clients tested on multiple Web browsers. It is a good idea to find out which features are available for each device because not all features are supported on all platforms.

Synchronization modes. There are two ways to synchronize PIM data: through a companion PC and directly to the PIM server. Devices that do not have wireless access typically synchronize to a companion PC, which has desktop software that communicates with the PIM server. Devices that have wireless PIM access must be able to communicate directly to the PIM server so that any changes made on the device are reflected in the enterprise groupware system.

Notifications and push. Most users want their mobile PIM solution to behave like their desktop system. This means they want to be notified when a new message arrives and to be able to access its contents immediately. Providing this capability in a wireless environment is not a straightforward task. First you must have a wireless network that supports an always-on connection, such as Mobitex or GPRS or any of the 3G networks. Second, you must have software that can provide the notification of new messages and download the messages automatically without user interaction. The RIM BlackBerry devices have become very popular largely because they have these capabilities.

Enterprise integration. For corporate users it is important that a mobile PIM solution can integrate with existing email and groupware applications. The most common products include Microsoft Exchange, Lotus Domino, and common Internet protocols such as SMTP, IMAP4, and POP3. In addition, having support for PIM standards such as SyncML, iCalendar, and vCard allows for integration to the new breed of standards-based PIM servers.

Minimal data transfer. Data compression is particularly important when transferring data over a wireless network. Transferring compressed data can help to improve performance and minimize costs. Along the same lines, it is critical to ensure that only the required data is sent. Smart client PIM applications need to send only new email messages and calendar entries to the remote client. Online PIM solutions should send the user a minimal set of information, such as who the message is from, the message subject, and a few lines of the message text, and then let the user decide if he or she wants to download the entire message.

Message filtering. Along with data compression, being able to filter messages on a set of defined criteria is a great way to limit the amount of data being transferred. Users may want to receive messages only from certain individuals, or that contain specific keywords, or that are under a certain size. In addition, they may want to remove any attachments unless they are specifically requested to be downloaded. Message filtering is an effective way to reduce the amount of data being downloaded while viewing only important messages. Any messages that do not pass the filter criteria can be viewed at a later time from a desktop environment.

Security. As with any corporate application, security is paramount. For mobile PIM solutions, several types of security are required. They include encryption of the client message store and of the communication stream between the client and the PIM server, user authentication against the enterprise servers, and a secure connection between the PIM server and the enterprise groupware servers.

Attachments. Several solutions are available for handling mobile email attachments. On more capable devices running Pocket PC, Palm OS, or Symbian OS, software is often available to view and edit common attachments. If this is not wanted, either because of the attachment size or relevance, other options have to be provided. Some solutions allow the attachment to be transformed into another format, such as HTML, that can be viewed on the device. The attachment content is then sent to the user for viewing. If the user then needs to forward the attachment to another party, most PIM servers will allow it to be forwarded from the server in its original format. Another solution is to send the attachment to a nearby resource, such as a printer or fax machine, for printing. The user is then able to read the contents and react accordingly.

Standardization Efforts

Currently, several efforts are underway to standardize PIM applications and server integration. The most comprehensive effort is SyncML, which provides a standard protocol for synchronizing all forms of PIM data. Other standards, such as iCalendar and vCard, are more narrowly focused on particular applications within a complete PIM suite.

SyncML

SyncML is an industry initiative to provide a standard protocol for synchronizing data between all types of devices (PDAs, smart phones, pagers, cell phones), over any network protocol, to any data store. Its foundation is based on the synchronization of PIM data, although it has expanded into mobile device management and enterprise database synchronization as well. Many of the leading PIM vendors have either incorporated support for SyncML into their products or have announced future plans for implementing it. At the same time, mobile operating system providers are also working to include SyncML support to their offerings, allowing for easier integration of third-party PIM applications.

NOTE A complete overview of SyncML and its role in data synchronization is provided in Chapter 10, "Enterprise Integration through Synchronization."

The main benefit of SyncML is interoperability: It enables multiple products and synchronization technologies to work together. It does not define the user interface for specific applications; rather, it provides a specification for application providers to adhere to. By doing so these vendors can be assured that their products will work with other SyncML-compliant products on the market. From the user's perspective, this makes the synchronization process smooth and simple. As more SyncML-compliant products are developed, users will be able to choose from a variety of product offerings that are all totally integrated with their corporate applications. For example, it will be possible to have a variety of PIM components from different vendors all synchronize to a single SyncML-compliant PIM server.

Device manufacturers will also benefit because they will have to support only one synchronization protocol, which will work with a wide range of applications and wireless networks. This will allow them to concentrate on creating innovative products and not on interoperability and integration issues. Finally, application developers will be able to create applications that can communicate with a diverse set of device and networked data using a single synchronization protocol.

At a protocol level, SyncML offers many compelling advantages, such as support for multiple transport protocols, support for synchronizing arbitrary sets of data, support for Internet standards, and support for both wireless and wireline networks. (Chapter 10 outlined each of the major advantages of SyncML and gave an overview of how SyncML works.)

Before all of these benefits can come to fruition, however, we need to see more SyncML products on the market. This is starting to happen, but many vendors still are not totally committed to the standard. They are waiting on the sidelines to see if SyncML lives up to its promises. This may happen sooner rather than later, as many of the large SyncML promoters, such as Nokia, Ericsson, IBM, and Symbian are including support for SyncML into their core product offerings. For most of these vendors, PIM products are the first to implement SyncML.

vCalendar/iCalendar

The vCalendar standard was created to promote interoperability between calendaring products. The goal was to provide a way to easily add entries into a personal calendar from any location, as well as to be able to schedule meetings with a group of people who all use different calendaring applications. For this to happen, a defined standard is necessary, one to which all of the calendar products adhere.

vCalendar/iCalendar are specifications that define the data format for exchanging calendaring and scheduling information between applications, such as personal information managers, groupware calendaring programs, word processors, and Web browsers. vCalendar, the older format of the two, has widespread support in PIM products. iCalendar, a newer and more robust version of vCalendar, is capable of transferring larger amounts of information. It will eventually replace vCalendar as the leading calendar specification.

Using vCalendar/iCalendar, you can distribute entries as email attachments, or make them available for download from a Web page. They hold information about events and to-do items that are stored in PIM applications. These entries can then be inserted into a compliant calendar client with minimal effort. In this way, calendar entries can be distributed and accessed by any product that has support for the standard. You will find that many of the PIM servers on the market have support for vCalendar and/or iCalendar.

vCard

vCard is similar in concept to vCalendar, except it stores contact information. It is a specification that defines how contact information data can be exchanged between software applications. vCard stores information such as your name, address (home, business, mailing), telephone numbers (work, home, cellular, pager, etc.), email address, and Internet URLs. vCards can also contain graphics and multimedia content such as photographs, company logos, and audio clips. They essentially can be thought of as an electronic business card that can be distributed via multiple channels, including as an email attachment or a Web page download. A vCard recipient can easily add the included information to his or her contact list or electronic address book with a single click.

Enterprise PIM Vendors

Wireless carriers often provide wireless Internet PIM access. In most cases, they license technology from one of the vendors listed in Table 16.2 to offer this functionality. The vendor used behind the scenes does not really matter from the end user's perspective, as long as the solution works. Consumers typically rely on their carrier to provide this service. For corporate use, most companies prefer to manage the solution in-house so they can control the entire implementation and provide access to their enterprise PIM data.

The companies listed in Table 16.2 all provide viable mobile PIM software solutions, although the features may vary significantly. A vendor usually offers either a smart client or an online PIM solution, and a few provide both. For more information about any particular vendor's offering, visit the accompanying Web site.

Table 16.2 Mobile PIM Vendors

VENDOR NAME	PRODUCT NAME	PRODUCT URL
Aether	ScoutExtend	http://www.aethersystems.com /webfiles/productsservices /aetherproducts/aetherscoutextend /default.asp
AvantGo	AvantGo Pylon	http://avantgo.com/products /businesses/workforce /productivity/pylon/pylon_pim _server.html
Extended Systems	XTNDConnect	www.extendedsystems.com /ESI/Products/default.htm

Table 16.2   *(Continued)*

VENDOR NAME	PRODUCT NAME	PRODUCT URL
FusionOne	Mobile Sync for PIM	www.fusionone.com/services/mobile_sync.htm
iAnywhere Solutions	Mail Anywhere Studio	www.ianywhere.com/product/mail_anywhere.html
IBM	Lotus Domino Everyplace	www.lotus.com/products/wireless.nsf
Infowave	Symmetry Pro	www.symmetrypro.com/learn.htm
Lotus	Lotus Mobile Notes	www.lotus.com/products/wireless.nsf
Microsoft	Microsoft Outlook Mobile Manager	www.microsoft.com/outlook/mobile/
Microsoft	Mobile Information Server 2002	www.microsoft.com/miserver/default.asp
Motorola	Mobile Office	www.motorola.com/internet/mobileoffice/product.htm
Palm	Tungsten Mobile Information Management Solution	www.palm.com/enterprise/products/mims/
PumaTech	Intellisync	www.pumatech.com/is_desktop_main.html
Research In Motion	RIM BlackBerry Software	www.blackberry.net/solutions/enterprise/index.shtml
Starfish	TrueSync	http://www.starfish.com/solutions/solutions.html
Synchrologic	Email Accelerator	www.synchrologic.com/about/about_realsync_server.html
Time Information Systems	Time IS Synchronization	www.timeis.com
Openwave	Mobile Email	www.openwave.com/products/messaging_suite/mobile_email/index.html
Wireless Knowledge and Server	Workstyle Desktop	www.wirelessknowledge.com/products_and_solutions/desktop_edition/index.asp

Mobile Device Management

As enterprises deploy mobile applications to increasing numbers of employees, a need arises for an effective way to manage both the applications and the devices themselves. For a small number of remote users, it is possible for IT staff to configure each device individually. As the number grows, mobile device management software quickly becomes necessary. Mobile device management (MDM) platforms focus on providing an intuitive way for system administrators to deploy and manage software applications, usually to a wide variety of devices running on an even wider number of wireless networks. This technology is often provided as an out-of-the-box solution that can be used with existing applications and security architectures.

Architecture

Figure 16.2 illustrates the most common architecture for device management platforms. As the diagram shows, both client and server components are required for a complete solution. The management server is responsible for the overall device management, software distribution, and administration facilities, while the client component carries out much of the on-device configuration and maintenance. This is accomplished by deploying mobile agent software to the remote devices. This agent can then be configured to carry out a variety of tasks locally on the device, often without requiring a network connection. Depending on the solution being used, there may be client agents for Windows-based laptops, Windows CE/Pocket PC devices, Palm OS, Symbian OS, RIM BlackBerry, and Java clients. Usually, the client functionality varies depending on the platform being used. Laptops have the most complete set of features, and RIM devices usually have the least.

The devices can connect to the server directly using wireless or wireline (very often dial-up) connections or by using a companion PC as a proxy server. The server itself usually stores device statistics in a relational database. The server may also interact with other management software such as Microsoft Systems Management Server (SMS) or directory services such as LDAP or Active Directory.

What Mobile Device Management Software Solves

Traditional LAN-based application management systems have a hard time functioning in mobile and wireless environments where low bandwidth and unreliable network connectivity are common properties. The introduction of new mobile operating systems and wireless protocols do not improve the situation. The cost of managing these devices can surpass the cost of the devices themselves.

Mobile device management software has been designed to overcome these issues in order to reduce the overall cost of a mobile solution. These platforms also provide other capabilities that are well suited for a mobile environment such as remote administration and backup and recovery.

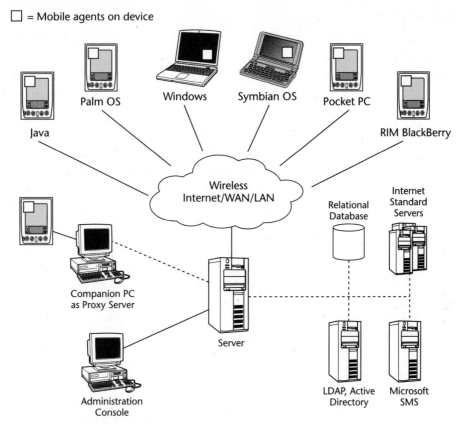

☐ = Mobile agents on device

Figure 16.2 Mobile device management architecture.

Software Distribution

Software distribution is difficult in a networked environment. It becomes even more challenging when mobility is added to the mix. Device management platforms provide the capability to distribute software over a variety of wireless network protocols such as CDPD, Mobitex, GPRS, and 802.11b. In addition, they work with LAN and dial-up configurations and often over a serial port. The software distribution is controlled from a central location making it possible for a single administrator to configure devices in multiple locations using several types of devices. It is even possible to schedule future software installation and activation for a controlled application rollout.

Additionally, remote software management allows administrators to ensure that applications are installed correctly, thereby reducing user error. Both push and pull mechanisms are typically provided. In addition, many of the mobile device packages provide the following software distribution benefits:

- Task scheduling
- Automated upgrades
- File collection and retrieval
- Remote command execution and control

Application and Asset Management

As application updates are deployed and users install personal software on their devices, a need arises for a way to manage that software. Snapshots of device configurations can be downloaded to a server, where they can be analyzed to ensure that remote users are following corporate policies with regard to information content and software licensing. It is also possible to remove applications from the device to prevent further data access. This is especially important in the case of a lost device or an unauthorized user.

Hardware tracking is also possible. Corporations can keep track of the number and type of devices that are deployed remotely. With this information they can make educated decisions on additional hardware purchases such as modems, memory cards, and other peripherals. In addition, the cost of future device upgrades can be predicted more accurately.

Remote Administration

Remote administration is a requirement for any sizable mobile environment. It is not feasible to continually ship devices back to corporate headquarters, where IT staff configure the applications on site. At the same time, mobile device users should not have to be concerned with hardware and software maintenance issues. Remote device administration solves this problem.

Administration functionality may include, but is not limited to, running custom scripts to delete temporary files, running maintenance applications (scan disk, defrag, virus scans), and monitoring available memory. In some cases, it is also possible to remotely control a device for troubleshooting purposes.

System Backup and Restoration

Backup and restoration capabilities are required when a device is lost or malfunctions. By archiving vital mobile device information, corporations can feel confident that important information will not be lost. Data backup can occur transparently so that remote users do not have to be concerned with performing manual backups. When the device comes back online, current data can be downloaded and new software can be installed to replace what was lost.

Core Mobile Device Management Software Features

When evaluating mobile device solutions, certain core features deserve consideration. The platform capabilities discussed in the previous section are common to nearly all vendor solutions. The features described in this section will vary between vendor offerings. Usually only a subset of these features are required for any particular deployment, so during an evaluation you will want to give extra consideration to the components that are relevant to your situation.

Server Administration

Every device management platform has an administration console for managing and scheduling data and software transfers, monitoring systems activity, and viewing client inventory. This console is the main interface into the system, and its importance cannot be overstated. Each vendor provides a different feature set. Some may focus on ease of use by providing advanced wizards to set up common tasks, while others may feel that providing administration from remote sites is important. Another core administration feature is the ability to authorize users for different tasks. For example, you probably would not want to give helpdesk staff the same rights as the system administrator. Finally, the types of client software may make a difference. Very often, both an Internet interface and a Windows desktop client are available.

Mobile Agents/Operating System Support

For on-device functionality, mobile agent software is required. The agent software is responsible for the most common tasks such as application installation, software configuration, and data transfer management. Device management vendors list support for Windows, Windows CE, Palm OS, Symbian OS, RIM, and Java clients, but it is important to realize that the level of functionality varies dramatically between client types. Windows and Windows CE clients usually have the most complete feature sets, while RIM clients are usually quite limited. Keep this in mind when making both device and management platform decisions.

Before the mobile agents can be used, they have to be deployed to each mobile device. There are a variety of ways to accomplish this. Three approaches are common: provide an install tool in an email message; make agents downloadable from a Web page; or activate them automatically during a system login. A fourth option is to install the agent using CompactFlash cards or a companion PC.

Software is also installed on a network server. This software usually stores the device information in a relational database, commonly MS SQL Server, Sybase Adaptive Server Anywhere, or Oracle. Find out what the software requirements are, including whether it has to be installed on a dedicated machine, its hardware requirements, and which enterprise systems it can integrate with.

Extensibility

Though the standard device management functions provide 90 percent of what you will require, there must be a way to perform custom tasks to meet the other 10 percent. This is typically accomplished by enabling the administrator to create and execute custom scripts. This allows for the platform to extend beyond the typical uses, and be useful for situations that may not have been foreseen.

Connectivity Options

There are two main ways for mobile agents to connect to the device management server: directly or via a companion PC. For direct connectivity, the main network protocols (both wireless and wireline) should be supported. These include LAN and dial-up connectivity over TCP/IP or HTTP, 802.11, CDPD, CDMA, GSM, TDMA, GPRS, and Mobitex. If the connection is being made via a companion PC, then you must understand how easy or difficult it is to install and maintain the proxy server that resides on the PC.

Support for Standards

Device management platforms support a variety of standards to help make the systems easier to use and configure. For file and data transfer, support for HTTP/HTTPS provides connectivity through corporate firewalls from virtually anywhere on the Internet. FTP support can provide similar benefits. For data access and reporting, many systems store the captured data in relational databases that can be accessed using ODBC, and queried using SQL. This makes it possible to use reporting packages such as Crystal Reports to analyze the device data. Support for the Desktop Management Interface (DMI) and Windows Management Interface (WMI) is beneficial for collecting hardware and system information, while integration with the Lightweight Directory Access Protocol (LDAP) and Active Directory are useful for accessing user, group, and resource information. Finally, as discussed later in this chapter, the SyncML Initiative has released a standard for device management called SyncML DM.

Enterprise Integration

Any new technology that is introduced to an enterprise should integrate well with existing systems. Mobile device management platforms are no exception. They need to work in conjunction with existing Web servers, firewalls, relational databases, and other system management software such as Microsoft SMS. In addition, being able to integrate with directory servers such as LDAP or Active Directory will allow for anyone to locate resources such as individuals, users, lists, and groups.

Security

Mobile device management solutions address two areas of security. The first is the security of the data and applications themselves. This is accomplished by encrypting the data communication stream and by using digital certificates for authentication. The second is by enforcing corporate security policies. Device Management software can ensure that security software such as firewalls, virtual private networks (VPNs), and virus databases are installed and configured correctly. If a user removes or alters the software in any way, the mobile agent can transparently reinstall or reconfigure it, thereby ensuring the security infrastructure remains intact. Finally, in the case of a lost or stolen device, the device management software can automatically remove any confidential applications or data as soon as the device connects to the Internet. This level of security is not possible without implementing device management software.

Scalability

Many techniques can be used to provide scalable solutions. Device management platforms provide load balancing and failover capabilities, in addition to various deployment strategies for distributed environments. At least two servers operating in a cluster are required to take advantage of these features.

Load balancing ensures that the users are distributed evenly across the servers within a cluster, while failover allows a system to keep operating as long as any server in the cluster is still active. For deploying applications across multiple geographies, it is possible to distribute management servers in each location to handle that region's set of users. These remote servers can then communicate with one another during system updates.

Optimized Communication

Unlike corporate LANs, wireless networks are inherently unreliable. Mobile device management platforms include several features that address this. Because it is best to transfer as little data as possible over these networks, only changed data is transmitted, and it is compressed prior to transmission. Also, during transmission, bandwidth "throttling" can detect the bandwidth and scale back the transmission rate during a package download. This makes it possible to distribute applications and data in the background. Larger files can be segmented into smaller chunks before transmission and then reassembled once all the pieces have been sent. This increases the chances that an application will be deployed without interruption. When a connection is dropped during transfer, these platforms provide a way to restart the download at the exact place where it was stopped. (Note: This feature is a requirement for wireless environments, because dropped connections are common.)

Self-Healing

A snapshot of the correct client configuration can be taken and compared to the current configuration at defined intervals. If an error is detected, the correct image can be reinstalled to prevent any potential problems. The status can then be sent back to the server for logging purposes. This process can be applied to the mobile operating system itself. Self-healing, as it is called, enables the maintenance of critical applications without intervention by the user or system administrator.

Logging and Reporting

All the information about a mobile environment can be logged in to a database for reporting purposes. This may include application usage patterns, software and hardware inventory, device statistics, and error information. Once this information is captured, many useful reports can be created. For example, for licensing purposes, you may need to know how many users have a certain software program installed; conversely, you may want to know how many users do not have a set of required software on their devices. Along the same lines, you can view application usage statistics to see if the applications are being used to their full potential. When an error occurs, the logs are an invaluable set of information for finding and correcting the problem. But note that the logging and reporting capabilities vary between platforms, so it is a good idea to choose a solution that provides the information that is most important to your organization.

The Role of SyncML DM

In February 2002, the SyncML Initiative (now part of the Open Mobile Alliance) released the SyncML Device Management (SyncML DM) Protocol specification. SyncML DM defines a protocol for creating interoperable, device- and network-independent management platforms. Many of the concepts are similar to those specified in the SyncML synchronization protocol released in mid-2001. (Refer back to Chapter 10 for more on the SyncML synchronization specification.)

The SyncML DM Protocol consists of two parts: a setup phase and a management phase. The setup phase is responsible for authentication and device information exchange. SyncML DM uses the SyncML authentication framework, with some security extensions defined in the SyncML Device Management Security specification. It is mandatory that the client and server be authenticated to each other before any data is transferred. This can be accomplished using existing transport-level authentication mechanisms, if they use a form of strong authentication; if not, the SyncML DM Protocol-level authentication must be used. Ideally, the underlying transport protocol will permit authentication to occur only once per session. If this is not possible, then each request has to be authenticated. The setup phase only occurs once during a session. Once it is completed, the management phase takes over.

The management phase is responsible for the remaining communication between the client and the server. This involves one or more content packages being sent from the server to the client. If the package requires a response, then the client will send a package back to the server with its response. The response package starts a new proto-col iteration. The server can continue to send new management operation packages and continue the session with new protocols iterations as many times as required. The information in each package is defined using XML. Unless specified, the management commands can be executed in any order. There is no timeout between package itera-tions, because operations consume an unpredictable amount of time. Either the client or the server can abort a session at any time for reasons such as server shutdown, client powerdown, or user interaction from the client.

A SyncML DM Protocol iteration can be initiated either by the client or the server. Where both the client and the server are on and listening for connections, the setup phase initiates an iteration as just defined. Unfortunately, many devices cannot contin-uously listen for incoming requests from a management server or do not want to keep a port open for security reasons. In these situations, a push message such as an alert or notification can be sent to the device to cause the client to initiate a session back to the management server. This technique provides a way to "wake up" the device to start a management session. Once the session is initiated, the setup and management phases occur.

NOTE For complete information on the SyncML DM specification, including common usage scenarios and example code, visit the SyncML Initiative Web site at www.syncml.org, or the Open Mobile Alliance Web site at www.openmobilealliance.org.

The SyncML Initiative predicts that the DM protocol will first be used by wireless carriers, followed by corporations. The carriers will provide services such as automatic backup of phone books, and device preferences. They will then add more advanced features such as over-the-air provisioning of new applications. Without device man-agement capabilities, a user would have to go to a store and have the phone memory "reflashed" to install a new application. Corporations are expected to use SyncML DM for more traditional device management features such as software distribution and asset tracking. The main benefit for enterprise deployments is interoperability. SyncML DM will enable a single client to synchronize with multiple information sources. With proprietary solutions, the client can only communicate with the corre-sponding server software, usually at a single location. In addition, with a standard pro-tocol, the mobile agent software can be preinstalled on devices. This will remove one of the challenges with current device management software. It is expected that, as the standard evolves, many of the current device management vendors will add SyncML DM support to their enterprise offerings.

MDM Vendors

Table 16.3 lists the companies that provide mobile device management platforms with many of the core features discussed in this chapter. Note that these companies are regarded as having the leading offerings in this field. Over time, more companies will surely develop products in this area. At the same time, consolidation of the market is expected.

Table 16.3   Mobile Device Management Vendors

VENDOR	PRODUCT NAME	PRODUCT URL
Extended Systems	XTNDConnect	www.extendedsystems.com /ESI/Products/default.htm
iAnywhere Solutions	Manage Anywhere Studio	www.ianywhere.com /products/manage_anywhere .html
Microsoft	Systems Management Server (SMS) 2003	www.microsoft.com /smserver/default.asp
Mobile Automation	Mobile Automation 2000	www.mobileautomation.com
Novell	ZENworks for Handhelds	www.novell.com/products /zenworks
Synchrologic	iMobile Systems Management	www.synchrologic.com /about/about_imobile _software_distribution.html
Tivoli Software	Tivoli Configuration Manager	http://www.tivoli.com /products/
XcelleNet	Afaria	www.xcellenet.com/public /products/afaria/afaria.asp

Summary

Mobile information management (MIM) comprises both personal information management (PIM) and mobile device management (MDM). PIM applications can add tremendous value to remote workers, while MDM capabilities are important for system administrators. Currently, most PIM and MDM offerings come from different vendors, but the trend is for these solutions to be provided by a single vendor. All of the mobile operating systems include some level of PIM support for individual users. Several vendors produce complete mobile PIM solutions, for those who require integration with enterprise groupware products. For mobile device management, vendor platforms provide many features for software distribution, asset management, remote administration, and backup and recovery. Using a device management solution can dramatically reduce the administration costs for remote users.

In the next chapter we will learn about location-based services, an exciting technology that looks to revitalize the market for both consumer and corporate mobile applications.

Helpful Links

These Web links will provide more information on subjects related to this chapter. Vendor URLs were included in the body of the text:

iCalendar Specification (RFC 2445)   http://xml.resource.org/public /rfc/html/rfc2445.html

SyncML Initiative
  (a component of the
  Open Mobile Alliance)
www.syncml.org
www.openmobilealliance.org

Symbian and SyncML   www.symbian.com/technology /standard-syncml.html

vCard and vCalendar Web site   www.imc.org/pdi/

CHAPTER 17

Location-Based Services

Adding location information to mobile applications adds an important dimension to many solutions. Location aware applications increase user efficiency by providing customized access to data based not only on user preferences, but also on the user's current position. This takes content personalization to a new level, providing tremendous benefits to both consumer and enterprise applications.

This chapter gives a complete overview of location-based services (LBS). This includes an examination of location positioning technology and examples of how location information can be used in both consumer and corporate applications. The chapter also previews the standardization efforts that are underway for location information. We begin by discussing why location is important and what forces are driving the market.

Location-Based Services: What, Why, and When

Industry analysts, service providers, and consumers all agree on one thing: that location-based services are valuable for the mobile user. Studies conducted to determine which mobile features are in greatest demand indicate that location information is high on the wish list of the average mobile user. The reasons people cite for wanting location capabilities range from safety concerns to m-commerce; but in all cases, they believe that having LBS on their mobile devices is a must.

Before we delve into why this may be the case, let's first take a look at what location-based services are: They are applications that take advantage of location information to provide a timely service. These services can range from finding the nearest gas station to obtaining instructions on how to get from point A to point B. Or, in many cases, they provide information about a user's current location to emergency personnel, such as the 911 emergency service. Other LBS uses, such as receiving an alert on your phone about a sale as you walk by a particular store, though promised, are still some time away from being available to the mobile user. Before we get too far ahead of ourselves, let's look at the three generations of location-based services:

First generation. These applications require the users to manually input location information into their devices. Input can be in the form of a postal or zip code or possibly a city name or street address. Based on this information, the application can provide customized content such as driving directions, nearby restaurant or store locations, or weather information.

Second generation. These applications can determine location information without the assistance of the mobile user. The location is usually accurate to within a few kilometers (1 to 2 miles), similar to the accuracy level in first-generation applications. They also provide similar services to those in first-generation applications.

Third generation. These applications can obtain more accurate location information and can initiate services based on location. Third-generation applications can provide timely updates about nearby brick-and-mortar services, asset-tracking information, and street-level mapping and routing. In general third-generation applications are feature-rich and user-friendly, although some groups are voicing concerns over privacy.

As you may be aware, currently the market is in the second generation of LBS, and working its way toward third-generation services. What is driving the market to implement these new types of LBS? There is no single answer to this question. Many factors are driving the new breed of LBS, factors that depend on both the region and type of user being addressed. In the United States, one of the strongest drivers of LBS is the U.S. Federal Communication Commission (FCC) mandate for the Enhanced 911 (E911) service (more on this shortly). In Europe and Asia, m-commerce is the driving factor. And in all regions, LBS can provide tremendous advantages for corporate applications such as fleet management and asset tracking.

NOTE There are those who argue that there is no need for automatic positioning technology because many location services are not based on a user's current position, but rather on a location that the user will be at in the future. For example, a user might want to find a nice restaurant near a customer site for a dinner meeting, but he or she is nowhere near the particular location when the information is required. This is definitely true in some cases, but there are just as many situations when knowing the user's current position is critical.

Let's get back to the aforementioned E911 service requirements for wireless devices, as mandated by the U.S. FCC. To date, there are two phases of implementation:

Phase I. This involves providing information about the cellular tower where the call originated. Wireless networks are divided into cells, so knowing the cell that the user is currently in provides some idea of the user's location, usually accurate to within a few kilometers (1 to 2 miles). Also, the user's callback number is usually provided so the person can be contacted if more information is required. All wireless carriers meet the Phase I requirements.

Phase II. This involves providing much more accurate position information to the 911 emergency services. Carriers are required to provide location information accurate to within a 125-meter (400 feet) radius at least 67 percent of the time. Phase II was originally scheduled to be completed by October 1, 2001, but providing this level of accuracy has proven to be difficult. Consequently, most carriers have been granted extensions to this date since both the time required and cost of upgrading their networks is much higher than anticipated. Some analysts estimate that the total cost of the Phase II upgrade will cost U.S. wireless carriers more than $3 billion! Full deployment (covering 95 percent of users) is expected in 2005.

Prior to Phase I implementation, calling 911 from a wireless phone was somewhat ineffective. The operators not only had to find out what the emergency was, but also the location of the problem. This was especially difficult because usually the caller was not at a fixed location, but rather on the side of the road somewhere, in the woods, or some other unfamiliar location. Obtaining accurate location information is just as important as finding out what the emergency is. With Phase I implementations in place, they can at least get a general idea of the location and work from there. Phase II will provide the level of position detail that is required to execute effective emergency services. (More on the mobile positioning technologies used for Phase I and Phase II implementations is provided later in this chapter.)

NOTE If you are interested in learning more about the FCC E911 mandate, or want to find out about carrier's compliance plans, visit www.fcc.gov/911/enhanced/.

Fortunately, the benefit of all of this infrastructure development applies to more than emergency services. Once the technology is in place to provide accurate positioning information, it can be used by LBS providers for both consumer and corporate applications. This may prove to be a big plus for m-commerce, as it would have taken much longer for these technologies to be put in place if carriers based their decisions on market demand and return on investment. Imagine how many applications would be required to generate the $3 billion required to add these capabilities to the wireless networks.

It is expected that once these positioning technologies are in place, they will be used for a variety of applications. Once you can locate a customer to within a few hundred meters, there are many useful and exciting applications that can be developed that take advantage of the location information. That said, the cost of implementing these types of solutions is still high, and until vendors are convinced they will see a return on that investment, location services will not become readily available.

Location Applications

Now that we are familiar with the concepts behind LBS, it is time to discuss where location services are being used and what value they bring to consumer, enterprise, and government applications. The adoption of these applications is expected to grow as consumers and corporations realize the value of all that they offer. But before this can happen, LBS have to be accurate, fast, and easy to use. Many mobile operators are looking at LBS to help increase customer loyalty and satisfaction, as well as the average revenue per user.

Let us take a look at some of the most common uses of location information for each application category. Some of these applications use the user's current location to provide a service, while others provide a user with the location of an item or place they are interested in finding. The following are common uses of LBS:

Emergency services. In an emergency, users can contact a call center and have the current location tracked so that assistance can be deployed. This is the foundation of E911, but it can also be used for roadside emergency ("Help, my car broke down").

Traffic information. LBS can help detect your location in traffic flow or pinpoint areas of traffic congestion. Then it can provide a better route to avoid upcoming traffic.

Navigation. This category applies to many types of applications. The navigation service can answer a question as simple as "How to I get from Point A to Point B?" or those more complex, such as "What is the ideal delivery route for product distribution?" Later in this chapter we describe how Geographic Information Systems (GIS) help to provide this kind of service (see *What Is a GIS?*).

Field service management. This service makes it possible to track the location of field service agents so the closest agent can be sent to a pending service call. This type of location service is particularly valuable to emergency service vehicles such as police cars, ambulances, and fire trucks.

Fleet management. Using LBS it is possible to track the locations of an entire delivery fleet at any given time. There are several areas where this is useful: for example, to track the location of assets within each vehicle and predict arrival times for delivery vehicles.

Asset tracking. This is similar to fleet management, but it enables tracking the location of any asset, which may include delivery packages, mobile devices, or any other item of value.

Wireless advertising. Users may be able to subscribe to services that will notify them of relevant product information when they enter a defined zone. For example, as you approach a shopping mall, you could be notified if your favorite store is having a sale. This type of LBS requires very accurate location information, and will most likely not be seen in the near future.

Find-it services. This helps a mobile user find a particular feature that is within a specified range of his or her current location; for example, nearby restaurants, gas stations, golf courses, or any other retailer or venue. These LBS can also be tied to booking or purchasing applications to facilitate m-commerce.

Automatic vehicle location. A common LBS being deployed today is vehicle-tracking. In this way, stolen vehicles can easily be located and recovered.

Mapping. Many vehicles now come equipped with mapping services that provide information about the user's current location as well as a detailed map of nearby facilities. Very often, navigation services are also offered in conjunction with the mapping capability.

Weather information. Current conditions and weather forecasts can be supplied based on the user's current location. Note, however, that this information will only be as accurate as the nearest weather station.

This is just a partial list of where location-based services are being deployed for all types of applications, and signals only the beginning of the deployment of these services. As location technologies become more accurate, and development platforms become easier to use, expect to see even more LBS being deployed for both consumer and corporate markets.

Mobile Positioning Techniques

Before location-based solutions can be created, we first have to be able to get the location of the mobile user. Many technologies are in place that can provide this information. Deciding which technology to use will usually be based on a combination of accuracy and cost. In most cases, as the required level of accuracy increases, so does the cost. This cost is usually shared between the mobile user and the wireless carrier. Developers typically have to rely on the information provided to them from the handset and the carrier, precluding their ability to directly influence the accuracy of the location information.

In most cases, the location accuracy depends on the type of positioning technology being used. There are network-based solutions that can be implemented by wireless carriers to provide position information for both new and legacy handsets. These solutions are quite cost-effective, although their accuracy is often not ideal; they range from several hundred meters (or yards) to several kilometers (or miles) depending on the solution. Handset-based solutions can dramatically improve the accuracy, although they introduce significant costs for both the handset manufacturers and network operators. With these solutions, it is possible to get location information that is within meters (5 to 10 feet) of the user's position. In many cases, a mixed solution is the best approach, with the handsets and networks working together to provide a solution with acceptable accuracy at a reasonable cost.

The objective of all positioning technologies is to capture the location of a mobile device and convert it into a meaningful X,Y coordinate. This section describes the leading methods for accomplishing this task.

Network-Based Solutions

One way to locate a mobile user is to use the fixed-base stations that comprise the wireless carrier's network. Each of these stations contains radio intercept equipment that can receive a signal from any active phone. By taking the signals from one or more base stations, the location of the mobile user can be determined. In general, the more base stations that are used, the more accurate the location information will be.

Network solutions can work with existing handsets, making them an ideal first step in providing location information. Carriers are expected to use network-based solutions to meet the FCC E911 Phase II requirements for legacy handsets.

Cell Identity

Cell identity is the most simplistic and cost-effective way to provide position information. It simply determines which cell of a wireless network the device is using and reports its location. Since the base station for each cell is in a fixed location, the cell identity can easily be translated into a location for the mobile user. The downside of this approach is that the user's precise location within the cell is unknown. This method typically provides location information accurate within a kilometer or two (about one mile) which may be acceptable for getting a general idea of where the user is located, but does not provide terribly useful information for emergency services or tracking, let alone targeted advertising or driving instructions.

Fortunately, there are ways to improve the accuracy of cell identity. Some cells are divided into sections, thereby reducing the total area of the possible location. This can often reduce the area cross-section by two-thirds. For example, if the total area of one cell is four square kilometers, the location of a user is limited to that area. If, however, that cell can be identified in sections one-third the size, then the area where that particular user is located can be reduced to under one-and-a-half square kilometers.

To get an even more accurate reading on the location, a technique called timing advance (TA) can be used. TA provides a way to find out how far a user is away from the base station, thereby dramatically reducing the possible locations for that user. The information is not exact, but it does improve the overall accuracy of using cell identity for determining a user's position. Unfortunately, TA information is not easy to obtain without access to a mobile positioning center (MPC). An MPC can provide more detailed position information using a defined API. Developers can write applications that talk to the MPC to get the TA information along with cell identity. Figure 17.1 shows the accuracy for the various positioning techniques using cell identity and timing advance. The shaded areas in the diagrams represent the possible locations for the mobile user.

When these positioning methods are combined, they are commonly referred to as cell global identity-timing advance or CGI-TA. This approach can yield results that are accurate within 100 to 200 meters (100 to 200 yards), quite impressive for such a simple technique that does not require handset or network upgrades. It is also worthwhile to note that the accuracy of CGI-TA is better in cities than in rural areas due to the higher density of base stations in populated areas.

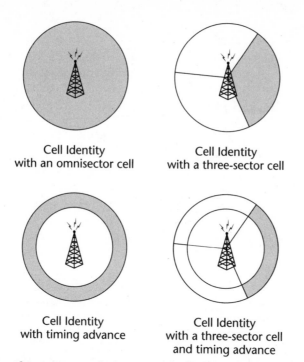

Cell Identity
with an omnisector cell

Cell Identity
with a three-sector cell

Cell Identity
with timing advance

Cell Identity
with a three-sector cell
and timing advance

Figure 17.1   Cell Identity and timing advance positioning areas.

Time of Arrival (TOA)

Even when timing advance information is available, CGI-TA does not provide accurate-enough information for most location-based services. Using a time of arrival (TOA) approach—also known as time difference of arrival (TDOA) or uplink time of arrival (ULTOA)—can dramatically improve the location accuracy. Rather than using one base station to determine the location, TOA uses information gathered from three or more base stations. It works by having the phone send out a signal that is received by all of the base stations within range. Each station then measures the amount of time it took to receive the signal from the time it was sent (T1, T2, T3). These time differences have to be very accurate, requiring all of the base stations to be synchronized. This also requires using either GPS systems for synchronization or an atom clock, both of which are costly solutions.

Since the signal moves at a fixed speed, the distance of the device to the base station can be determined. The distance from a single base station does not help a lot as it is not possible to know in which direction the mobile user is moving. By using the information from three base stations, it is possible to triangulate the coordinates of the user relative to the base stations, as shown in Figure 17.2. Since the base stations are in a fixed location, the relative coordinates can be translated into absolute coordinates that can be used to create LBS.

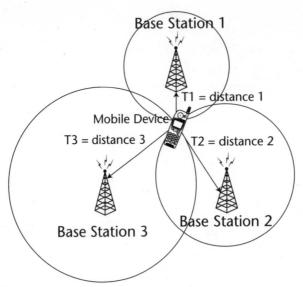

Figure 17.2   Using time of arrival to determine location.

TOA technology does not require any changes to the handset itself. This makes it a potential solution for meeting the E911 Phase II requirements for legacy handsets. The accuracy of this solution is fairly decent, coming in at around 50 meters (50 yards) in urban areas and 150 meters (150 yards) in rural settings.

> **NOTE** TOA solutions are more practical on CDMA/CDMA2000 networks, as they are already synchronized from the start and do not require using GPS or atom clocks.

Angle of Arrival (AOA)

Angle of arrival (AOA) works in a similar fashion to TOA, but instead of using the time it takes for a signal to reach three base stations, it uses the angle at which a device's signal arrives at the station. By comparing the angle-of-arrival data among multiple base stations (at least three), the relative location of a device can be triangulated. On its own, AOA is not commonly used, and is rarely discussed with LBS. That said, some systems may use the angle of arrival along with the time of arrival to get an even more accurate location.

Handset-Based Solutions

When more accuracy is needed, handset-based solutions are required. In these solutions the handset participates in the position determination. The accuracy of these technologies allows for the introduction of the third generation of location-based services, where precise location information is required. The two handset-based systems described next use similar ways of calculating location, with one major difference: E-OTD relies on base stations and GPS uses satellites.

Enhanced Observed Time Difference (E-OTD)

Enhanced observed time difference (E-OTD) technology works in a similar way to time of arrival, but the handset makes the time measurements instead of the base stations. E-OTD relies on measuring the time at which signals from the base station arrive at two geographically dispersed locations: the mobile device and a fixed measuring location called the location measuring unit (LMU). For accurate triangulation, at least three base stations have to participate in the calculation.

For this method to work, the participating base stations have to transmit a very accurate clock time to the mobile device. Using this approach, all of the signals have to be sent at the same time because the mobile user may be moving during the measurements. This is where the LMUs come into play. They provide an accurate timing source for the measurements, ensuring the accuracy of the data. Once the measurements are taken, the E-OTD-enabled handset records the time differences from the three base stations. The distance between the mobile device and the base stations can then be calculated by comparing the time differences between the timing measurements. The time difference can be translated into a distance because the signals move at a fixed speed. Once the measurements are taken, the relative location can be calculated on the network or on the handset itself. The handset performs the measurements using a software solution; but in order to perform the calculations, a hardware upgrade is required as well. Once the relative location is determined, it can be translated into an absolute position since the base station coordinates are known. Figure 17.3 depicts the E-OTD positioning system.

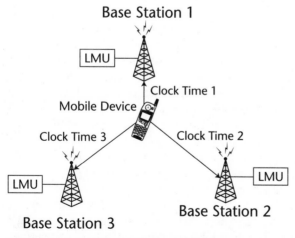

Figure 17.3   E-OTD positioning architecture.

E-OTD provides an accurate and cost-effective way to determine a mobile position. The results are typically accurate within 50 to 100 meters (50 to 100 yards), meeting the FCC E911 Phase II guidelines. As a result, E-OTD has become the de facto standard among U.S. GSM carriers for E911 Phase II implementation.

GPS and A-GPS

Global Positioning System (GPS) is the most popular positioning technology being used today. It uses 24 global satellites that orbit the Earth to send signals to a GPS-enabled receiver. The receiver can communicate with three or four satellites at any single point in time. For this to work, however, there has to be a line of sight between the receiver and the satellites, precluding the use of GPSs inside buildings. Once the receiver obtains the positioning measurements, it can either calculate the location coordinates directly on the device or send the measurement results back to the network server for processing. Similar to E-OTD, the calculations are quite complex and require adequate processing power. If the calculations are to take place on the device, the device manufacturer has to include the appropriate hardware, adding to the device cost. In many product offerings, the GPS receiver is a separate unit that can be connected to a mobile device using a cable hookup or wireless technology such as Bluetooth. This means the GPS unit can include the required hardware, without having a direct impact on the form factor and power consumption of the device. Over the past several years, the size, power consumption, and cost of GPS chipsets have fallen, leading to widespread use of this technology in the mobile environment.

GPS works similar to other triangulation-based positioning technologies. The satellites constantly broadcast signals that can be read by GPS-enabled devices. It is unimportant to the satellite how many devices are receiving the signal because the communication goes in only one direction. The device measures the amount of time it takes for the satellite signals to reach it. This measurement is taken from three distinct satellites to provide precise location information. Mathematically, four measurements are required, but three usually provide sufficient information to give an accurate result. The speed of the signal is known, enabling the GPS to determine the distance from the satellite. As you can imagine, it is very important that these time measurements be incredibly precise. A time calculation that is off by one-thousandth of a second can result is a location variation of over 300 kilometers (200 miles)! For this reason, the GPS receivers use atomic clocks on each satellite to ensure the time is correct. Once the distances have been determined, triangulation calculations are executed to determine the absolute location coordinates. GPS produces very accurate results, typically to within 5 to 40 meters (5 to 40 yards)of the actual location. GPS also provides three-dimensional location information, latitude, longitude, and elevation.

Even though GPS-based systems provide accurate information, they are not terribly versatile. As already mentioned, in order to get a reading, the GPS receiver has to have a line of sight to the satellites. This is a significant limitation for the mobile workforce. Many applications require use inside of buildings or vehicles, making it difficult for GPS to provide the required service. In some cases, this is addressed by incorporating a second positioning technology, typically network-based, along with GPS. For example, a cell identity solution could be used as a backup where a line of sight is not available. A second limitation is the time required to obtain the location information. With standard GPS configurations, this time ranges between 20 to 40 seconds, a delay that could have a negative impact of the usability of the application.

One solution to both the line-of-sight and time delay issues is network-assisted GPS or A-GPS. A-GPS uses modified handsets that receive the GPS signals and then send those readings to a network server. The server uses network-based GPS receivers to help the handset measure the GPS data. The network GPS receivers are placed around the network several hundred kilometers apart. They regularly collect GPS satellite data and provide this data to the handsets, enabling them to make timing measurements without having to decode the actual satellite messages. This makes a substantial difference in the time it takes to get the location information. Using A-GPS, the time is typically between one and eight seconds.

To alleviate the line-of-sight restriction, the handsets send the measurement information to network servers, so that complex computations can be done away from the device. This extra computing power allows for multipath mitigation and signal processing techniques to locate devices indoors and other locations that are challenging for conventional GPS. Figure 17.4 demonstrates the A-GPS architecture.

GPS receivers are commonly used in conjunction with a Geographic Information System (GIS; described in the next section) to provide a complete topographic view of different areas in the country. One of the most common GPS applications is for vehicle positioning and tracking services. These services require more than just position to make them functional. That is where a Geographic Information System becomes useful.

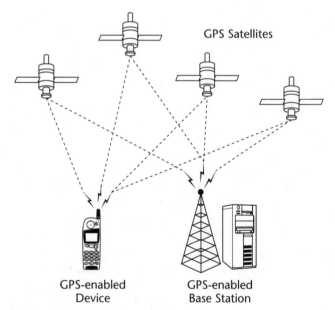

Figure 17.4   A-GPS architecture.

What Is a GIS?

Mobile positioning is only one aspect of providing location-based services. Having the location coordinates obviously is essential, but what is done with them is just as important. A Geographic Information System (GIS) is mapping software that relates the location information with other pertinent information to give it meaning and value. This is typically accomplished by providing multidimensional mapping of information such as building locations, street layouts, population densities, and a plethora of other information. A full GIS consists of hardware, software, data, and trained users who have knowledge of how to perform analysis on the information provided by the GIS.

A GIS is helpful for two primary purposes:

Finding a feature. This refers to supplying information on where something is or what something may be. This can include finding the closest restaurant or gas station or determining the best route to get to a certain location.

Finding patterns. This capability is very relevant for business analysis, and not necessarily only used from mobile devices. Businesses are often more interested in distribution of information rather than individual features. For example, when deciding on a location to hold a seminar, a salesperson may want to determine the area where the majority of his or her customers work.

Clearly, a GIS is useful for mobile applications, but it offers benefits that go well beyond what is required in a mobile environment. For example, using a GIS, users can decide what information is and is not relevant to them, and formulate their queries based on their personal criteria. Unlike a paper map, a GIS allows for in-depth analysis and problem solving that can make marketing, sales, and planning much more successful. The following are some common GIS uses:

Finding what is nearby. This is the most common use for mobile users. Given a specific location, The GIS finds institutions within a defined radius. This may include entertainment venues, medical facilities, restaurants, or gas stations. Users might also use the GIS to locate vendors that sell a specific item they want. This promotes m-commerce by matching buyers with sellers. The results can be provided using a map of the surrounding area or the destination addresses.

Routing information. This is another common use for mobile users. Once users have an idea of where they want to go, a GIS can help provide directions on how to get there. Once again, this can be provided graphically using a map or with step-by-step instructions. For mobile applications, it is often helpful to provide routing information in conjunction with search services.

Information alerts. Users may want to be notified when information that is relevant to them becomes available based on their location. For example, a commuter might want to know if he or she is entering a section of the highway that has traffic congestion, or a shopper might want to be notified if his or her favorite store is having a sale on a certain item.

Mapping densities. For business analysis, knowing population densities can be extremely useful. This allows users to find out where high concentrations of a certain population may be. Densities are typically mapped based on a standard area unit, such as hectares or square kilometers (or miles), making it easy to see distributions. Examples of density mapping may include the location of crime incidents for police to determine where additional patrolling is required, or of customers to help determine ideal field delivery routes.

Mapping quantities. People map quantities to find out where the most or least of a feature may be. This information could, for example, be used to determine where to locate a new business or service. For example, let's say you are interested in opening a laundromat: It would be prudent to determine how many other laundromats are in the area and what the population base is. This information could help you choose a location that would be receptive to your service. This type of mapping can also be useful for urban planning and environmental studies; for example for city planners who are trying to determine where to build more parks.

A GIS can provide information and insight to both mobile users and people at fixed locations. This information uses the location coordinates provided by one of the positioning technologies to give details that are relevant to the user at that specific moment. Many of the location-based services discussed earlier in this chapter would benefit from the information provided from a GIS.

LBS Development

The mobile location industry epitomizes an emerging market: The environment is complex and many different technologies and methodologies can yield the same results; furthermore, there are no precedents or best practices yet established, making the LBS landscape unfriendly to developers.

Currently, vendors typically have a proprietary API for accessing their position information. In general, two types of API are being used:

Network-based API. The positioning information is available on the network server. This information can be accessed by any party that has rights to access the server.

Handset-based API. The positioning information is generated on the device and is directly accessible by applications running on the device.

The majority of positioning solutions provide a network API for developers. Access to the position information is provided by a mobile positioning center (MPC). The same API is used regardless of the positioning technology, thereby providing a layer of abstraction for developers. The interface to the MPC is typically provided using HTTP. MPCs are a step in the right direction for developers, but they do not quite solve the problem. There are still some challenges to face, such as the lack of standardization and roaming among location services offered by disparate mobile operators. Until these issues are resolved, many vendors will continue to hold off on their LBS efforts.

Standards in the location market are required to resolve interoperability issues between positioning technologies and network operators. Without a defined standard, the rollout of LBS will continue to lag behind expectations. To address this need, in 2000, Nokia, Ericsson, and Motorola jointly formed the Location Interoperability Forum (LIF). The goal of LIF is to promote common and ubiquitous mobile location services (MLS). In 2002, LIF became a member of the Open Mobile Alliance, to make location-based services part of future OMA standardization efforts.

A common API for development across vendor platforms and location technologies will help drive the LBS market. It will give developers a way to design location-based services that are both network-protocol- and positioning-technology-independent. Other issues the LIF addresses include privacy, roaming, and charging.

Other standardization efforts include the Open GIS Consortium (www.opengis.org) and the Magic Service Initiative (www.magicservicesforum.org).

LBS Vendors

The list of vendors involved in the location-based services field is extensive due to the mix of components required for a complete solution. Figure 17.5 shows how these various providers work together to provide a complete solution. At the base level are the position technology vendors, which provide either hardware or software solutions for determining the user's location. The infrastructure providers use the location information to provide services to carriers and content providers. The location-based applications use all of the aforementioned technology to target both the consumer and enterprise markets.

Each of these categories is explained briefly in this section, followed by a list of some of the technology vendors, along with their Web sites in Table 17.1. Keep in mind that the LBS market will continue to change as demand for these services increases and as the technology matures. New vendors will come to market with products, and mergers between existing vendors will likely happen.

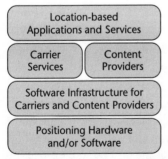

Figure 17.5 Location technology layout.

Position Determination Software Vendors. These vendors provide software solutions to determine a mobile user's position. The software is typically installed on network servers, although in some cases it is deployed to wireless handsets.

Position Determination Hardware Vendors. These vendors provide hardware chipsets or receivers for position determination.

Location Infrastructure Providers. These vendors provide software platforms that use position information for creating location-based services. In this category you will also find GIS vendors and mapping software vendors. Corporations can use these products to create a variety of location-based applications.

Content Providers. These vendors provide location-based services to both the consumer and corporate markets. The number of vendors in this category will surely increase as LBS grow in popularity. (Note that many of the network infrastructure vendors, such as Motorola, Ericsson, Nortel, and Nokia, provide LBS content.)

Development Platforms. These vendors provide platforms targeted at developers for creating location-based services. Many of these platform vendors have partnered with other location vendors to provide integrated solutions.

Table 17.1   LBS Vendors

VENDOR	URL
@Road	www.@road.com
AirBiquity	www.airbiquity.com
Cambridge Positioning Systems	www.cursor-system.com
Cell-Loc	www.cell-loc.com
Earthlink	www.earthlink.com
Ericsson	www.ericsson.com
ESRI	www.esri.com
Garmin	www.garmin.com
Global Dining	www.globaldining.com
Global Locate	www.globallocate.com
Grayson Wireless	www.grayson.com

(continues)

Table 17.1   LBS Vendors *(Continued)*

VENDOR	URL
IBM	www.ibm.com
iPlanet	www.iplanet.com
Kivera	www.kivera.com
LoJack	www.lojack.com
MapInfo	www.mapinfo.com
Maporama	www.maporama.com
Maptuit	www.maptuit.com
Motorola	www.motorola.com
ObjectFx	www.objectfx.com
Oracle	www.oracle.com
ParthusCeva	www.parthusceva.com
Pharos	www.pharosgps.com
QUALCOMM	www.qualcomm.com
SignalSoft	www.signalsoft.com
SiRF Technology	www.sirf.com
Sun Microsystems	www.sun.com
Telcontar	www.telcontar.com
Telespatial	www.telespatial.com
Televigation	www.televigation.com
TruePosition	www.trueposition.com
TCS	www.telecomsys.com
Trimble	www.trimble.com
Vindigo	www.vindigo.com
Webraska	www.webraska.com

Summary

Location-based services are expected to be major contributors to the success of mobile computing. They provide value for consumer, corporate, and government markets by allowing a user's current position to be used within mobile applications that provide routing, asset tracking, information retrieval, emergency services, and a significant number of other features. At the heart of LBS are the location positioning technologies, which vary significantly in accuracy and cost. As these positioning technologies become more accurate and less expensive, expect to see the LBS market expand significantly.

Developers should not be concerned with the underlying technology being used when creating location-based services, as long as the location information is accurate enough to satisfy the application requirements, and as long as the time required to determine the location does not detract from the application's usefulness. To aid with development, a number of vendors offer LBS components as part of their mobile platforms.

The next chapter focuses on four technologies that are expected to grow in importance in the mobile industry: Web Services, Binary Runtime Environment for Wireless (BREW), Speech Application Language Tags (SALT), and M-Services.

Helpful Links

The following Web links can be used to learn more on subjects related to those described in this chapter:

FCC E911 Information	www.fcc.gov/911/enhanced
Cambridge Positioning Systems – E-OTD information source	www.cursor-system.com
GPS Information from Trimble	www.trimble.com/gps/why.html
Location Interoperability Forum (LIF) (Component of Open Mobile Alliance)	www.openmobilealliance.org
General LBS Information	www.mobileinfo.com /LocationBasedServices www.jlocationservices.com
Wireless Developer Network LBS Channel	www.wirelessdevnet.com/channels/lbs/

Other Useful Technologies

This chapter focuses on technologies that, while promising, have not yet reached their market potential. Each of the technologies discussed here has been developed to improve upon previous technologies in the same space; in fact, the goal of each is to become the standard in its field.

The goal of this chapter is to introduce each of these technologies and explain how it relates to mobile computing. Thereafter, we will examine the reasons why it is expected that they will be able to survive in the very competitive mobile industry. The chapter starts by taking a look at Web services from a mobile perspective. Of the technologies included in this chapter, Web services are clearly the leader in market acceptance and standardization. The others, such as BREW, SALT and M-Services, are still working to gain meaningful vendor and developer acceptance.

Web Services

The W3C definition for Web services is as follows:

"A Web service is a software application identified by a URI, whose interfaces and bindings are capable of being defined, described, and discovered as XML artifacts. A Web service supports direct interactions with other software agents using XML-based messages exchanged via internet-based protocols." (http://www.w3.org/TR/2002/WD-wsa-reqs-20020819).

Essentially, Web services are segments of business logic that are made accessible over the Internet. They can be incorporated into any application that can communicate with XML over HTTP. They may be incredibly simple, such as validating a user, or intricately complex, such as executing a business transaction across many systems in multiple organizations. In short, Web services will change the way Internet applications are assembled. This is quite a profound statement to make about a relatively new technology, but it is not at all unreasonable for the simple reason that Web services provide a standardized way for corporations to both publish their own applications and access other prebuilt applications over the Internet; and these tasks can all take place without requiring the participating parties to develop any formal relationship.

It is interesting to note that the idea behind Web services is not unique. Many other technologies provide similar capabilities (CORBA, Java RMI, and DCOM) but all in a more proprietary manner. Each is capable of providing remote access to business logic. But Web services are succeeding where others have failed because they provide the same services in a simple, standardized way. Web services use XML for communication, rather than the proprietary binary formats used by other technologies. In addition, Web services are both platform- and language-neutral; they can be implemented using any piece of middleware, based on J2EE, .NET, CORBA, DCOM, or others. These technologies all allow for distributed application development using industry specified protocols. Earlier in this book we have discussed the various components of both J2EE and .NET, but we have not as of yet defined either CORBA (Common Object Request Broker Architecture) or DCOM (Distributed Component Object Model). CORBA is a language neutral distributed object model specified by a consortium of over 500 vendors through the Object Management Group (OMG). It allows applications at different locations, by different vendors, to communicate with one another, typically over the Internet using Internet Inter-ORB Protocol (IIOP). Notably absent from the OMG member list is Microsoft, who instead of supporting CORBA, created their own distributed component model in DCOM. DCOM, which is based on COM, is a protocol that enables software components to communicate directly over a network in a reliable and efficient manner. Web services simply provide the interface into this logic using HTTP and XML. More specifically, Web services are a combination of several technologies, including SOAP, WSDL, and UDDI (all of which will be described shortly).

All of the major industry players agree on the Web services concept, so it almost certainly will become the de facto method for accessing application logic on the Internet. Initially, it is expected that corporations will make Web services available internally, via corporate intranets. This will enable them to become familiar with the technology in a controlled, secure environment. Over time, these same services will be made available to other parties over the Internet. Eventually, we will have a complete library of Web services applications available for use to our applications, both wireline and wireless.

Before we explore how Web services fit into a mobile environment, some background on the core Web services technologies is in order.

Web Services Technologies

In order to understand how Web services work as a whole, we need to understand each of the core technologies involved in using them. Over the past three years, the following technologies have come to the forefront for implementing Web services:

- Simple Object Access Protocol (SOAP)
- Web Services Description Language (WSDL)
- Universal Description, Discovery, and Integration Service (UDDI)

For universal integration of Web services, these technologies have to be adopted by all major Web services vendors. It is likely that other technologies will either complement, or possibly replace, these technologies as Web services evolve.

The following subsections offer a brief description of each of these technologies. Then we will examine how they work together to define the complete Web services platform.

Simple Object Access Protocol (SOAP)

The Simple Object Access Protocol, SOAP, is a specification for the exchange of structured XML documents between parties in a distributed environment. The data exchange can occur over standard Internet technologies such as HTTP, SMTP, or FTP. While SOAP is fundamentally a stateless one-way message exchange, applications often use it to create the more complex request/response and request/multiple response interactions commonly used in Internet applications.

SOAP consists of the following three parts:

SOAP envelope. Defines an overall framework for expressing what is in a message, who should deal with it, and whether it is optional or mandatory.

SOAP encoding rules. Defines a serialization mechanism that can be used to exchange instances of application-defined data types.

SOAP RPC representation. Defines a convention that can be used to represent remote procedure calls and responses.

The fundamental benefit of using SOAP is interoperability. Current middleware technologies have difficulty communicating with one another. By using XML for data exchange, SOAP enables heterogeneous technologies to communicate with each other. For example, a Java client can communicate with a .NET component exposed using SOAP; or in the same manner, a .NET client can communicate with an EJB using SOAP. This is extremely powerful in the current fragmented middleware environment. Many of the current middleware environments provide a way to translate their native function calls into a SOAP message. For example, Microsoft provides tools to translate COM function calls to SOAP, and many J2EE vendors provide tools to translate Java function calls to SOAP.

IBM, Microsoft, Lotus, DevelopMentor, and UserLand created the SOAP specification in 2000 and submitted it to the W3C for standardization. The XML Protocols Working Group is responsible for all future specifications.

Web Services Description Language (WSDL)

The Web Services Description Language (WSDL) is an XML-based specification that provides a standard way to describe Web services over different protocols, or encodings. WSDL is used to describe what a Web service can do, where it resides, and the parameters required to invoke it. Even though it is designed to work over any protocol, the specification currently only describes its usage in conjunction with SOAP, HTTP, and Multi-Purpose Internet Mail Extensions (MIME). This makes it well suited for our discussions of Web services, since those are the common protocols used.

In order to interact with a Web service you have to have knowledge of its input and output parameters, its function structure, and its protocol bindings. Service providers use WSDL to provide this information to clients who want to invoke the service. In this way, disparate clients can automatically understand how to interact with a Web service wherever it may be located. This is a key aspect to the ubiquitous nature of Web services. Without a standard way to describe the interface, Web services cannot be accessed from remote parties using disparate client technology.

Similar to SOAP, the WSDL specification is now the responsibility of the W3C.

Universal Description, Discovery, and Integration (UDDI)

The Universal Description, Discovery, and Integration (UDDI) service is a mechanism clients use to dynamically find Web services. It provides a platform-independent, open framework for describing services, discovering businesses, and integrating business services using the Internet. More simply, UDDI can be thought of as the "yellow pages" of Web services: It is where you look to find available Web services by searching for names, categories, or other specific characteristics of a Web service.

The UDDI specification lets two companies query each other's capabilities and for each to describe its own capabilities. Businesses can submit their services to be included in the UDDI business registry. Each entry is an XML file that describes a business and the services it offers. There are three parts to an entry in the UDDI directory, as follows:

White pages. Includes name, address, contacts, telephone numbers, and other known identifiers.

Yellow pages. Includes industrial categories based on standard taxonomies such as the Standard Industrial Classification.

Green pages. Includes technical information about Web services that are provided by a given business. These pages should provide enough detail so that a third party can write an application to use the Web service.

UDDI takes advantage of standard technologies and protocols such as XML, HTTP, Domain Name System (DNS) protocols, and SOAP.

Bringing Them Together

In isolation, no one of these technologies is a major breakthrough; but when brought together in a cohesive manner, they establish a strong foundation for the next generation of business integration. All three of these specifications (SOAP, WSDL, UDDI) are based on Internet standards and use XML as a core language. This commonality is behind the success of Web services, and marks the first time a standard solution has been made available for integrating dynamic business services.

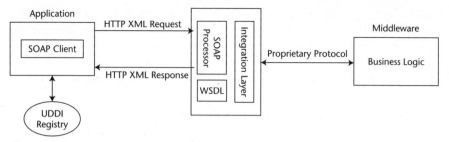

Figure 18.1 Web service using SOAP, WSDL, and UDDI.

Figure 18.1 shows how these three technologies interact within a Web service architecture. The client application has a need to find a particular service on the Internet. The client queries the UDDI registry for this service by using an identifier such as its name or category. Once the Web service is located, the client obtains information about the location of the WSDL document. This document contains the information on what the service provides and how it can be called. The client then creates a SOAP message that adheres to the information provided by the WSDL document. The SOAP message is then sent to the host of the Web service, where it can then be processed.

Another specification that warrants discussion is Electronic Business using eXtensible Markup Language, or ebXML. ebXML is a standard specification that provides guidelines for electronic business. It takes advantage of technologies such as SOAP, UDDI, and WSDL to provide a real-world view at how Web services can be implemented for business transactions. In general, ebXML is largely complementary to these Web services technologies, although it does offer alternative technologies to both WSDL and UDDI in the form of Collaboration Protocol Profile (CPP) and the ebXML Registry Service, respectively. When it comes to a comparison between the standards, Web Services provides the low-level technologies for enterprise integration, while ebXML provides a real-world implementation of Web services for electronic business transactions. The complete ebXML Technical Architecture Specification can be found at www.ebxml.org.

.NET versus J2EE

The topic of Web services always generates debates as to the relative merits of J2EE versus .NET. Based on the strengths that each platform offers, a conclusion will not be reached anytime soon. That said, the vendors promoting J2EE platforms, and Microsoft promoting the .NET platform, do agree that Web services form the cornerstone of Internet applications; hence, they are scrambling to make sure the core Web services technologies are supported in their offerings. Some observers feel that Microsoft has the lead because of its strong support for XML standards, such as SOAP, which is native to its platforms. Others feel J2EE will be dominant because of the strong industry support for this platform from software leaders such as Sun, Oracle, BEA, and Sybase, all of which have J2EE Web services-enabled platforms. More likely, there will be no clear winner; probably both .NET and J2EE will continue to attract developers to their respective platforms, meaning that many corporations will end up deploying a combination of .NET and J2EE servers for some time to come.

> **NOTE** J2EE and .NET were defined earlier, in Chapter 11, "Thin Client Overview."

Because both have strong Web services support, it is important to understand the differences between the .NET and J2EE offerings. The following are the main areas where the two platforms differ:

Core technologies supported. Both J2EE and .NET have support for core Web services technologies such as SOAP, WSDL, and UDDI, although they provide access in different ways. The .NET platform was built with Web services in mind and therefore has native support for SOAP, WSDL, and UDDI. In contrast, J2EE had to be retrofitted by adding Java APIs to access these technologies. These APIs are not difficult to use with current versions of J2EE, and future J2EE versions will support Web services technologies natively, so this eventually will be a moot point.

Programming languages. This is an obvious distinction: J2EE is based on the Java programming language, while .NET enables the use of many programming languages. In both cases the programming languages are compiled into binaries that can be executed by a virtual machine. The Java Virtual Machines (JVM) require Java binaries, while the Microsoft common language runtime can use Microsoft languages such as Visual Basic, C#, and a variety of others.

Supported platforms. Another obvious difference: J2EE platforms can be deployed on nearly all server platforms, including Solaris, Windows, Linux, AIX, and many others, whereas .NET currently can be deployed only on Windows-based platforms. For organizations that are not solely Windows-based, this could be a deciding factor on which platform to implement.

Scalability. Along with the server platforms comes scalability. J2EE is proven to be very scalable for large deployments. High-end .NET deployments have yet to be seen on a large scale. By the same token, .NET has been shown to have better performance than J2EE, especially for dynamic Web sites.

Ease of use. This may cause some disagreement, but the .NET platform is easier to develop on. Ease of use is a core .NET design goal. Since a single vendor provides all aspects of the platform, they work very well together. Implementing a complete J2EE solution can be quite complex. While there are many great Java development tools on the market, they cannot change the fact that J2EE is inherently complex.

In spite of their differences, however, these two platforms may find common ground in Web services. That said, true interoperability will only be achieved if vendors of both platforms work together to achieve a common specification. The most logical way for this to happen is through the Web Services Interoperability Organization (WS-I). The goal of WS-I is to promote Web services interoperability across platforms, applications, and programming languages. To accomplish this, WS-I is providing some key deliverables to help developers create interoperable Web services. These include implementation guidelines, sample implementations, and a set of tools for creating, monitoring, and detecting errors in Web services applications. The WS-I has adopted specifications such as SOAP, WSDL, and UDDI from other standards bodies, including the W3C and IETF, for usage in its Web services profiles. And WS-I encourages all organizations interested in promoting Web services interoperability to join the WS-I. At the time of this writing, more than 140 companies had accepted the invitation to become WS-I community members.

Web Services in a Mobile Environment

By their nature, Web services are ideal for mobile applications. They provide access to business logic and enterprise data using standard Web-based protocols such as SOAP, WSDL, and UDDI over an HTTP network connection. This allows Web services to play a role in the development of both smart client and thin client wireless applications. Since most devices do not possess the processing power for complex transactions, these calculations have to occur in a different location. Even when devices have the required capabilities, creating the applications and distributing them to mobile devices that are running different operating systems is not a trivial task. The same is true for implementing security and administration.

Having a way to offload these capabilities to a centralized server—or, better yet, to license them from a third party—could mean tremendous benefits for mobile deployments. Instead of having to create all of the content in-house and then finding a way to make it available to mobile applications, corporations can take advantage of Web services to access the required functionality. In the near term, it is likely that corporations will gradually implement logic as Web services for internal usage. As the standard becomes more prevalent, these same Web services will be made available to outside parties. For example, rather than creating a custom authentication routine, services such as Microsoft Passport could be used to authenticate users. Or credit card clearance centers could allow for credit card purchases to be approved using Web services. The possibilities are nearly endless.

To make Web services easily accessible from Windows CE-based devices, Microsoft has enabled Web services on Windows CE .NET. The first step was to provide native support for SOAP in Windows CE, then to provide in the .NET compact framework a set of services and APIs that enable application developers to quickly and easily offer Web services support for their mobile applications.

The same ideas are being incorporated into J2ME. With support for XML, HTTP, and the Java APIs for Web services, J2ME applications can take advantage of Web services. This makes possible a whole new set of applications for the wide range of Java-enabled devices on the market.

For applications on more limited devices, such as Web-enabled phones, the integration with Web services may have to occur on a server, since the client does not have the necessary processing power to parse the SOAP message and consume a Web service. In this architecture, a server could consume the Web service, and relay the relevant information to the mobile device using a standard wireless markup language such as WML or XHTML Mobile Profile. It is expected that this will be the most common methodology for limited devices to access Web services in the near future.

The great part about both of these architectures is that the result of the complex logic on the server is an XML document or a wireless markup language, which is easily consumed by a mobile client application. This means that even the simplest client-side applications can access subsets of critical data. This data can then be used within a client-side application, and possibly be stored in a local data store or simply be displayed using a microbrowser. In either case, Web services provide an important interface to complex enterprise logic and data from a variety of mobile clients.

Binary Runtime Environment for Wireless (BREW)

The Binary Runtime Environment for Wireless (BREW) gives developers a mechanism for creating portable applications that will work on Code Division Multiple Access (CDMA)-based handsets. BREW was developed by QUALCOMM, which also created CDMA technology. It is a lightweight framework that runs between the applications and the chip operating system software. BREW manages all of the telephony functions on the device, thereby allowing application developers to create advanced wireless applications without having to program to the device's system-level interface. Figure 18.2 shows the BREW architecture and the relationship between the BREW API platform and the mobile applications. Later in this section we will also discuss how J2ME and BREW can coexist.

> **NOTE** Even though BREW currently supports only QUALCOMM's CDMA chipsets, the platform is air-interface-independent, meaning that handsets using other wireless technologies can be BREW-enabled in the future.

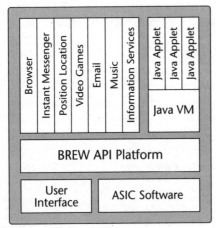

Figure 18.2 BREW architecture.

BREW Benefits

Historically, mobile phones have been closed systems; that is, the software that shipped with the device was all that was available to the user. There was no capacity to add new applications to the device or even to update existing applications. If you were unhappy with the calendar application, for example, there was really nothing you could do about it. The same was true for the phone book, email clients, and micro-browsers. And you could not even think about adding a custom application like an MP3 player or a video game. Those restrictions were very limiting for both the consumer and the wireless operator. This is where BREW comes in: By providing a standard application layer, consumers, developers, wireless operators, and handset manufacturers all benefit.

> **Consumers.** The benefit for consumers is quite clear-cut: BREW allows for consumers to download new applications over the air, which means they can personalize their devices. BREW will promote a whole new range of applications for the consumer, ranging from instant messaging to location-based services. The consumer will have the final say in what is deployed to their device; moreover, they will only have to pay for applications they are interested in.

> **Developers.** BREW makes it possible for developers to create new applications for mobile phones. They no longer have to establish a relationship with wireless handset manufacturers or carriers or have to own a physical device prototype. Instead, developers can use C/C++ or other high-level languages to create their applications for a wide range of BREW-enabled devices. QUALCOMM has also established a TRUE BREW testing process to ensure application compatibility and quality.

> **Handset manufacturers.** Once handset manufacturers integrate BREW into their devices, they will be able to release new products faster. Application testing will no longer have to be included in the factory integration tasks because they can be deployed to the device at a later time. In addition, handset manufacturers will be able to use a wide range of third-party applications, allowing them to spend more time on handset design and development.

> **Wireless operators.** Wireless operators benefit from the capability to provide a broader range of applications for handsets running on their networks. This can help drive mobile phone adoption, as well as increase airtime usage. In many cases, it is the wireless operator that provisions the applications, thereby taking a percentage of the application revenue.

On the development side, a BREW SDK is available on Windows platforms. It includes a phone emulator that supports phone customization by OEMs or developers who are looking to customize the device. It also has sample applications and tools to aid with development. Although any language can be used, the language of choice for creating BREW applications is C/C++, which will run faster because it is native and does not require bytecode interpretation. The BREW environment can be used with other mobile operating systems such as Palm OS, Windows CE, and Symbian OS. It will complement these environments by enabling over-the-air provisioning of applications and management of telephony functions in the devices. BREW can act as the common denominator for the entire range of mobile devices, from mobile phones to high-end PDAs. As good as all this sounds, however, it is still too early to tell if this type of integration will prove to be successful, as many of the mobile operating system vendors are just starting to add these capabilities natively to the operating system, or by using J2ME.

Relationship between BREW and J2ME

At first glance, it might appear as though BREW and J2ME were direct competitors. Each technology provides an abstract layer over the native device APIs. They both provide a platform-neutral API that makes it possible to deploy applications on any compliant device; and they both provide over-the-air provisioning of applications. These similarities often lead developers to believe that they have to choose one platform or the other. This is not the case. BREW enables Java Virtual Machine (JVM) vendors to create JVMs that are BREW-compatible. This will allow the JVM to be deployed to a BREW-enabled device and, therefore, to execute Java applications. The benefit is that BREW-enabled devices will be able to run both BREW *and* J2ME applications. Figure 18.2, shown earlier, depicts this relationship.

To support this scenario, Hewlett-Packard has ported its MicroChaiVM to the BREW platform. Other VM vendors, such as IBM and Insignia Solutions, have announced similar plans. All that said, in truth, even with JVM support, it is unlikely that J2ME and BREW will exist in harmony. In most cases, handset manufacturers and wireless operators support one technology or the other, but rarely both. J2ME has a strong lead in terms of the number of developers, as well as the potential market opportunity, and J2ME applications are not limited to particular devices or wireless networks. Any device with a JVM can execute J2ME applications. On the other hand, as noted previously, BREW is available only for CDMA chipsets. This may not be a limiting factor, though, as close to 100 million CDMA handsets are sold each year. For comparison purposes, this is more than double the total number of PDAs that are sold. So even though the market for BREW applications is smaller than that for J2ME applications, it is still sizable.

Speech Application Language Tags (SALT)

The Speech Application Language Tags (SALT) technology was created in 2001 by Cisco, Comverse, Intel, Microsoft, Phillips, and Speechworks. These companies, along with many others, have formed the SALT Forum to oversee future development and standardization of the SALT specification. Their goal is to provide a royalty-free, platform-independent standard for creating multimodal and telephony-enabled applications that can be accessed from PCs, telephones, and PDAs. SALT is a technology for extending common markup languages (including HTML, XHTML, cHTML, WML, etc.) with a spoken dialog interface for Internet applications. It is designed for both voice-only applications and multimodal applications that combine voice and visual displays.

SALT Elements

SALT is a lightweight specification consisting of a small number of XML elements, with associated attributes and DOM object properties, events, and methods. SALT has four main top-level elements, as follows:

<listen...> . Used for speech recognition and information input.

<prompt...>. Used for prompt playing and speech synthesis configuration.

<dtmf...>. Used for configuration and control of DTMF input.

<smex...>. Used for general-purpose communication with the host platform.

In addition to these top-level elements there are also <grammar>, <bind>, and <record> elements for specifying grammars, processing recognition results, and recording audio input. In total, SALT has only around 10 elements, making it a very simple language. In most cases, these elements are used in conjunction with ECMAScript to allow for a programmatic approach to application development. This simplicity makes SALT easy to use in conjunction with other markup languages such as HTML or WML.

There are two main usage scenarios for SALT:

Multimodal. SALT can be used to extend existing Internet applications with speech for both input and output. For example, a user can request a set of information using voice and have the response returned visually using HTML. Conversely, a user can enter a request using HTML and have the response spoken back to them.

Voice only. SALT can be used for applications without a visual display to control the application flow, input, and output. The interaction occurs over any telephone (wireline or wireless), allowing for universal access to Internet content.

The SALT specification is designed to benefit a variety of users. It gives end users more options for interacting with applications. SALT-enabled applications allow for speech, text, or graphical interfaces on their own or in combination with each other. Since it integrates into existing markup languages, developers can continue to use the tools and technologies they are comfortable with while adding advanced speech interfaces. These speech interfaces can help reduce overall application costs while reducing application complexity. In addition, using SALT, businesses can continue to use their existing Web infrastructure and expertise.

Competition between SALT and VoiceXML

VoiceXML and SALT are competitors in the speech application market. VoiceXML is an established specification with a growing user base. Its focus has been to replace or augment existing interactive voice response (IVR) systems, whose software and hardware are proprietary, resulting in higher costs and vendor lock-in. VoiceXML is changing that by taking a standard approach for creating cross-platform, advanced voice applications. SALT, on the other hand, is more focused on multimodal applications, where voice and text are combined to provide more effective interfaces to Internet applications.

The promoters of SALT argue that VoiceXML is inflexible and does not work well with existing Web development tools and server platforms. They are planning to address these failings with SALT, which, because it extends existing markup languages, can be used to add voice to Internet applications in a simplified manner. Microsoft is also spending significant resources on supporting SALT throughout its platform. Visual Studio. NET, ASP.NET, Internet Explorer, and PocketIE will all add support for SALT in the near future. This should increase SALT adoption, as both Visual Studio .NET and ASP.NET are widely used for Internet application development.

Not surprisingly, VoiceXML promoters contend there is no need for SALT, citing the fact that VoiceXML has been around since 1999 and has a strong industry following, proven by the many VoiceXML-based applications that have been successfully deployed. SALT, of course, is still in its conception stage, without any proven market success.

In rebuttal is the fact that SALT can be used with existing speech recognition and text-to-speech technology, which will enable many of the existing VoiceXML vendors to add SALT support to their solutions. This will help SALT to gain some market momentum without requiring companies to put forth a lot of effort. Still, SALT lags considerably behind VoiceXML in terms of both market adoption and maturity of the specification. Whereas the VoiceXML Forum has nearly 400 member companies, has released the second version of its specification, and is recognized as an official standard by the W3C, SALT has just over 50 member companies and only in August 2002 submitted SALT version 1.0 to the W3C Multimodal Interaction Working Group and Voice Browser Working Group for standardization.

The point here is, regardless whether VoiceXML or SALT becomes the dominant voice technology, clearly, an increasing amount of Internet content will become available through a voice interface. This will enable the over 1 billion wireline and wireless telephones to access the Internet through a simple call. (For a complete overview of VoiceXML technology see Chapter 15, "Voice Applications with VoiceXML.")

M-Services

By now you are no doubt familiar with the shortcomings of the Wireless Application Protocol (WAP). Some of these shortcomings—low bandwidth and high latency—have to be addressed by network operators. Others—usability and richness—have to be addressed by the application developers themselves. The latter had proven to be a difficult task until Mobile Services (M-Services) were introduced by the GSM Association (GSMA) in June 2001. With the introduction of M-Services, developers now have a standard, cross-platform way to create sophisticated and user-friendly wireless Internet applications. But for M-Services to be successful, network operators, handset manufacturers, and application developers have to cooperate on the guideline implementations.

M-Services: What Are They?

M-Services are guidelines for creating wireless Internet applications. They are not an industry standard, as the GSMA is not a standards body; rather they are requirements that members of the GSMA adhere to for developing and deploying wireless data applications. M-Services provide a way for wireless operators, handset manufacturers, and application developers to provide value-added services that are attractive to consumers—similar to what is provided by NTT DoCoMo with the i-Mode service in Japan. Unlike previous solutions that are largely browser-based, M-Services address the entire development and deployment platform. In doing so, they incorporate several elements, many of which are existing standards:

- WAP June 2000 Conformance Release
- WAP 2.0 (for WML2/XHTML Basic and provisioning)
- Graphical user interface (GUI)
- Download of media objects
- Multimedia Message Service (MMS) and, optionally, email
- Enhanced Message Service (EMS)
- SIM Application Toolkit
- SyncML for vCard and vCalendar

NOTE When M-Services were released, many misinterpreted them to be intended as a replacement for WAP. This is not true. The M-Services guidelines complement WAP; they define areas that the WAP specification does not address, such as application user interfaces, navigation, and content downloads. For more information on WAP and wireless Internet applications, see Chapter 11, " Thin Client Overview."

Not all of these features are immediately required in M-Services. Rather, M-Services are being rolled out in two phases over a two-year period. Phase I was introduced in June 2001; its objective was to deploy the first set of M-Services by the end of 2001. The requirements for Phase I included handset support for GPRS and WAP provisioning; support for WAP 1.2.1 features, such as WAP Push, WTLS 2, and UAProf; support for Enhanced Message Service (EMS) and long messaging; and downloadable object support for ringtones and wallpapers. Phase II, introduced in February 2002, is intended to complete the picture. Two implementation dates were defined for Phase II: September 2002 and April 2003. Building on Phase I, Phase II requires additional support in many of the same areas, with a focus on the user interface, downloadable objects, and messaging. The following subsections give a more in-depth explanation of what is required in these categories.

User Interface and Navigation

The first, and in many cases most important, set of requirements applies to the user interface and navigation of M-Services applications. These requirements include a new suite of graphical components, as well as a specification for navigation between screens. This will address one of the major problems of WAP, where the UI varies from phone to phone, making development incredibly difficult.

The following are some improved navigation and UI features aimed at making the wireless browsing experience more similar to a traditional desktop:

- A set of new components for more advanced applications. These components include pop-up menus, radio buttons, check boxes, push buttons, text boxes, tables, and inline hypertext links.

- Support for WAP components such as UAProf, WAP Push, WTLS Class 2, and Provisioning agent. It is also recommended that WML2 or XHTML Basic be used.

- Use of soft keys to support icons. In addition, soft key labels must appear near the physical key, out of the text flow.

- The use of a title bar to let users know what part of the application they are currently in. The title bar must scroll with the content of the page.

- Push buttons must have the label inside of the button, rather than beside it. This will help distinguish them from text entry fields.

This list is not exhaustive. Many other user interface requirements will make wireless Internet applications much more approachable. In order for these applications to be successful, handset manufacturers have to produce M-Services-compatible devices, with support for over-the-air provisioning, color, graphics and multimedia. By 2002, the first wave of M-Services-enabled devices had come to market, and many more are expected in 2003 and beyond.

Downloadable Objects

The ability to download larger and richer content is important to the developers of M-Services. Though users can download content such as ringtones, wallpaper, and screensavers using SMS, anything more advanced, such as digital pictures, Java MIDlets, or multimedia files, were too large for these transport mechanisms. M-Services solves this by adopting Download Fun (DF), a structured system for managing content downloads. It allows wireless operators to control application content distribution and provides an opportunity for them to share in the deployment revenues.

Many types of content are suitable for download to a wireless phone. The more common types are ringtones, wallpaper, screensavers, pictures (JPEG, GIF, WBMP, PNG), games, Java applications (that are compliant with CLDC 1.0 and MIDP 1.0 or later), audio clips, video clips, vCard and vCalendar entries, and bookmarks. A Download Fun server is required to support these downloadable objects. The M-Services Phase II requirements specify which formats are mandatory, recommended, or optional. Most of the formats listed here are mandatory.

Messaging

In Phase I, M-Services require support for Long SMS (640 characters in length) EMS; in Phase II, for MMS. Support for these advanced messaging protocols is essential if M-Services are to be universally accepted by both consumers and corporations. In order to take advantage of these new messaging protocols, faster networks (2.5G and 3G), as well as handsets with color and graphics support, are required.

Support for email is optional in M-Services.

Who Benefits from M-Services?

The primary beneficiary of M-Services is the consumer. In fact, providing effective services for the end user is the driving force behind the M-Services guidelines. When applications based on the M-Services guidelines are deployed, consumers will finally have sophisticated, user-friendly services available to them from their wireless devices. These applications have interfaces similar to standard Web applications, making them much more approachable to the average consumer.

But to be successful in creating effective consumer services, the other parties involved in the creation and deployment of M-Services have to work together—the network operators, handset manufacturers, and developers. Fortunately, these parties will also benefit from M-Services, giving them the incentive to bring these solutions to market quickly.

- **Network operators.** The operator will form the relationship with the consumer. They will take control of application management and provisioning, giving them new revenue streams by selling M-Services. In addition, M-Services will help drive wireless data usage, thereby increasing operator revenues.

- **Application developers.** Second only to consumers, developers benefit most from M-Services, primarily because M-Services improve upon WAP in two main areas: increased application usability and a standard client interface. M-Services help developers create applications that are more attractive to users and that will work equally well on any mobile handset. Anyone who has developed WAP applications knows that both of these were difficult to achieve before the advent of M-Services.

- **Handset manufacturers.** The compelling set of M-Services applications will lead consumers to upgrade their devices to support the latest applications. By using mobile phones for more than just voice calls, consumers will realize more value from their device, hence will be more inclined to spend more on a handset. In addition, because M-Services guidelines describe how content should be displayed and how user interactions are to be handled, true application interoperability between handsets becomes possible.

M-Services have the potential to revive the floundering WAP industry. Consumers have become frustrated by the lack of compelling applications available on wireless phones, and developers have had a hard time addressing the issue because of the fragmented WAP market. Thanks to M-Services, network operators, handset manufacturers, and developers have a standard platform to aim for.

NOTE It is important to point out that M-Services require the network operator to support GPRS. Other network protocols, including GSM, are not supported. The GSMA does not see this as a limitation since close to 70 percent of the world's digital subscribers are represented by the GSMA. As 3G technologies are rolled out, they expect this number to grow to 80 percent.

RECOMMENDED READING If you are planning to implement M-Services applications, be sure to visit www.gsmworld.com and download the M-Services Phase I Guidelines and M-Services Phase II Requirements.

Summary

Of the technologies discussed in this chapter, Web services are by far the most likely to achieve widespread industry adoption, because they provide access to enterprise data and logic using standard Internet protocols. The core technologies on which Web services are based include SOAP, WSDL, and UDDI. Both smart client and thin client wireless applications can benefit from Web services. The combination of server-based functionality and client-side processing can be especially powerful in smart client applications.

BREW offers a simplified way for developers to create a wide range of wireless applications for CDMA-based handsets. With it, developers can use a higher-level API, rather than developing applications at the device system level. BREW takes a similar approach to J2ME, but does not necessarily rule out using J2ME along side it.

SALT is a technology for integrating voice interaction into Internet applications. It consists of simple XML elements that can be used in conjunction with existing markup languages such as HTML and WML. The SALT Forum was founded in 2001 and has submitted SALT version 1.0 to the W3C for standardization.

M-Services are guidelines that address many of the limitations inherent to wireless Internet applications. It is complementary to, not a replacement for, WAP. By the end of 2003, we will see complete M-Services Phase II implementations that will provide sophisticated wireless Internet applications with support for advanced user interfaces with multimedia content, downloadable objects, and enhanced messaging.

These up-and-coming technologies may not have an immediate impact on your mobile solutions, but they eventually will be factors in your decision making process. Keeping an eye on future technologies and trends will allow you to make important decisions today for the applications of tomorrow.

Helpful Links

The following Web links can be used to find more information on subjects related to those covered in this chapter:

BREW	www.qualcomm.com/brew/
M-Services Initiative	www.gsmworld.com/technology /services/index.shtml
SALT Forum	www.saltforum.org
W3C Web services activity	www.w3.org/2002/ws/
SOAP	www.w3.org/TR/SOAP/ http://www.w3.org/TR/2002 /WD-soap12-af-20020924/
WSDL	www.w3.org/TR/wsdl.html
UDDI	www.uddi.org/about.html
Web Services Interoperability Organization	www.ws-i.org
ebXML	www.ebxml.org

Index